PLD Based Design with VHDL

Vaibhav Taraate

PLD Based Design with VHDL

RTL Design, Synthesis and Implementation

Springer

Vaibbhav Taraate
Pune, Maharashtra
India

ISBN 978-981-10-3294-3 ISBN 978-981-10-3296-7 (eBook)
DOI 10.1007/978-981-10-3296-7

Library of Congress Control Number: 2016958476

Printed on acid-free paper

This Springer imprint is published by Springer Nature
The registered company is Springer Nature Singapore Pte Ltd.
The registered company address is: 152 Beach Road, #22-06/08 Gateway East, Singapore 189721, Singapore

Dedicated to my Spiritual Guru, my Teachers and to my Parents

Preface

In the present decade, the complexity of the ASIC and FPGA design has grown rapidly. Due to that there is need of the intelligent and complex devices, and hence the FPGA prototyping area has evolved during this decade.

Major FPGA vendors such as XILINX and Altera (Intel FPGA) have come up with the complex FPGAs which are required for design and realization of the system on chip (SOC). During this decade, the era of miniaturization has lot many challenges. The major challenges are to design and deliver the intelligent products for lesser cost, high speed, less area, and less power.

Under such circumstances for the idea or product feasibility, the complex FPGAs are used and the complexity of FPGA architecture has grown in the past decade. Even the multiple FPGA designs are used to validate the complex SOCs. For easy understanding of the FPGA designs and ASIC prototyping using FPGAs, this book is organized. This book covers the design for the lower gate count to higher gate count designs. Even this book is written in such a way that it can give information about the VHDL, synthesis, FPGAs, and ASIC prototyping.

Chapter 1 of this book discusses the evolution of the logic design, need of HDL, and differences between the VHDL and other higher level languages, and even this chapter describes about the different modeling styles using VHDL.

Chapter 2 of this book describes about the basic combinational elements and their use in the design. Even this chapter describes how to write synthesizable RTL using the VHDL constructs. This chapter is useful for the beginners to understand about the basic VHDL constructs and the synthesis outcome of few low gate count designs.

Chapter 3 discusses the key VHDL constructs such as processes, signals, and variables, when else, with select, if-then-else and case. Even this chapter covers the practical scenarios and use of these constructs.

Chapter 4 describes the how to write an efficient RTL using VHDL. Even this chapter covers the design for the combinational logic such as multibit adders, multiplexers, decoders, and encoders. The synthesis for the RTL design using VHDL is covered with the detailed explanation and practical scenarios.

Chapter 5 covers the sequential design scenarios and the RTL using VHDL for the latches and flip-flops. Even this chapter covers the BCD counters, binary counters, gray counters, ring counters, Johnson counters and the RTL design and synthesis for the same. This chapter has information about the timing parameters and timing analysis for the synchronous sequential designs. This chapter even gives information about the basics of asynchronous and multiple clock domain designs and the issues like metastability and how to overcome those during design cycle.

Chapter 6 covers the PLD-based designs and the detail practical-oriented examples and scenarios for the design using SPLDs, CPLDs, and FPGAs. This chapter covers the XILINX and ALTERA (Intel) FPGA architectures and their use in the design and prototyping. The vendor-specific design guidelines are covered in this chapter.

Chapter 7 covers the VHDL constructs and the use of VHDL for the verification and simulation of the design. This chapter is useful to understand the test benches and how to simulate the design for early detection of bugs. Even this chapter covers the practical issues in the design verification using practical scenarios and examples.

Chapter 8 covers the design and coding guidelines for the PLD-based designs. How to use the VHDL for the efficient design is explained in detail with the practical scenarios and synthesizable VHDL constructs. This chapter covers techniques such as grouping, parallel and concurrent logic, logic duplications, and resource sharing. Even this chapter covers the low-power basics as clock gating and clock enabling.

Chapter 9 discusses the finite state machines (FSMs) using the VHDL. The Moore and Mealy machines and their use to code the sequence detectors and counters are described in this chapter. Even the FSM synthesis issues and how to improve the design performance are discussed with the practical scenarios. Even this chapter covers the FSM synthesis guidelines and FSM optimization techniques used while prototyping ASICs using the complex FPGAs.

Chapter 10 covers the complex designs such as multipliers, barrel shifters, arbiters and the processor logic as ALU, and the other basic protocols. This chapter is useful to understand the synthesis issues in the complex designs and how to overcome those using the techniques described in Chap. 7.

Chapter 11 covers VIVADO based design flow and case study using VIVADO for the design implementation. The case study of FIFO is covered in this chapter.

Chapters 1–11 are organized in such a way that it covers the small gate count RTL using VHDL to the complex design using VHDL with the meaningful scenarios. This book is useful for the beginners, RTL design engineers, and professionals. I hope that this book can give you the excellent understanding of VHDL constructs and use of VHDL in ASIC prototyping!

Pune, India Vaibhav Taraate

The original version of the book was revised:
The incorrect informations in chapters
have been corrected. The erratum
to this book is available at
DOI 10.1007/978-981-10-3296-7_12

Acknowledgements

This book is possible due to direct and indirect contribution of many people. While writing this book, I got the great help from many people. I am thankful to all my students to whom I have taught this subject over past almost 15 years.

I am thankful to my wife Somi for her sacrifices during the period of writing this book. Although Somi was hospitalized for three months after returning from the hospital, she has helped me in finding the grammatical mistakes and even corrected initial proofs of this book.

I am very much thankful to my dearest friend Ishita Thaker (Ish) for her indirect help and motivation while writing this book.

I am very much thankful to dearest Kaju for the great wishes and prayers.

Especially I am thankful to my Son Siddesh and my daughter Kajal for understanding me during this period and for helping me with few suggestions for the representation of diagrams.

This book would not have been possible without the help of Swati Meherishi and Aparajita Singh at Springer.

I am thankful to all the Springer staff, especially Praveen V for the review of this book and for the easy-to-understand outline of this book.

Finally, in advance, I am thankful to all the readers and buyers for buying and enjoying this book!

Contents

About the Author

Vaibbhav Taraate is an entrepreneur and mentor at "Semiconductor Training @ Rs. 1." He holds a BE (electronics) degree from Shivaji University, Kolhapur, in 1995 and secured a gold medal for standing first in all engineering branches. He has completed his M. Tech (aerospace control and guidance) in 1999 from IIT Bombay. He has over 15 years of experience in semicustom ASIC and FPGA design, primarily using HDL languages such as Verilog and VHDL. He has worked with few multinational corporations as consultant, senior design engineer, and technical manager. His areas of expertise include RTL design using VHDL, RTL design using Verilog, complex FPGA-based design, low-power design, synthesis/optimization, static timing analysis, system design using microprocessors, high-speed VLSI designs, and architecture design of complex SOCs.

Chapter 1
Introduction to HDL

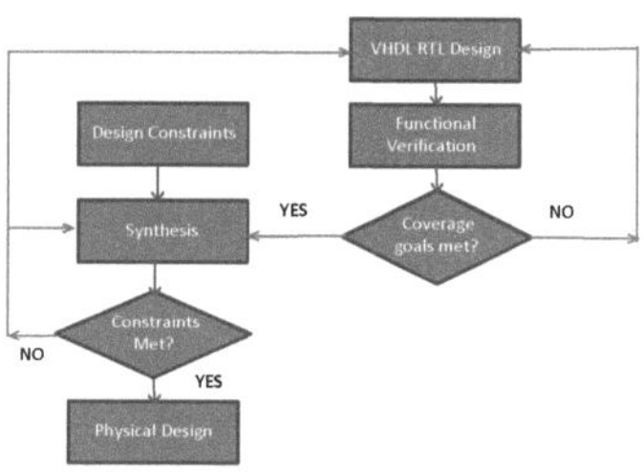

"Imagination is very important than Knowledge."
--- Albert Einstein

The HDL design engineer should have the Knowledge of hardware and should apply it to solve the practical problems with imagination.

Learn to understand and imagine description using efficient VHDL constructs!

Abstract This chapter discusses the digital logic design evolution and the basic ASIC design flow. The chapter describes the necessity of ASIC SOC prototype. The comparison of ASIC and FPGA implementation is described in this chapter. The chapter even discusses the need of HDL and VHDL different modeling styles using the small gate count example. This chapter is useful to the HDL beginners to understand about the difference between high-level language and HDL modeling styles.

Keywords ASIC · FPGA · HDL · Prototype · C · C++ · Concurrent · Sequential · ABEL · Data flow · Logic capacity · RTL · Simulation · Verification · Implementation · Structural model · Data-flow model · Behavior model · ASIC prototyping

---

The original version of this chapter was revised: The incorrect information have been corrected. The erratum to this chapter is available at 10.1007/978-981-10-3296-7_12

V. Taraate, *PLD Based Design with VHDL*,
DOI 10.1007/978-981-10-3296-7_1

1.1 History of HDL

The invention of CMOS logic during 1963 has made integration of logic cells very easy and it was predicted by Intel's cofounder Gordon Moore that the density of the logic cells for the same silicon area will get doubled for every 18–24 months. What we call as Moore's law!

How Moore's prediction was right that experience engineers can get with the complex VLSI-based ASIC chip designs. In the present decade, the chip area has shrunk enough, and process technology node on which foundries are working is 14 nm and chip has billions of cells for small silicon die size. With the evolutions in the design and manufacturing technologies, most of the designs are implemented using **V**$_{\mathbf{HSIC}}$**HDL**: **V**ery **H**igh **S**peed **I**ntegrated **C**ircuit **H**ardware **D**escription **L**anguage or using Verilog. We are focusing on the VHDL as hardware description language. The evolution in the EDA industry has opened up new efficient pathways for the design engineers to complete the milestones in less time.

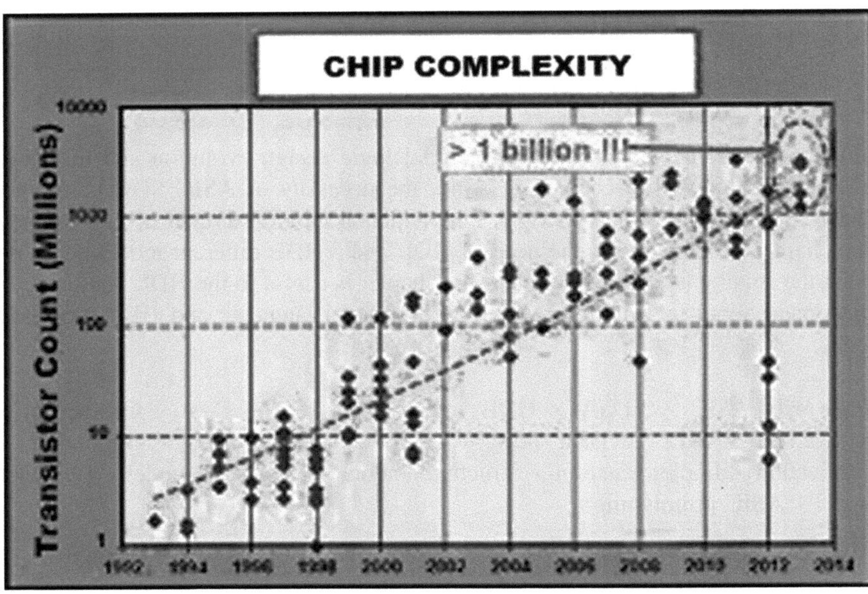

Table 1.1 Hardware description language and evolution

HDL	Description	Application	Standard
AHDL	Analog hardware description language	Open source and used for analog verification	1980
Verilog-AMS	Verilog for analog and mixed signals	Open standard and used for the mix of digital and analog simulation	Derived from IEEE 1364
VHDL-AMS	VHDL for analog and mixed signals	Standard language for both the analog and digital mixed signal simulations	IEEE 1076.1-2007
ABEL	Advanced Boolean expression language	Used for the PLD-based design	None
System C	The high-level abstraction language for the hardware designs	It uses the C++ classes for higher level behavioral and transaction-level modeling(TLM) for describing hardware at system level	IEEE 1666-2011
System Verilog	Superset of Verilog	Used to address the system-level design and verification	IEEE 1800-2012
Verilog	Widely used hardware description language (HDL)	Used for the design and verification of digital logic	IEEE 1364-2005
VHDL	Very-high-speed integrated circuit (HSIC) hardware description language (HDL)	Use for the design and verification of digital logic	IEEE 1076-2008

Table 1.1 describes the various hardware description languages (HDLs) and their standard with the description.

1.2 System and Logic Design Abstractions

As shown in Fig. 1.1, most of the designs have various abstraction levels. The design approach can be top-down or bottom-up. The implementation team takes decision about the right approach depending on the design complexity and the

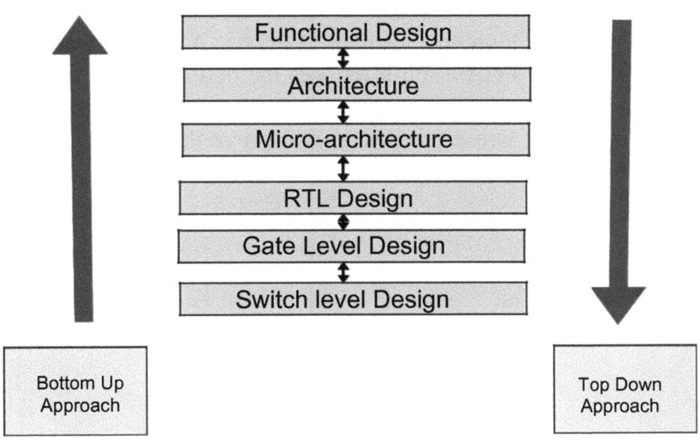

Fig. 1.1 Design abstractions

availability of design resources. Most of the complex designs are using the top-down approach instead of bottom-up approach.

The design is described as functional model initially, and the architecture and microarchitecture of the design are described by understanding the functional design specifications. Architecture design involves the estimation of the memory processor logic and throughput with associative glue logic and functional design requirements. Architecture design is in the form of functional blocks and represents the functionality of design in the block diagram form.

The microarchitecture is a detailed representation of every architecture block, and it describes the block and sub-block level details, interface and pin connections, and hierarchical design details. The information about synchronous or asynchronous designs and clock and reset trees is also described in the microarchitecture document.

RTL stands for register transfer level. RTL design uses microarchitecture as reference design document and can be efficiently coded using VHDL for the required design functionality. The efficient design and coding guidelines at this stage play an important role and efficient RTL can reduce the overall time requirement during the implementation phase. The outcome of RTL design is gate-level netlist. Gate-level netlist is output from the RTL design stage after performing RTL synthesis and it is a representation of the functional design in the form of combinational and sequential logic cells.

Finally, the switch-level design is the abstraction used at the layout to represent the design in the form of CMOS switches—PMOS, NMOS.

1.3 ASIC Prototyping

ASIC prototyping is also called as FPGA prototyping or SOC prototyping. ASIC is an application-specific integrated circuit, FPGA is field programmable gate array, and SOC is system-on-a-chip. If we consider the past one decade, then due to availability of high logic density FPGAs the ASIC prototyping using FPGA area have has been evolved. The main goal is to validate the firmware, software, and hardware of SOC using high-end available FPGAs. ASIC designs can be prototyped by using suitable FPGAs, and it reduces the delivery time, budget, and even the product launch targets in the market. For million logic gate SOCs, the ASIC prototyping using FPGA is used to design and prove the prototype and it reduces the risk while manufacturing of ASICs.

As the process node has shrunk to 14 nm and even will shrink to less than 10 nm, the complexity of design, the design risk, and the development time has increased. The main challenge for every organization is to develop the lower cost products with improved design functionality in small silicon area. In such scenario, the designers are facing the development and verification challenges. Under such circumstances, the high-end FPGAs can be used to prototype the ASIC functionality and it reduces the overall risk. The verified and implemented design on high-end FPGAs can be resynthesized using standard cell ASIC using the same RTL, constraints, and scripts. There are many EDA tools available to port an FPGA prototype on structured ASICs. This really reduces the overall risk in ASIC design and saves money and time to market for the product.

Following are key advantages of ASIC prototyping using FPGAs

1. The shrinking process node and chip geometries involve the investment in millions of dollars in the early stage of design. Using FPGAs, the investment risk reduces.
2. Due to the uncontrolled market conditions, there is risk involved in the design and development of products. FPGA prototype reduces such risk as the product specifications and design can be validated depending on the functional requirements or changes.
3. FPGA prototyping is efficient as the bugs, those were not detected in simulation, can be addressed and covered during prototyping.
4. Full-system verification using FPGA prototype can detect the functional bugs in the early stage of design cycle.
5. FPGA prototyping saves the millions of dollar of EDA tool cost and even it saves the millions of dollar engineering efforts before ASIC tape-out.
6. As design using FPGA can be migrated using the EDA tool chains onto the ASICs, it saves the time to market the product with intended functionality.
7. Multiple IPs can be integrated and design functionality can be verified and tested and that speed up the design cycle.
8. Most of the cases the hardware software portioning is visualized at higher abstraction level. The hardware software codesign can be evaluated at the hardware level and it is more important milestone in overall design cycle. So the ASIC prototyping can be useful in tweaking of the architecture. If there is additional design overhead in the hardware, then the design architecture can be changed by pushing few blocks in software and vice versa. This will give the more efficient architecture and design.

Table 1.2 Comparison of FPGA with ASIC implementation

	FPGA	Hard copy	Structured ASIC	Standard cell ASIC
NRE, mask and EDA tools	Up to a few thousand US$, so the overall cost is low	Couple of hundred thousand US$ for FPGA conversion and masks. So the overall cost is moderate	A couple of hundred thousand US$ for interconnect/meta-one masks so the overall cost is moderate	A million US$ depending on the design functionality. So the cost is high
Unit price	High	Medium-low	Medium-low	Low
Time to volume	Immediate	Almost around 8–10 weeks. The additional conversion time may require for other structured ASIC products	Almost around 8–10 weeks. The additional conversion time may require for other structured ASIC products	Almost around 18 weeks + conversion time of another 18 weeks
Engineering resources and cost	Minimum	Minimal from developers but other structured product may require the additional engineering resources	Nominal but for the other structured ASIC products may require the additional engagement of the resources	High as most of the work requires the development from scratch and requires good support from the backend team
FPGA prototype correlation	Same device	For hard copy-structured ASIC: Nearly identical—Same logic elements, process, analog components, and packages	It depends upon the type of IP used and the functionality. Same RTL but potentially different libraries, process, analog, and packages	Same RTL but potentially different libraries, process, analog, and packages

The Table 1.2 gives information about the pros and cons of FPGA and ASIC.

There is always confusion between the prototype and the migration. The ASIC prototyping is basically the design or validation of idea to check for the early functional and feasibility of new designs. The design migration from ASIC to FPGA involves the flow from RTL design to implementation and may be useful in the upgradation of design with additional features.

Following are the key points need to be considered during ASIC prototyping and design migration using high-end FPGA.

1. Use the universal prototype board as it saves the time of almost four months to twelve months for the high-speed prototyping board development.

2. Choose the FPGA device depending on the functionality and gate count. It is not possible to fit whole ASIC into single FPGA even if we use the high-end families of ALTERA or XILINX FPGAs. So the practical solution is use of multiple FPGAs. But the real issue is the design partitioning and the inter-communication between multiple FPGAs. If the design is well defined and partitioned properly, then the manual partitioning into multiple FPGA can give the efficient results. If the design has complex functionality, then the use of automatic partitioning can play efficient role and can create the efficient prototype.
3. As the design library for ASIC and FPGA is totally different, the key challenge is to map the primitives. So it is essential to map the directly instantiated primitives during synthesis and during the implementation level. That is at the post-synthesis, all the primitives from ASIC library need to be remapped for getting the FPGA prototype.
4. High-end FPGA may have 1000–1500 pins and if one FPGA is used for prototype, then there are limited issues in the pin assignment and pin interface. But if IO pins required more than the pins available in one FPGA, then the real issue is due to multiple FPGA interfaces and connectivity. The issue can be resolved by using the partitioning with the signal multiplexing. This will ensure the efficient design partitioning and efficient design prototype.
5. Implementation of single clock domain design prototype is easy using FPGAs. But if the design has more than one clock that is multiple clock domains, then it is quite difficult to use the clock gating and other clock-generation techniques during prototype. So the migration of ASIC design into FPGA needs more efforts and sophisticated solutions. One of the efficient solutions is to convert the designs into smaller design units clocked by the global clock source.
6. The memory models used in the FPGA are different as compared to ASIC. So it is essential to use the proper strategy during memory mapping. Most of the time, the synthesized memory models required are not available. Under such scenario, the best possible solution is to use the prototyping board with the required specific memory device.
7. The full functional testing and debugging is one of the main challenges in the ASIC prototyping. During this phase, it is essential to use the debugging platform which can give the visibility of the results such as speed and functional testing results.

The ASIC prototyping is achieved by using industry's standard leading tools such as Design Compiler FPGA. The design compiler is industry's leading EDA tool which is used to get best optimal synthesis result and best timing for the FPGA

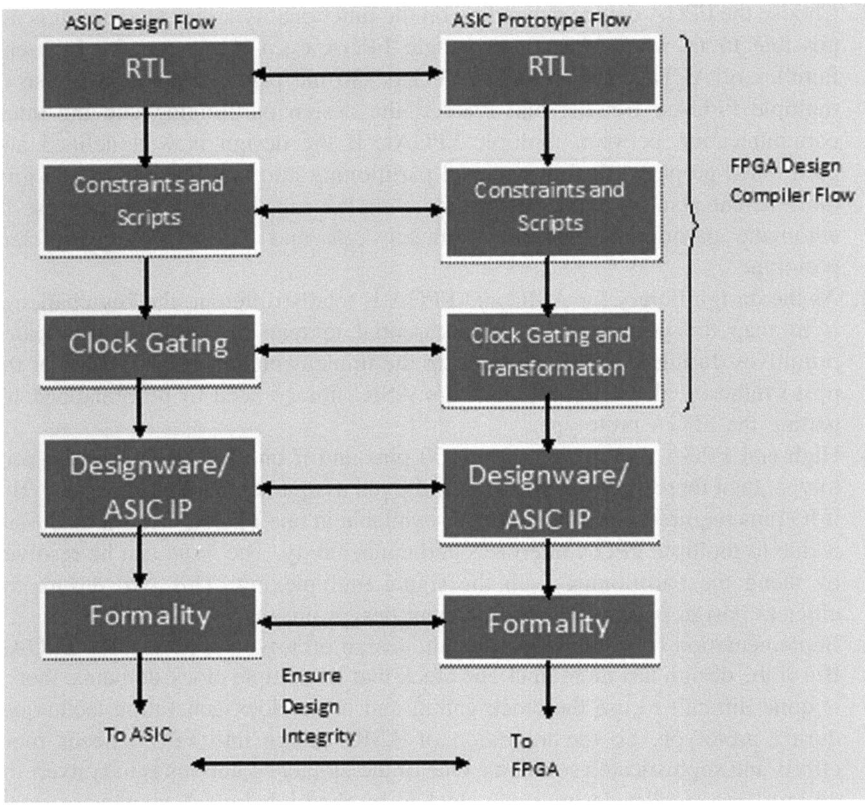

Fig. 1.2 ASIC prototype flow

synthesis. The basic flow for the ASIC prototyping is shown in Fig. 1.2, and in the subsequent chapters, we will discuss the FPGA based designs and key steps, to achieve the efficient ASIC prototype using XILINX/Altera FPGA.

1.4 Integrated Circuit Design and Methodologies

With the evolution of VLSI design technology, the designs are becoming more and more complex and SOC-based design is feasible in shorter design cycle time. The demand of the customers is to avail the product in the shorter span of time is possible due to efficient design flow. The design needs to be evolved from specification stage to final layout. The use of EDA tools with the suitable features has

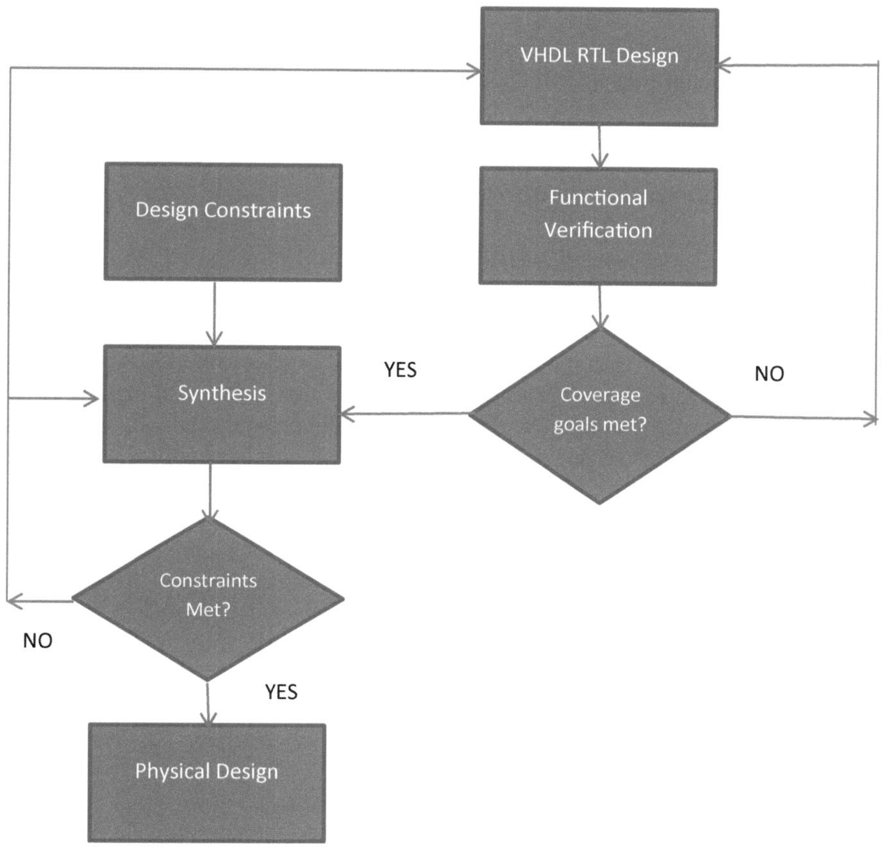

Fig. 1.3 Simulation and synthesis flow

made it possible to have the bug-free designs with proven functionality. The design flow is shown in Fig. 1.3, and it consists of the three major phases to generate the gate-level netlist.

1.4.1 RTL Coding

Functional design is described in the document form using the architecture and microarchitecture. Architecture and microarchitecture design is the functional representation of the design in the block and sub-block levels. This design document includes the block level interfaces, timing and logic blocks. The RTL design using VHDL uses the microarchitecture document as reference document to code the design. RTL designer uses the suitable design and coding guidelines while implementing the RTL design. An efficient RTL design always plays important role during implementation cycle. During this, designer describes the block-level and top-level functionality using an efficient VHDL RTL.

1.4.2 Functional Verification

After completion of an efficient VHDL RTL for the given design specifications; the design functionality is verified by using industry standard simulator. Pre-synthesis simulation is without any delays and during this the focus is to verify the functionality of design. But common practice in the industry is to verify the functionality by writing the testbench. The testbench forces the stimulus of signals to the design and monitors the output from the design. In the present scenario, automation in the verification flow and new verification methodologies has evolved and used to verify the complex design functionality in the shorter span of time using the proper resources. The role of verification engineer is to test the functional mismatches between the expected output and actual output. If functional mismatch is found during simulation, then it needs to be rectified before moving to the synthesis step. Functional verification is iterative process unless and until design meets the required functionality.

1.4.3 Synthesis

When the functional requirements of the design are met, the next step is synthesis to perform the RTL synthesis for the design. Synthesis tool uses the RTL VHDL code, design constraints, and libraries as inputs to generate the gate-level netlist as an output. Synthesis is iterative process until the design constraints are met. The primary design constraints are area, speed, and power. If the design constraints are not met then the synthesis tool performs more optimization on the RTL design. After the optimization if it has observed that the constraints are not met then it becomes compulsory to modify RTL code or tweak the microarchitecture. The synthesizer tool generates the area, speed, and power reports and gate-level netlist as an output.

1.4.4 Physical Design

It involves the floor planning of design, power planning, place and route, clock tree synthesis, post-layout verification, static timing analysis, and generation of GDSII for an ASIC design. This step is out of scope for the subsequent discussions!

1.5 Programming Language Verses HDL

Most of the engineers have familiarity with the programming languages such as C and C++. The most important point is to understand the differences between the programming language and the HDL. Table 1.3 illustrates the key differences between the programming language and HDL.

1.5.1 VHDL Evolution and Popularity

Very-high-speed integrated circuit hardware description language used to describe the hardware is also called as the programing language. It is used to describe the hardware for the programmable logic devices and the integrated circuit designs. The design automation flow using VHDL RTL plays crucial role while implementing the designs for high-end PLDs and ASICs.

To document the behavior of the ASICs, the VHDL was introduced by US Department of Defense. The initial version of VHDL was named as IEEE 1076-1987 standards and has wide variety of the data types. But this was not

Table 1.3 Programming language verses HDL

Parameters	Programming language (C or C++)	HDL
Instructions	Understands only sequential constructs	Understands both the sequential and concurrent constructs
Description style	Description of program is always behavioral model. To code the behavior, programmer uses analytical, algorithmic, or logical thinking!	Description using HDL is register transfer level (RTL). To describe the functionality of electronic circuit, the designer should have knowledge and understanding of the hardware circuits
Resources and usage	While writing program in C or C++, the programmer will never consider the use of resources or area. Even most of the time programmer does not care about the use of memory and the speed for the program	While describing the electronic circuits using the HDLs, the designer needs to consider the area, speed, and power requirements. The use of memory and resources for the PLD-based designs is the important parameter needs to be understood by the designer
Application	Used as programming language to describe the functionality. The user is programmer. It is a mix of assembly and high-level language	Used to design an electronic circuit. The user is designer.
Time constructs	It does not support the notion of time	It supports the time constructs and the notion of time
Flow constructs	It supports the data flow in the sequential manner	It supports the data and control flow

enough to describe the behavior of the hardware and later updated with the multivalued logic (nine-valued logic) using IEEE std_logic_1164.all package. The IEEE 1076-1993 standard has made the syntax more consistent to describe the behavior of the hardware functionality and concurrency. To resolve the restrictions on the port mapping rules, the minor changes carried out during year 2000–2002 and even the class structure of C++ introduced in the standard. During June 2006, the new standard for the VHDL was introduced and it is backward compatible with all the older standards. During February 2008, technical committee of Accellera approved VHDL 4.0 and it is called as VHDL-2008. During the same year Accellera released the IEEE standard 1076-2008 and the standard was published during year 2009.

Table 1.4 describes the various VHDL revisions and the relevant description for the respective revisions.

Following are the key reasons for which VHDL is popular in the semiconductor industry.

1. Used to describe the synthesizable logic designs and used for the simulation of the logic design.
2. VHDL is not case-sensitive language and it is easy to interpret in the context of logic design.
3. VHDL supports parallelism due to the concurrent constructs.
4. VHDL supports the sequential statements to describe the RTL designs.
5. VHDL supports the notion of time and file input and output handling and thus used for the simulation of the described design.
6. VHDL code is translated into the real digital logic using the gates and nets (wires) and very user friendly to design the PLD/ASIC-based designs.
7. VHDL supports the synthesizable and non-synthesizable constructs.
8. VHDL descriptions are described by using the electronic design automation (EDA) tools. The popular EDA tools used for PLD-based applications are Xilinx ISE series, Altera Quartus II and Mentor Graphics ModelSim or

Table 1.4 VHDL IEEE standard and revisions

Revision year	IEEE standard	Description
1987	IEEE 1076-1987	First standard for the language from the United States of Air Force
1993	IEEE 1076-1993	The most widely used version with the EDA tool support
2000	IEEE 1076-2000	Minor additions in the 1076-1993 standard and support for the protected type
2002	IEEE 1076-2002	Minor additions in the 1076-2000 standard and support for the buffer ports
2008	IEEE 1076-2008	The standard supports the use of external names

QuestaSim. The ASIC EDA tools are Synopsys DC, PT, and IC compilers and Cadence SOC Encounter.

VHDL Description consists of the following:

1. ***Library declaration IEEE***
2. ***Package declaration for the required IEEE library STD_LOGIC_1164.all using 'USE'***
3. ***Entity declaration to describe the input and output interface***
4. ***Architecture declaration to describe the functionality***
5. ***Component: The instance used to describe the logic functionality is called as component. The component is associated with the 'entity architecture' pair. For example for the half adder description: xor_gate, and_gate are treated as components.***
6. ***Configuration: to define the linkage between the entity and architecture and components. Configuration is used for binding of all the components specified in the architecture with the entity, and will be discussed in the subsequent chapters***
7. ***Package : It is basically subprogram or procedures for the reuse. Declared by using the keyword 'USE' with the package name. Package consists of the multiple objects and is visible for the architecture functional description***

Note The configuration, component declarations, and the packages will be used according to the design requirements and will be discussed in the subsequent chapters.

The template shown in Fig. 1.4 describes the VHDL code structure with the relevant and required explanation in the respective boxes.

As described in Table 1.5 the VHDL supports nine-valued logic using STD_LOGIC and used to model or to describe the digital logic designs. Table 1.5 describes the nine-valued logic and the description for the respective logic level.

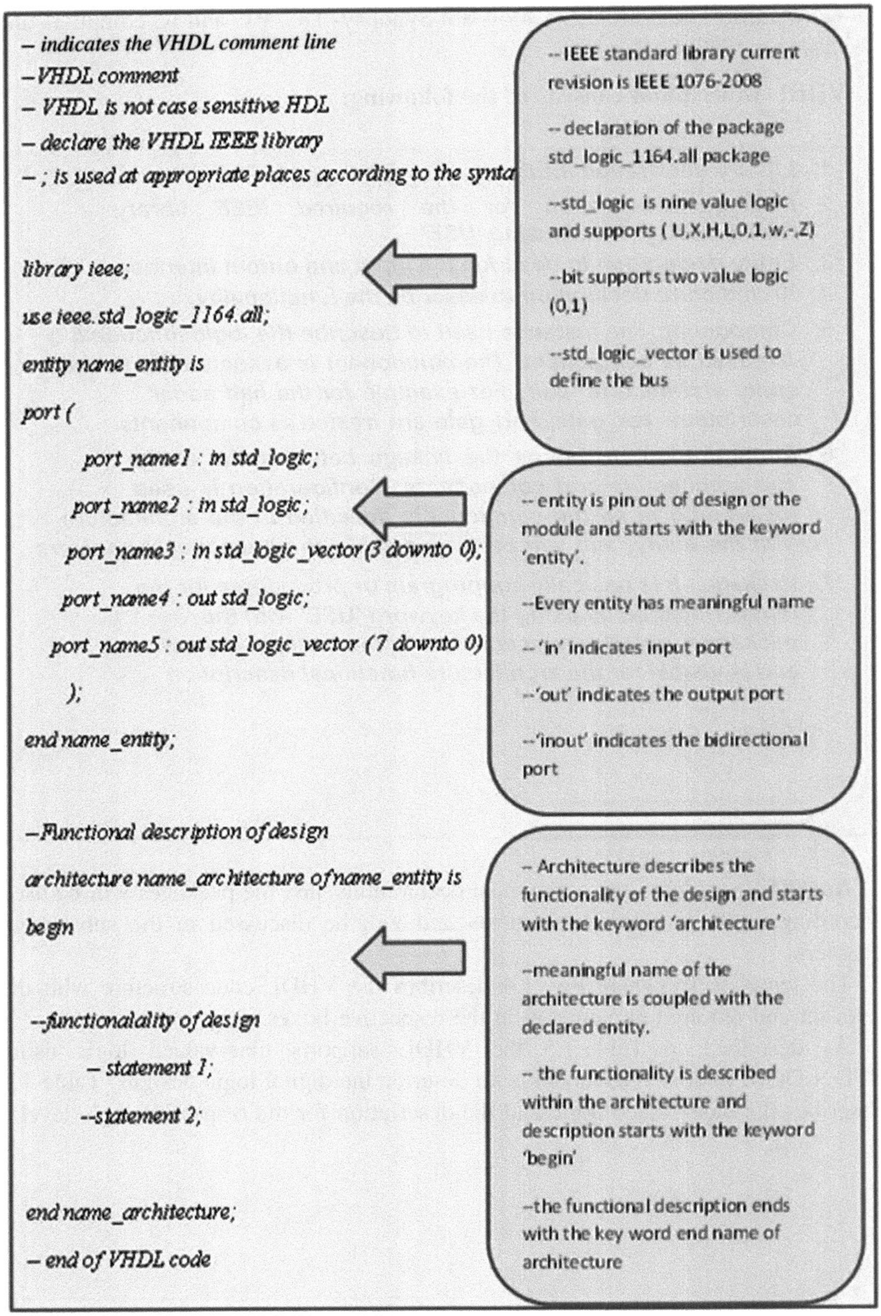

Fig. 1.4 VHDL code structure template

Table 1.5 Nine-valued logic

Character	Value description
'U'	Uninitialized value
'X'	Strong unknown value
'0'	Strong logic zero
'1'	Strong logic one
'Z'	High impedance
'W'	Weak unknown logic value
'L'	Weak logic zero
'H'	Weak logic one
'–'	Don't care

1.6 Design Description Using VHDL

In the practical scenarios, the VHDL is categorized into three different kinds of coding descriptions. The different styles of coding description are structural, behavioral, and synthesizable RTL. Figure 1.5 shows the truth table, schematic, and logic structure realization for half adder. The half-adder functionality is described by using the different modeling styles in this section.

1.6.1 Structural Design

Structural design defines a data structure of the design and it is described in the form of logic gates (logic components) using the proper net connectivity. Structural design is mainly the instantiation of different small complexity digital logic blocks or design components. It is basically design connection of small modules to realize moderate complex logic. Example 1.1 describes the structural code style for the half adder.

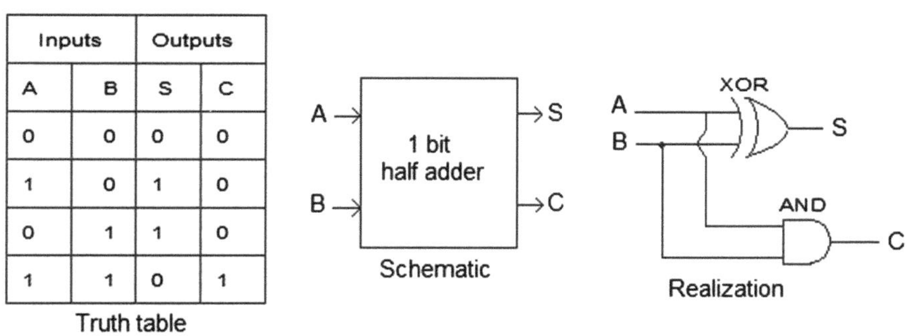

Fig. 1.5 Half-adder logic circuit

```
library ieee;
use ieee.std_logic_1164.all;

entity structural is
port(   a_in: in std_logic;
        b_in: in std_logic;
        sum_out: out std_logic;
       carry_out : out std_logic
);
end structural;

architecture arch_struct of structural is
begin
component XOR_gate
port (X, Y: in std_logic;
      Z: out std_logic));
end component;

component AND_gate
port (A, B: in std_logic;
         Y: out std_logic);
end component;
begin
U1: XOR_gate port map (a_in, b_in, sum_out);
U2: AND_gate port map (a_in,b_in,carry_out);
end arch_struct;
```

- Architecture of the structural is named as arch_struct
- Architecture describes the relationship between the inputs and outputs.
- The precompiled components 'XOR_gate', 'AND_gate' are used.
- Components are instantiated using the port mapping.
- Instance U1 generates the 'XOR_gate'
- Instance U2 generates the 'AND-gate'

Example 1.1 Structural style for the half adder

Note Structural style description is the digital logic in the form of components and their interconnections. Each component has its own ports and the directions. In this it is assumed that the pre compiled components XOR_gate, AND_gate are available in the work library.

1.6.2 Behavior Design

In the behavior style of VHDL, the functionality is coded from the truth table of the design. It is assumed that the design is black box with the inputs and outputs. The main intention of designer is to map the functionality at output according to the required set of inputs (Example 1.2).

```
library ieee;
use ieee.std_logic_1164.all;

entity behavioral is
port(   a_in: in std_logic;
        b_in: in std_logic;
        sum_out: out std_logic;
        carry_out : out std_logic
);
end behavioral;

architecture arch_behav of behavioral is
begin
 p1: process (a_in, b_in)
        begin
        if(a_in/=b_in) then
        sum_out <= '1';
        else
        sum_out <='0';
        end if;
        end process;
       p2: process (a_in, b_in)
        begin
        if(a_in='1'and b_in='1') then
        carry_out <= '1';
        else
        carry_out <='0';
        end if;
        end process;
end arch_behav;
```

- Architecture of the behavioral is named as arch_behav
- Architecture describes the relationship between the inputs and outputs.
- Two concurrent process blocks are described and named as 'p1', 'p2'.
- Process block 'p1' is used to describe the XOR logic functionality and generates the sum output.
- Process block 'p2' is used to describe the AND logic functionality and generates the carry output.

Example 1.2 Behavior style of the VHDL code for half adder

Note Behavior style of the VHDL description is the logic design in the form of behavior and not in the terms of the components or netlist.

1.6.3 Synthesizable RTL Design

Synthesizable RTL code is used in the practical environment to describe the functionality of design using synthesizable constructs. The RTL code style is high-level description of functionality using synthesizable constructs. The RTL coding style is treated as design description between the structural and behavioral

```
library ieee;
use ieee.std_logic_1164.all;

entity data_flow is
port(   a_in: in std_logic;
        b_in: in std_logic;
        sum_out: out std_logic;
       carry_out : out std_logic
);
end data_flow;

architecture arch_data of data_flow is
begin

  sum_out <= a_in XOR b_in;
 carry_out <= a_in AND b_in;

end arch_data;
```

- Entity is the pin out of the design named as data_flow and has four ports.
- Input ports are declared as 'a_in' and 'b_in'
- Output ports are declared as 'sum_out' and 'carry_out'

- Architecture of the data_flow is named as arch_data
- Architecture describes the relationship between the inputs and outputs.
- The concurrent assignments are used to describe the logic for the 'sum_out' and 'carry_out'
- Assignment executes for any change in the signal ('a_in' or 'b_in') on the right hand side of the assignments.
- Concurrent assignments are used to generate the parallel hardware and used to describe the small gate count dsigns

Example 1.3 Synthesizable RTL of VHDL code for half adder

model. For the combinational design, the design which results in the synthesizable netlist is treated as RTL description. For the sequential designs, the register-to-register timing path-based logic is treated as synthesizable RTL (Example 1.3).

Note RTL description or the data flow can be interchangeably used for the combinational logic designs. For the sequential logic description, the register-to-register (reg to reg) path results in the netlist and treated as register transfer level (RTL) description.

Although all above three representations generate the logic shown in Fig. 1.6, it is recommended to use the RTL design. RTL design is always synthesizable and uses all the synthesizable constructs. Many times for the small gate count (area) designs the data flow and RTL representations can be interchangeably used. Figure 1.6 describes the hardware inference for the half-adder using the XOR and AND logic gates.

1.7 Key VHDL Highlights and Constructs

VHDL has synthesizable and non-synthesizable constructs, the synthesizable constructs are used to describe the functionality of the design. This section discusses on the key VHDL highlights and frequently used VHDL constructs to describe the hardware.

1. VHDL is different from the software languages as it is used to describe the hardware. VHDL supports wide varieties of data types.
2. VHDL supports concurrent (parallel) execution of statements and even sequential execution of statements.
3. VHDL supports assignments to the signals and variables and these assignments will be discussed in the subsequent chapters.
4. VHDL supports the declaration of input, output, and bidirectional ports. VHDL supports file handling.
5. VHDL supports nine-valued logic using the STD_LOGIC.
6. VHDL supports the sequential execution of the statements inside the process block.

VHDL supports synthesizable constructs as well as non-synthesizable constructs. The template shown below describes key VHDL constructs used to describe most of the logic designs.

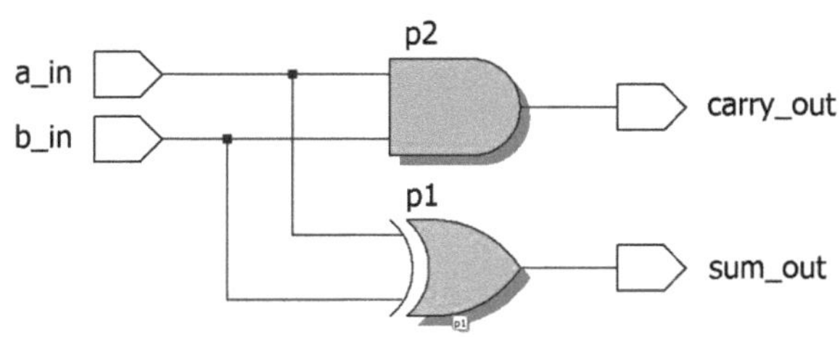

Fig. 1.6 Synthesized logic for half adder

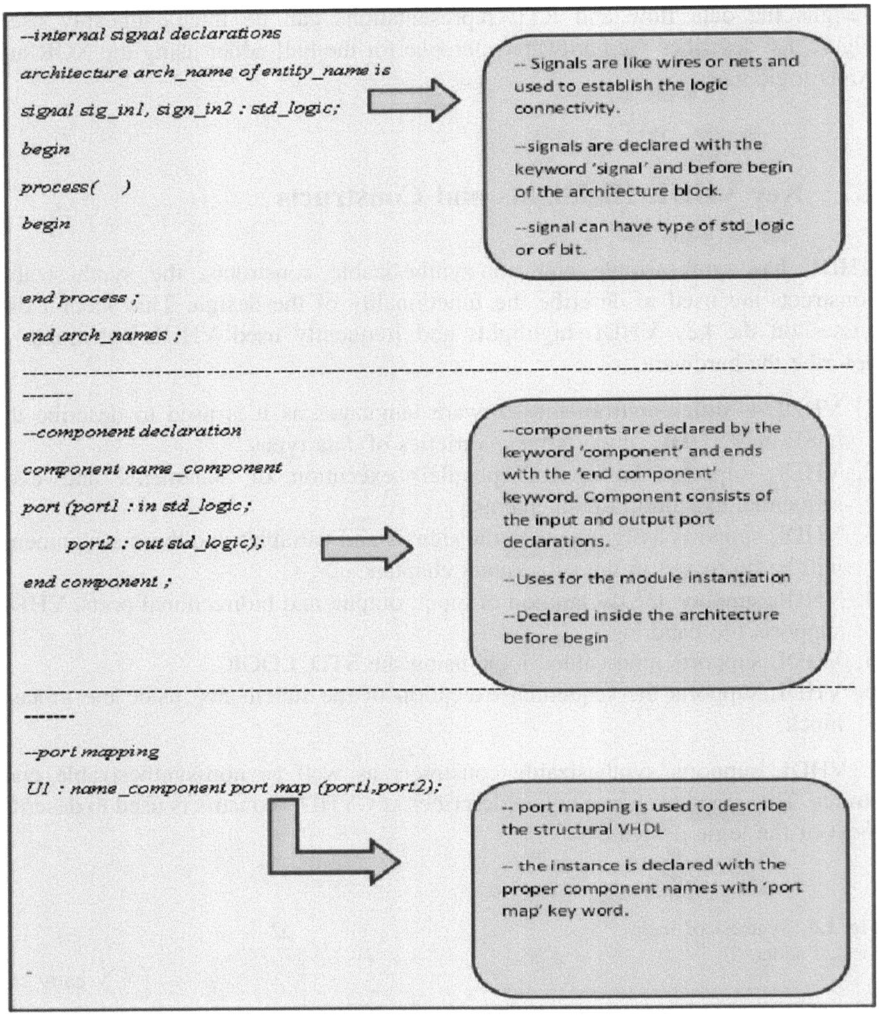
--internal signal declarations
architecture arch_name of entity_name is
signal sig_in1, sign_in2 : std_logic;
begin
process( )
begin
end process ;
end arch_names ;
-- Signals are like wires or nets and used to establish the logic connectivity.
--signals are declared with the keyword 'signal' and before begin of the architecture block.
--signal can have type of std_logic or of bit.
--component declaration
component name_component
port (port1 : in std_logic;
port2 : out std_logic);
end component ;
--components are declared by the keyword 'component' and ends with the 'end component' keyword. Component consists of the input and output port declarations.
--Uses for the module instantiation
--Declared inside the architecture before begin
--port mapping
U1 : name_component port map (port1,port2);
-- port mapping is used to describe the structural VHDL
-- the instance is declared with the proper component names with 'port map' key word.

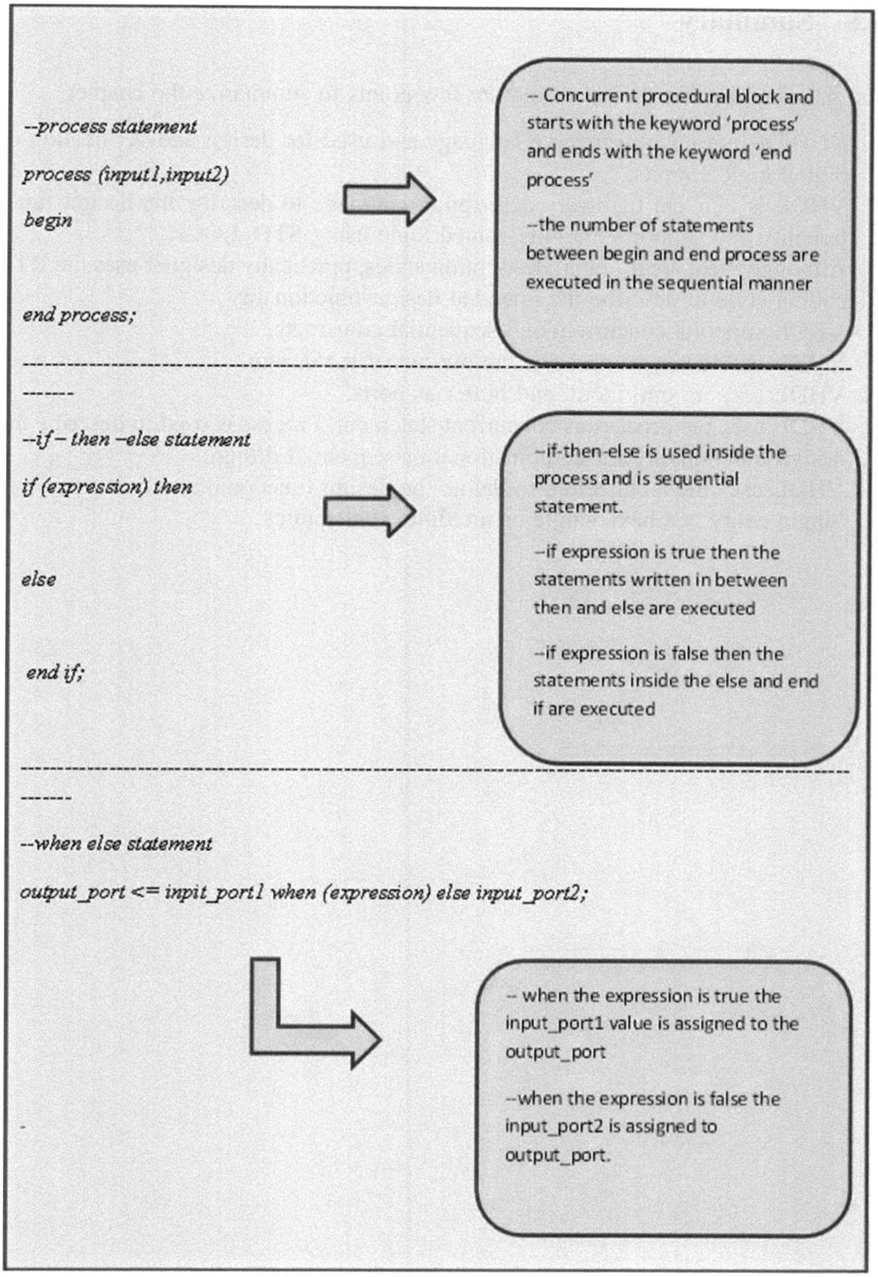
--process statement
process (input1,input2)
begin
end process;
-- Concurrent procedural block and starts with the keyword 'process' and ends with the keyword 'end process'
--the number of statements between begin and end process are executed in the sequential manner
--if – then –else statement
if (expression) then
else
end if;
--if-then-else is used inside the process and is sequential statement.
--if expression is true then the statements written in between then and else are executed
--if expression is false then the statements inside the else and end if are executed
--when else statement
output_port <= inpit_port1 when (expression) else input_port2;
-- when the expression is true the input_port1 value is assigned to the output_port
--when the expression is false the input_port2 is assigned to output_port.

1.8 Summary

As discussed earlier, the following are few points to summarize the chapter.

1. VHDL is not a case-sensitive language and used for design and verification of digital logic circuits.
2. VHDL is efficient hardware description language to describe the design functionality and supports the nine-valued logic using STD_LOGIC.
3. Although there are different description styles, practically designer uses the RTL coding style to describe the intended design functionality.
4. VHDL supports concurrent and sequential constructs.
5. VHDL uses entity to describe the pin-out of the design.
6. VHDL uses in, out, inout, and buffer as ports.
7. VHDL uses the process as concurrent statement. Process is used to describe the design functionality for combinational or sequential design.
8. VHDL uses the architecture to define the design functionality.
9. Single entity can have single or multiple architectures.

Chapter 2
Basic Logic Circuits and VHDL Description

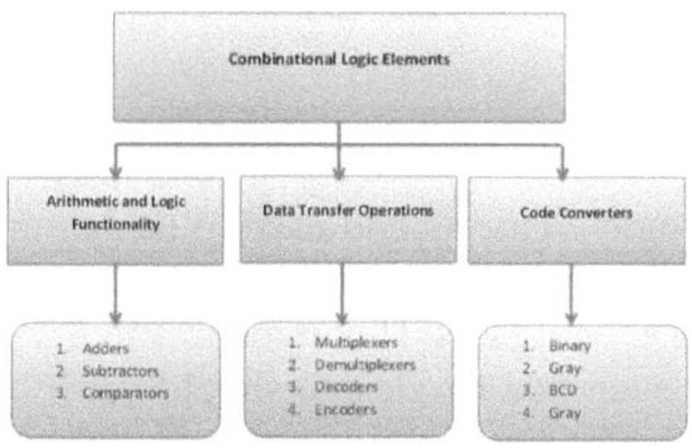

"We cannot solve our problems with the same thinking we used when we created them." ----- *Albert Einstein*

Like a C or C++ programmer don't apply the logic. Design the combinational logic by using the HDL

Learn the VHDL constructs and imagine the synthesizable designs and RTL designs using VHDL!

Abstract This chapter describes the overview of various combinational logic elements. The chapter is organized in such a way that reader will be able to understand the concept of synthesizable RTL for the logic gates and small gate count combinational designs using synthesizable VHDL constructs. This chapter describes the basic logic gates, adders, gray-to-binary and binary-to-gray code converters. This chapter also covers the key practical concepts while designing by using the combinational logic elements.

Keywords RTL · Synthesis · AND · NOT · OR · NOR · XOR · XNOR · Tri-state · Bus · Truth table · Combinational · Code converter · Adder · Gray · Binary · ALU · Critical path · Arithmetic operations · Logic minimization · Nine-valued logic · De Morgan's theorem

---

The original version of this chapter was revised: The incorrect information have been corrected. The erratum to this chapter is available at 10.1007/978-981-10-3296-7_12

V. Taraate, *PLD Based Design with VHDL*,
DOI 10.1007/978-981-10-3296-7_2

2.1 Introduction to Combinational Logic

Combinational logic is implemented by the logic gates, and in the combinational logic, output is the function of present input. The goal of designer is always to implement the logic using minimum number of logic gates or logic cells. Minimization techniques are K-map, Boolean algebra, Shannon's expansion theorems, and hyperplanes.

The conventional design technique using the Boolean algebra can be used for better understanding of the design functionality. The familiarity of the De Morgan's theorem and logic minimization technique can play an important role while coding for the design functionality. The De Morgan's theorem states that

1. Bubbled OR is equal to NAND.
2. Bubbled AND is equal to NOR.

The NOR and NAND gates are universal logic elements and used to design the digital circuit functionality. NOR and NAND can be used as universal logic cells. Figure 2.1 gives more explanation about the De Morgan's Theorem.

The thought process of designer should be such that the design should have the optimal performance with lesser area density. The area minimization techniques play an important role in the design of combinational logic or functions. In the present scenario, designs are very complex; the design functionality is described using the hardware description language as VHDL or Verilog. The subsequent section focuses on the use of VHDL RTL to describe the combinational design. Figure 2.2 illustrates the different types of combinational logic elements. This chapter discusses the basic combinational logic elements used to design the logic circuits.

The complex combinational logic circuits are discussed in the subsequent chapters.

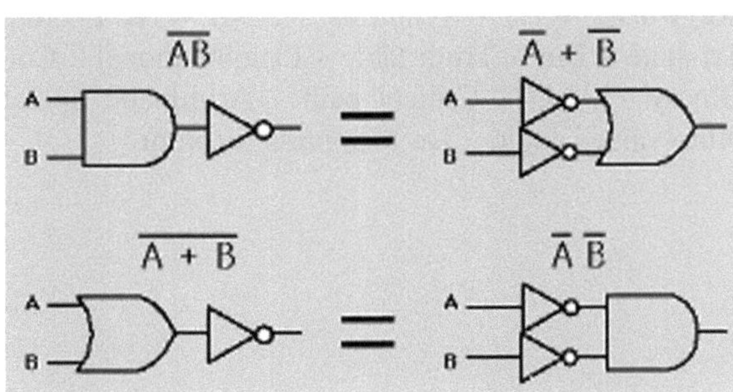

Fig. 2.1 De Morgan's theorems

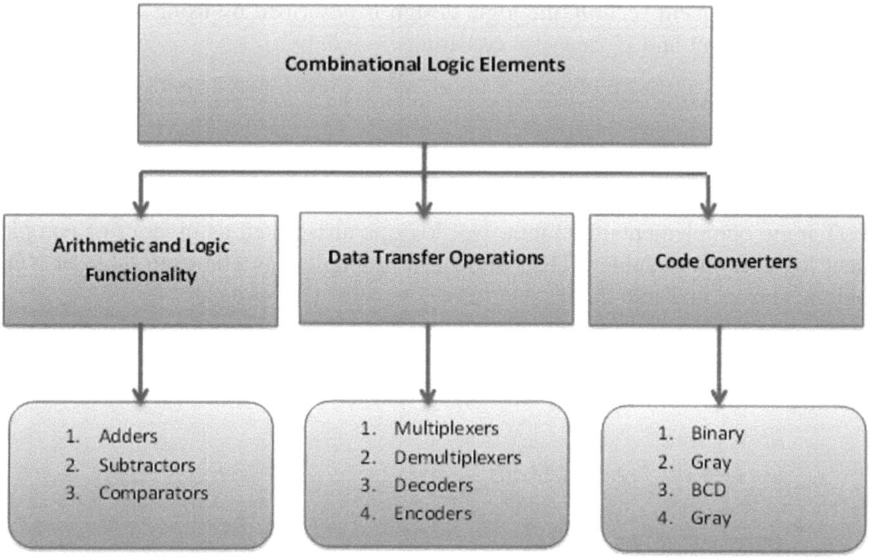

Fig. 2.2 Combinational logic elements

2.2 Logic Gates and Synthesizable RTL Using VHDL

This section discusses about the logic gates and the synthesizable VHDL RTL. In this section, the key VHDL constructs to describe the basic combinational logic gates are discussed.

To have a good understanding of VHDL, let us discuss on the key VHDL terminologies used to describe the combinational logic. Let us make our life simpler by understanding the logical operators as shown in Table 2.1.

Table 2.1 Logical operators

Logical operators	Operator description	VHDL description
NOT/not	Used as negation or to complement the input or signal	y_out <= NOT(a_in);
OR/or	To perform logical OR operation	y_out <= a_in OR b_in;
NOR/nor	To perform logical NOR operation	y_out <= a_in NOR b_in;
AND/and	To perform logical AND operation	y_out <= a_in AND b_in;
NAND/nand	To perform logical NAND operation	y_out <= a_in NAND b_in;
XOR/xor	To perform logical XOR operation	y_out <= a_in XOR b_in;
XNOR/xnor	To perform logical XNOR operation	y_out <= a_in XNOR b_in;

In the subsequent section, the logic design is described by using the concurrent process statement and if-then-else constructs.

2.2.1 NOT or Invert Logic

NOT logic complements the input. Not logic is also called as inverter or complement logic. Synthesizable RTL is shown in Example 2.1. The truth table of NOT logic is shown in Table 2.2.

```
-- invert or not logic using if-then-else

        library ieee;
        use ieee.std_logic_1164.all;

        entity not_logic_gate is
        port( a_in: in std_logic;
                    y_out: out std_logic
          );
        end not_logic_gate;

        architecture arch_not of not_logic_gate is
        begin

         process (a_in)
         begin
          if (a_in='0') then
                y_out<= '1';
                else
                y_out<= '0';
                end if;
                end process;
end arch_not;
```

> ➤ For the input 'a_in' is equl to zero the output 'y_out' is equal to logical one
> ➤ For the input 'a_in' is equl to one the output 'y_out' is equal to logical zero

Example 2.1 Synthesizable VHDL code for NOT logic

Note Operator (<=) is used for the port or signal assignment. On the other hand, concurrent construct 'process' is used to infer both combinational and sequential logic by using the required VHDL constructs.

Table 2.2 Truth table for NOT logic

a_in	y_out
0	1
1	0

The use of the STD_LOGIC is nine-valued logic, and for the NOT function it is described below.

Input	U	X	0	1	Z	W	L	H	–
Output	U	X	1	0	X	X	1	0	X

Synthesis result for the NOT logic is shown in Fig. 2.3; input port of not logic gate is named as 'a_in' and output as 'y_out'.

The implementation of NOT using universal NAND and NOR logic gate is shown in Fig. 2.4.

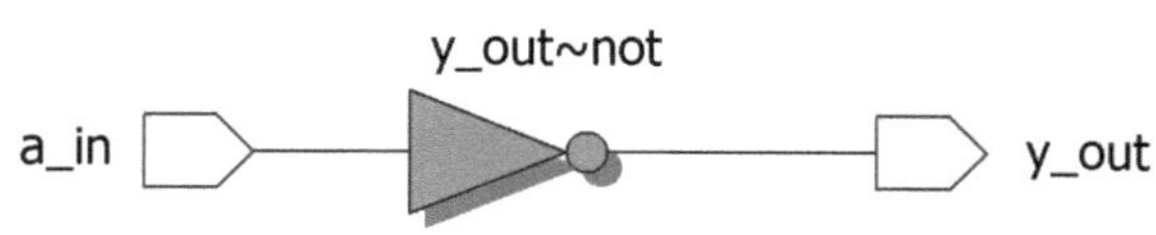

Fig. 2.3 Synthesis result for the NOT logic

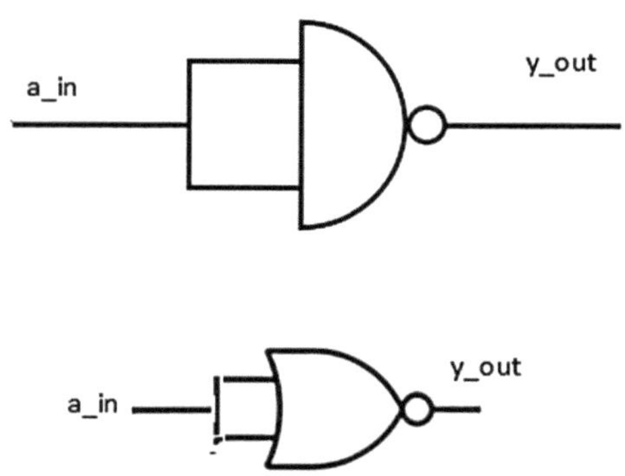

Fig. 2.4 NOT gate implementation using universal logic gates

2.2.2 Two-Input OR Logic

OR logic generates output as logical '1' when one of the inputs is logical '1'.

Synthesis result is shown in Example 2.2. The truth table of OR logic is shown in Table 2.3.

```
-- VHDL RTL for two input OR
library ieee;
use ieee.std_logic_1164.all;
entity or_logic_gate is
      port( a_in: in std_logic;
            b_in: in std_logic;
            y_out: out std_logic
            );
end or_logic_gate;

architecture arch_or of or_logic_gate is
begin
   process(a_in, b_in)
   begin
    if ((a_in='0') and  (b_in='0')) then
            y_out <= '0';
    else
            y_out <= '1';
   end if;
end process;
end arch_or;
```

- Entity is pin out of the design and named as or_logic_gate and has three ports
- Input ports are declared as 'a_in', 'b_in'.
- Output port is declared as 'y_out'

- Architecture of the design is named as arch_or
- Architecture describes the functionality of the design.
- The process is sensitive to inputs specified in the sensitivity list.
- For changes on the 'a-in' , 'b_in' the process gets invoked.
- For the inputs a_in='0' and b_in='0' the output y_out='0' otherwise output y_out='1'
- For the VHDL description using data flow replace the process block by using y_out<= a_in OR b_in;

Example 2.2 Synthesizable VHDL code for two-input OR logic

Note While describing the design functionality, make sure that all the input ports are listed in the sensitivity list. Missing required signal from sensitivity list will create simulation and synthesis mismatch and will be discussed in the subsequent chapters.

Table 2.3 Truth table for two-input OR logic

a_in	b_in	y_out
0	0	0
0	1	1
1	0	1
1	1	1

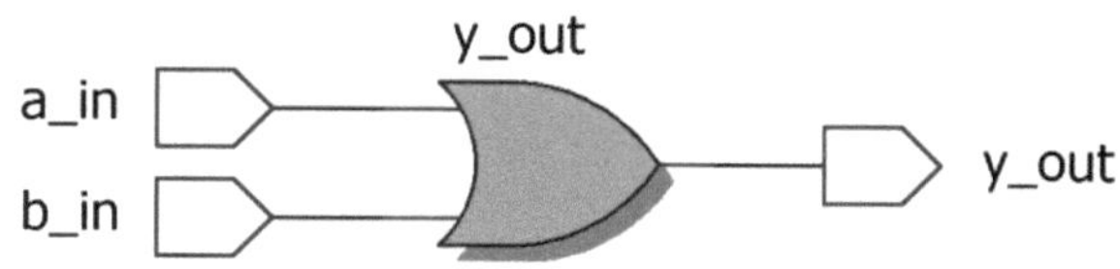

Fig. 2.5 Synthesis result for two-input OR logic

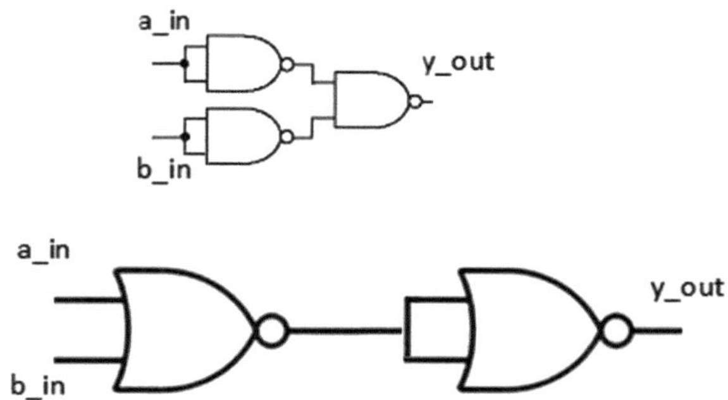

Fig. 2.6 OR gate implementation using universal gates

Synthesis result for the OR logic is shown in Fig. 2.5, input ports of OR logic gate are named as 'a_in' and 'b_in' and output as 'y_out'.

Using universal logic gates the implementation of OR gate is shown in Fig. 2.6.

2.2.3 *Two-Input NOR Logic*

NOR logic is opposite or complement of the OR logic. Synthesizable RTL is shown in Example 2.3. The truth table of NOR logic is shown in Table 2.4.

Synthesis result for the NOR logic is shown in Fig. 2.7; input ports of NOR logic gates are named as 'a_in' and 'b_in' and output as 'y_out'.

Two-input NOR gate implementation using universal logic gates is shown in Fig. 2.8.

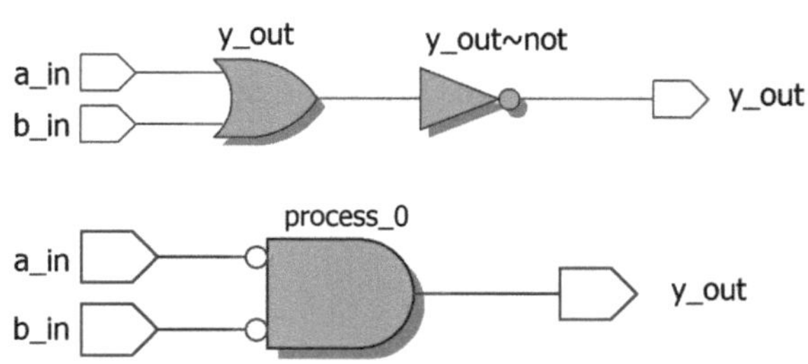

Fig. 2.7 Synthesized two-input NOR logic

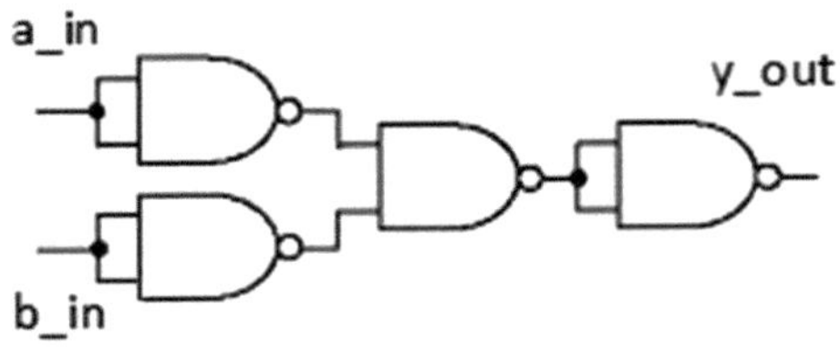

Fig. 2.8 NOR gate implementation using universal gates

```
-- VHDL RTL for two input NOR
library ieee;
use ieee.std_logic_1164.all;
entity nor_logic_gate is
      port ( a_in: in std_logic;
            b_in: in std_logic;
            y_out: out std_logic
             );
end  nor_logic_gate;
```

- Entity is pin out of the design and named as nor_logic_gate and has three ports
- Input ports are declared as 'a_in', 'b_in'.
- Output port is declared as 'y_out'

```
architecture arch_nor of nor_logic_gate is
begin
   process(a_in, b_in)
   begin
    if ((a_in='0') and (b_in='0')) then
             y_out <= '1';
    else
             y_out <= '0';
   end if;
end process;
end arch_nor;
```

- Architecture of the design is named as arch_nor
- Architecture describes the functionality of the design.
- The process is sensitive to inputs specified in the sensitivity list.
- For changes on the 'a-in' , 'b_in' the process gets invoked.
- For the inputs a_in='0' and b_in='0' the output y_out='1' otherwise output y_out='0'
- For the VHDL description using data flow replace the process block by using y_out<= a_in NOR b_in;

Example 2.3 Synthesizable VHDL code for NOR logic

Table 2.4 Truth table for two-input NOR logic

a_in	b_in	y_out
0	0	1
0	1	0
1	0	0
1	1	0

2.2.4 *Two-Input AND Logic*

AND logic generates an output as logical '1' when both the inputs 'a_in' and 'b_in' are logical '1'. Synthesizable RTL is shown in Example 2.4. The truth table of AND logic is shown in Table 2.5.

```
-- VHDL RTL for two input AND
library ieee;

use ieee.std_logic_1164.all;
entity and_logic_gate is
    port ( a_in: in std_logic;
           b_in: in std_logic;
           y_out: out std_logic
         );
end  and_logic_gate;

architecture arch_and  of and_logic_gate is
begin
  process(a_in, b_in)
  begin
   if ((a_in='1') and (b_in='1')) then
          y_out <= '1';
   else
          y_out <= '0';
  end if;
end process;
end arch_and;
```

- Entity is pin out of the design and named as and_logic_gate and has three ports
- Input ports are declared as 'a_in', 'b_in'.
- Output port is declared as 'y_out'

- Architecture of the design is named as arch_and
- Architecture describes the functionality of the design.
- The process is sensitive to inputs specified in the sensitivity list.
- For changes on the 'a-in' , 'b_in' the process gets invoked.
- For the inputs a_in='1' and b_in='1' the output y_out='1' otherwise output y_out='0'
- For the VHDL description using data flow replace the process block by using y_out<= a_in AND b_in;

Example 2.4 Synthesizable VHDL code for AND logic

Note AND gate is visualized as a series of two switches and used in programmable logic devices (PLD) as one of the elements to realize the required logic. Programmable AND plane can be created by using the AND logic gates with programmable inputs.

Table 2.5 Truth table for two-input AND logic

a_in	b_in	y_out
0	0	0
0	1	0
1	0	0
1	1	1

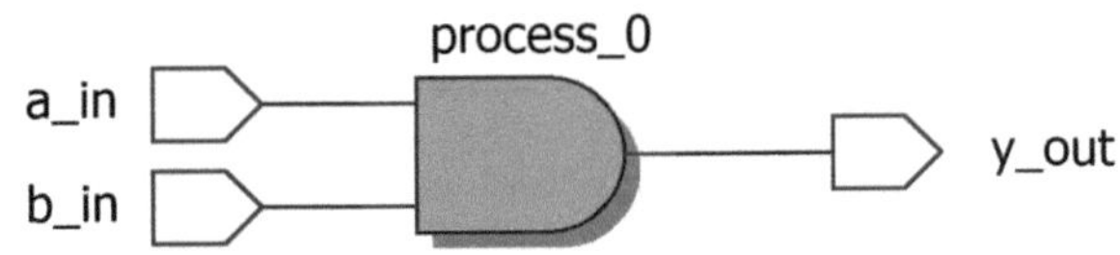

Fig. 2.9 Synthesis result for two-input AND logic

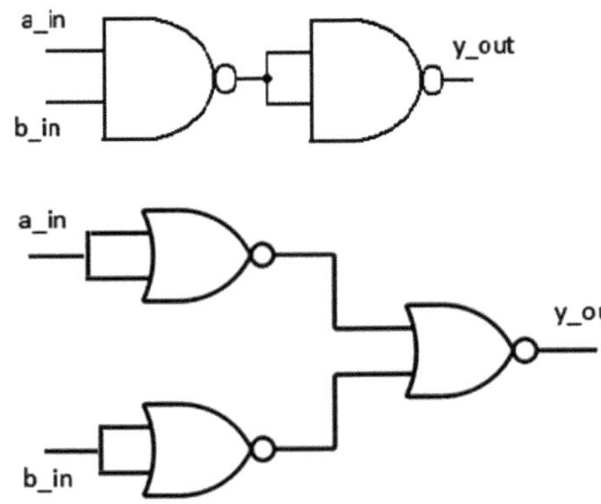

Fig. 2.10 AND gate implementation using universal gates

Synthesized two-input AND logic is shown in Fig. 2.9; input ports of AND logic gate are named as 'a_in' and 'b_in' and output as 'y_out'.

Two-input AND gate implementation using minimum number of universal gates is shown in Fig. 2.10.

2.2.5 Two-Input NAND Logic

NAND logic is opposite or complement of the AND logic. Synthesizable RTL is shown in Example 2.5. The truth table of NAND logic is shown in Table 2.6

Table 2.6 Truth table for two-input NAND logic

a_in	b_in	y_out
0	0	1
0	1	1
1	0	1
1	1	0

```
-- VHDL RTL for two input NAND
library ieee;

use ieee.std_logic_1164.all;
entity nand_logic_gate is
    port ( a_in: in std_logic;
           b_in: in std_logic;
           y_out: out std_logic
         );
end nand_logic_gate;


architecture arch_nand of nand_logic_gate is
begin
  process(a_in, b_in)
  begin
   if ((a_in='1') and (b_in='1')) then
         y_out <= '0';
   else
         y_out <= '1';
  end if;
end process;
end arch_nand;
```

- Entity is pin out of the design and named as nand_logic_gate and has three ports
- Input ports are declared as 'a_in', 'b_in'.
- Output port is declared as 'y_out'

- Architecture of the design is named as arch_nand
- Architecture describes the functionality of the design.
- The process is sensitive to inputs specified in the sensitivity list.
- For changes on the 'a-in' , 'b_in' the process gets invoked.
- For the inputs a_in='1' and b_in='1' the output y_out='0' otherwise output y_out='1'
- For the VHDL description using data flow replace the process block by using y_out<= a_in NAND b_in;

Example 2.5 Synthesizable VHDL RTL for two-input NAND logic

Note NAND logic is also treated as universal logic. By using NAND logic, all possible logic functions can be realized. NAND logic is used to implement the storage elements like latches or flip-flops and also to realize combinational functions.

Synthesis result for the NAND logic is shown in Fig. 2.11; input ports of NAND logic gate are named as 'a_in' and 'b_in' and output as 'y_out'.

As stated earlier, NAND is universal gate, and implementation of NAND using NOR is shown in Fig. 2.12.

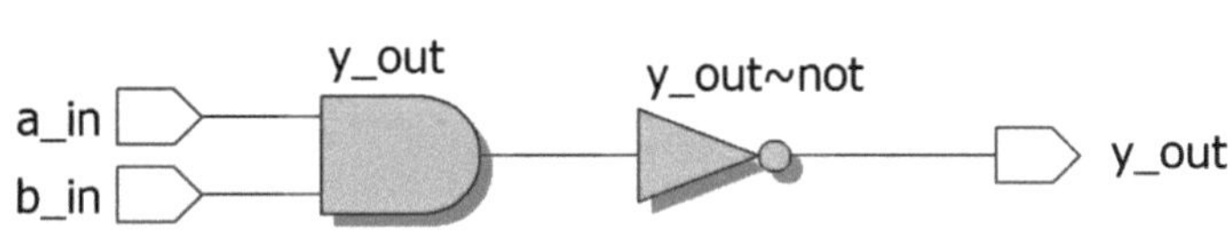

Fig. 2.11 Synthesis result for the result for the two-input NAND logic

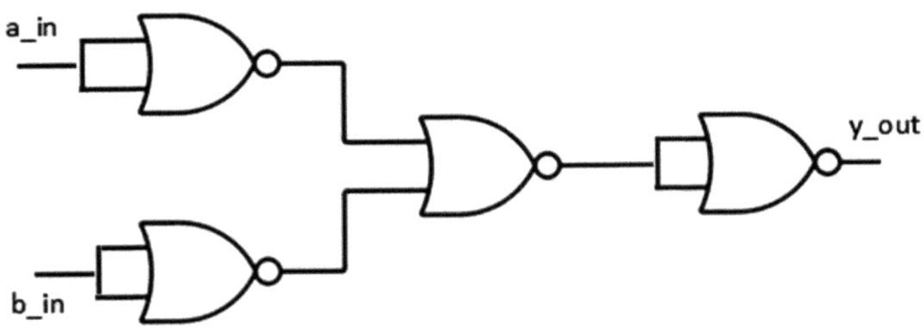

Fig. 2.12 Implementation of NAND using universal gates

2.2.6 Two-Input XOR Logic

Two-input XOR is called as exclusive OR logic and generates output as logical '1' when both inputs are not equal. Synthesizable RTL is shown in Example 2.6. The truth table of XOR logic is shown in Table 2.7.

```
-- VHDL RTL for two input XOR

library ieee;
use ieee.std_logic_1164.all;

entity xor_logic_gate is
port( a_in: in std_logic;
          b_in: in std_logic;
          y_out: out std_logic
);
end xor_logic_gate;

architecture arch_xor of xor_logic_gate is
begin

   process(a_in, b_in)
   begin

    if (a_in/=b_in) then
             y_out <= '1';
    else
             y_out <= '0';
   end if;

end process;

end arch_xor;
```

- Entity is the pin out of the design named as xor_logic_gate that has three ports.
- Input ports are declared as 'a_in' and 'b_in'
- Output port is declared as 'y_out'

- Architecture of the xor_logic_gate is named as arch_xor
- Architecture describes the relationship between the inputs and output.
- The procedural block 'Process' is always sensitive to the changes specified in the sensitivity list.
- For any changes in the inputs 'a_in' or b_in the process gets invoked.
- For the inputs, a_in is not equal to b_in output y_out = '1' otherwise output y_out = '0'
- For data flow model replace process block by using *y_out <= a_in xor b_in;*

Example 2.6 Synthesizable VHDL code for two-input XOR logic

Note XOR gate can be implemented by using two-input NAND gates. The number of two-input NAND gates required to implement two-input XOR gate are equal to 4. XOR gates are used to implement arithmetic operations like addition and subtraction. The implementation using minimum number of NAND gates is shown in Fig. 2.13.

Table 2.7 Truth table for two-input XOR logic

a_in	b_in	y_out
0	0	0
0	1	1
1	0	1
1	1	0

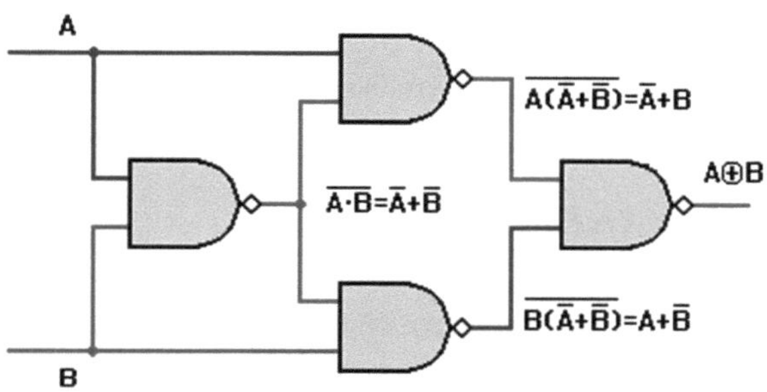

Fig. 2.13 XOR gate implementation using NAND

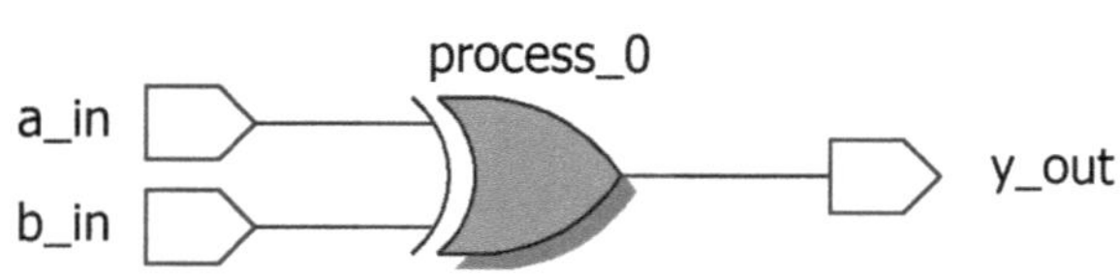

Fig. 2.14 Synthesis result for the two-input XOR logic

Synthesis result for the two-input XOR logic is shown in Fig. 2.14; input ports of XOR logic gate are named as 'a_in' and 'b_in' and output as 'y_out'.

If XOR cell or gate is not available in the library, then XOR logic is realized using AND-OR-Invert or by using minimum number of NAND gates.

XOR gate implementation using the universal gates is shown in Fig. 2.15.

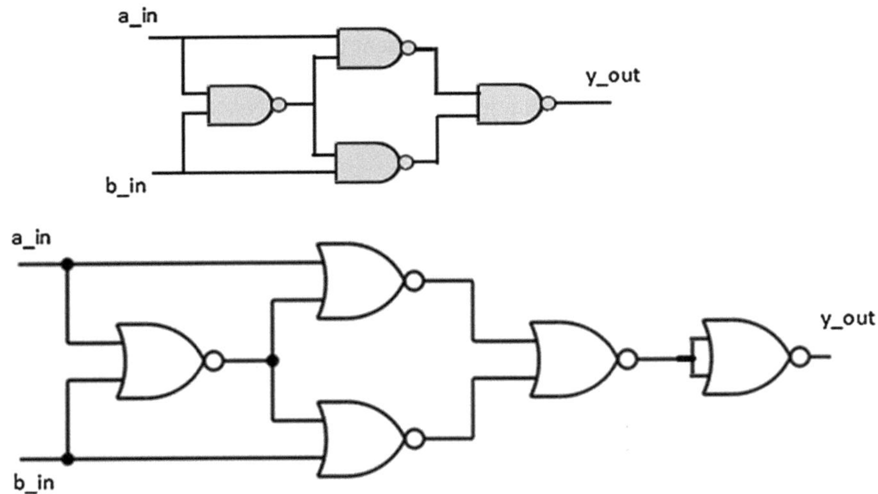

Fig. 2.15 XOR implementation using universal logic gates

2.2.7 Two-Input XNOR Logic

Two-input XNOR is called as exclusive NOR logic and generates output as logical '1' when two inputs are equal. XNOR is opposite or complement of XOR logic. Synthesizable RTL for XNOR is shown in Example 2.7. The truth table of XNOR logic is shown in Table 2.8.

Synthesis result for the XNOR logic is shown in Fig. 2.16; input ports of XNOR logic gate are named as 'a_in' and 'b_in' and output as 'y_out'.

If XNOR cell is not available in the library, then XNOR logic is realized by using Invert-AND-OR or by using minimum number of NAND or NOR gates.

The implementation of XNOR logic using universal gates is shown in Fig. 2.17.

In the practical scenario, the XOR and XNOR gates are used in the parity detection to detect for the even or odd parity. The subsequent chapter focuses on the complex designs and the synthesis. The even parity detector is shown in Fig. 2.18 and uses the XOR and XNOR gates to generate active high value at the output for even number of 1's in the input.

The odd parity checker to detect for odd number of 1's in the string is shown in Fig. 2.19. For odd number of 1's, it generates the active high output. As shown in figure, it uses three XOR gates.

```
-- VHDL RTL for two input XNOR

library ieee;
use ieee.std_logic_1164.all;

entity xnor_logic_gate is
 port( a_in: in std_logic;
          b_in: in std_logic;
          y_out: out std_logic
);
end xnor_logic_gate;

architecture arch_xnor of xnor_logic_gate is
begin

   process(a_in, b_in)
   begin

    if (a_in=b_in) then
            y_out <= '1';
    else
            y_out <= '0';
   end if;

end process;

end arch_xnor;
```

- Entity is the pin out of the design named as xnor_logic_gate that has three ports.
- Input ports are declared as 'a_in' and 'b_in'
- Output port is declared as 'y_out'

- Architecture of the xnor_logic_gate is named as arch_xnor
- Architecture describes the relationship between the inputs and output.
- The procedural block 'Process' is always sensitive to the changes specified in the sensitivity list.
- For any changes in the inputs 'a_in' or b_in the process gets invoked.
- For the inputs, a_in is equal to b_in output y_out= '1' otherwise output y_out = '0'
- For data flow model replace process block by using *y_out <= a_in xnor b_in;*

Example 2.7 Synthesizable VHDL code for XNOR logic

Table 2.8 Truth table for XNOR logic

a_in	b_in	y_out
0	0	1
0	1	0
1	0	0
1	1	1

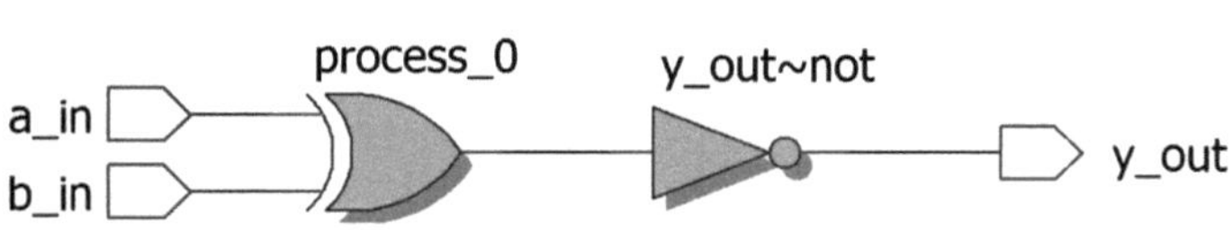

Fig. 2.16 Synthesis result for the XNOR logic

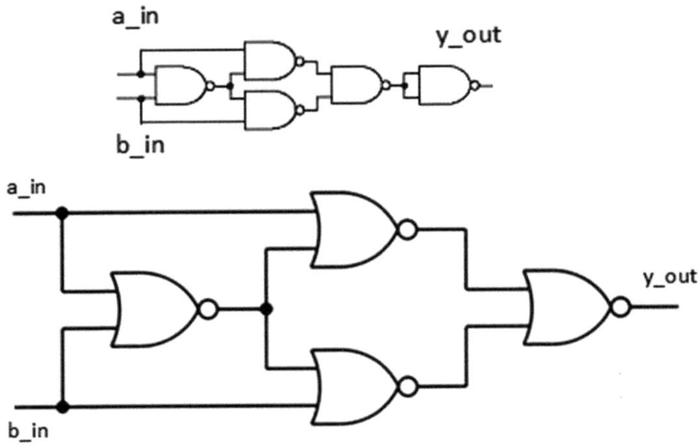

Fig. 2.17 XNOR implementation using universal logic gates

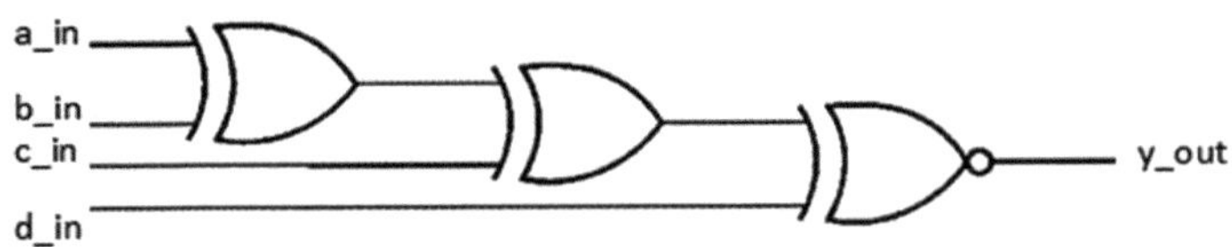

Fig. 2.18 Even-parity checker

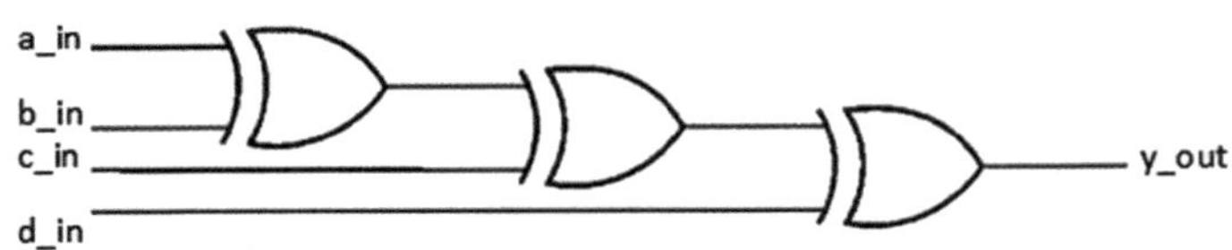

Fig. 2.19 Odd-parity checker

2.2.8 *Tri-State Logic*

Tri state has three logic states: logical '0', logical '1', and high impedance 'z'. Synthesizable RTL is shown in Example 2.8. The truth table of tri-state logic is shown in Table 2.9.

```
--Tri state logic

library ieee;
use ieee.std_logic_1164.all;
entity tri_state_bus is
port( data_in:      in std_logic_vector(3 downto 0);
      enable:       in std_logic;
      data_out:     out std_logic_vector(3 downto 0)
);
end tri_state_bus;
```

- Std_logic_vector is used to generate the bus.
- Bus width is defined by using '3 downto 0' and it is 4 bit wide.

```
architecture arch_tri_state_bus oftri_state_bus is
begin
 process(data_in, enable)
 begin
      if (enable='1') then
        data_out <= data_in;
      else
        data_out <= "ZZZZ";
      end if;
 end process;

end arch_tri_state_bus;
```

- The tri state logic functionality is described by using the 'if then else construct'
- Tri state logic generates the output 'data_out' = 'data_in' for the 'enable' equal to one.
- For the 'enable' equal to zero the output of the tri-state logic is forced to be zero.

Example 2.8 Synthesizable VHDL code for tri-state bus logic

Note Avoid use of tri-state logic while developing the RTL. Tri state is difficult to test. Instead of tri-state logic, it is recommended to use multiplexers to develop the logic with enable.

Table 2.9 Truth table for tri-state logic

Enable	data_in	data_out
1	0000	0000
1	1111	1111
0	xxxx	zzzz

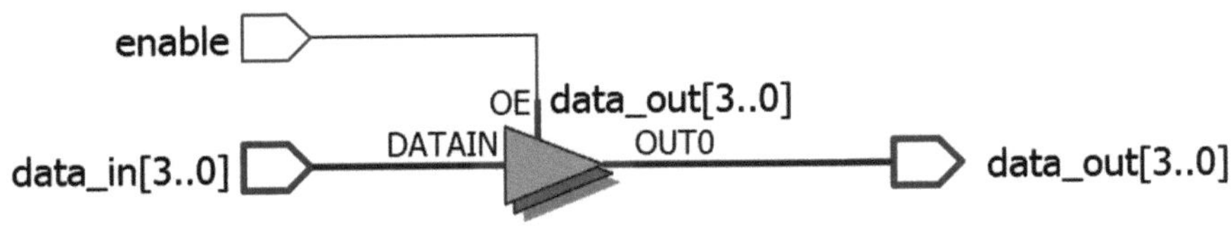

Fig. 2.20 Synthesis result for the tri-state logic

Synthesis result for the tri-state logic is shown in Fig. 2.20; input port of tri-state logic is named as 'data_in', enable input as 'enable', and output as 'data_out'.

2.3 Adder

Arithmetic operations like addition and subtraction play an important role in the efficient design of processor logic. Arithmetic and Logical Unit (ALU) of any processor is designed to perform the addition, subtraction, increment, and decrement operations. The arithmetic designs to be described by the RTL VHDL code to achieve the optimal area and to have less critical path. This section describes the important logic blocks to perform arithmetic operations with the synthesizable VHDL RTL description.

Adders are used to perform the binary addition of two binary numbers. Adders are used for signed or unsigned addition operations.

2.3.1 Half Adder

Half adder has two one-bit inputs 'a_in', 'b-in' and generates two one-bit outputs 'sum_out' and 'carry_out', where 'sum_out' is summation or addition output and 'carry_out' is carry output. Table 2.10 is the truth table for half adder, and RTL is described in Example 2.9.

Table 2.10 Truth table for half adder

a_in	b_in	sum_out	carry_out
0	0	0	0
0	1	1	0
1	0	1	0
1	1	0	1

```
--half adder using the logical operators.

library ieee;

use ieee.std_logic_1164.all;

entity half_adder is
port( a_in: in std_logic;
        b_in: in std_logic;
        sum_out: out std_logic;
        carry_out: out std_logic
);
end half_adder;

architecture arch_half_adder of half_adder is
 begin

        sum_out <= a_in xor b_in;
        carry_out <= a_in and b_in;

end arch_half_adder;
```

- The half adder is described by using the XOR and AND logic gates.
- The XOR and AND are logic operators and generates the required 'sum_out' and 'carry_out' outputs.

Example 2.9 Synthesizable RTL code for half adder

Note Half adders are used as basic component to perform the addition. Full adder logic circuits are designed using the instantiation of half adders as components.

Synthesis result for the half adder is shown in Fig. 2.21; input ports of half adder are named as 'a_in' and 'b_in' and output as 'sum_out', 'carry_out'.

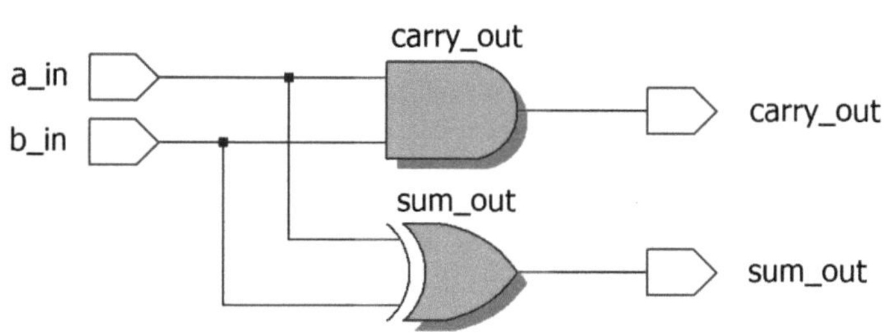

Fig. 2.21 Synthesis result for the half adder

2.3.2 *Full Adder*

Full adders are used to perform addition on three one-bit binary inputs.

Consider three, one-bit binary numbers named as 'a_in', 'b_in', 'c_in' and one-bit binary outputs as 'sum_out', 'carry_out'. Table 2.11 is the truth table for full adder and RTL is described in Example 2.10.

```
--full adder using the logical operators.

library ieee;
use ieee.std_logic_1164.all;

entity full_ader is
port( a_in: in std_logic;
      b_in: in std_logic;
      c_in : in std_logic;
      sum_out: out std_logic;
      carry_out: out std_logic
);
end full_adder;

architecture arch_full_adder of full_adder is

signal wire1,wire2,wire3 : std_logic;

begin
        wire1 <= a_in xor b_in;
        wire2 <= a_in and b_in;
        sum_out <= c_in xor wire1;
        wire3 <= wire1 and c_in;
        carry_out <= wire2 or wire3;

end arch_full_adder;
```

- The full adder is described by using the XOR and AND logic gates.
- The XOR and AND are logic operators and generates the required 'sum_out' and 'carry_out' outputs.
- Signals are used as data objects to establish the required connectivity.

Example 2.10 Synthesizable VHDL code for full adder

Note Full adder consumes more area, so it is highly recommended to implement the adder logic using multiplexers. Subtraction can be performed by using 2's complement addition.

Table 2.11 Truth table for full adder

c_in	a_in	b_in	sum_out	carry_out
0	0	0	0	0
0	0	1	1	0
0	1	0	1	0
0	1	1	0	1
1	0	0	1	0
1	0	1	0	1
1	1	0	0	1
1	1	1	1	1

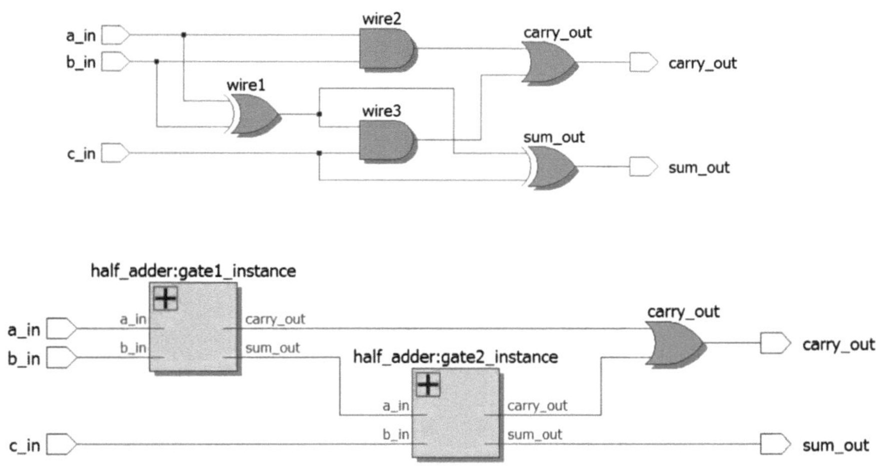

Fig. 2.22 Synthesized full adder

Synthesis result for the full adder is shown in Fig. 2.22; input ports of full adder are named as 'a_in', 'b_in', and 'c_in' and output as 'sum_out', 'carry_out'.

In the practical design scenarios, the multiplexers (MUX) can be used to implement the addition and subtraction operations. MUX is universal logic and discussed in the Chap. 4. The realization of the full adder is shown in the following figure. As shown in Fig. 2.23 by using 2:1 MUX, the logic is realized. The concept of using MUX to realize Boolean functions or logic gates is important to understand the PLD-based designs. Readers are encouraged to implement the logic of all the basic combinational elements using minimum number of 2:1 MUX.

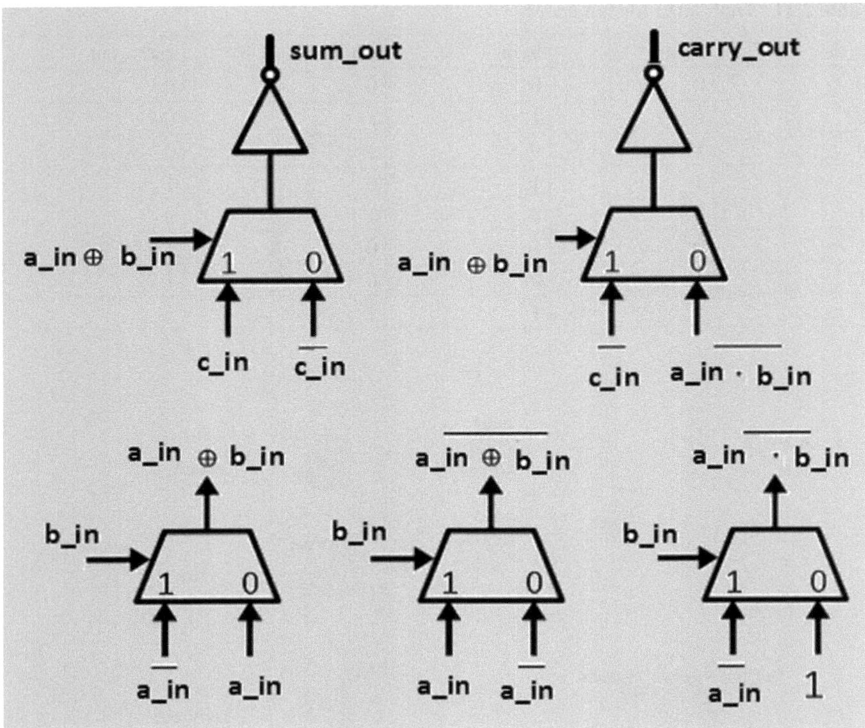

Fig. 2.23 Realization of full adder using MUX

2.4 Code Converters

This section deals with the commonly used code converters in the design. As name itself indicates, the code converters are used to convert the code from one number system to another number system. In the practical scenarios, binary-to-gray and gray-to-binary converters are used.

2.4.1 Binary-to-Gray Code Converter

Base of binary number system is 2, for any multibit binary number one or more than one bit changes at a time. In gray code, only one bit changes at a time. if we compare two successive gray codes.

The RTL description of 4-bit binary-to-gray code conversion is described in Example 2.11.

```
--Binary to Gray code converter
library ieee;
use ieee.std_logic_1164.all;
entity binary_to_gray is
port( data_in:      in std_logic_vector(3 downto 0);
      data_out: out std_logic_vector(3 downto 0)
  );
end binary_to_gray;
architecture arch_binary_to_gray of binary_to_gray is
begin
  data_out(3)<=data_in(3);
  data_out(2)<=data_in(3) xor data_in(2);
  data_out(1)<=data_in(2) xor data_in(1);
  data_out(0)<=data_in(1) xor data_in(0);
end arch_binary_to_gray;
```

- The 4 bit binary to gray converter is described by using the XOR logical operator
- 'data_in' is 4 bit binary input to the code converter
- 'data_out' is the 4 bit gray output from the code converter

Example 2.11 Synthesizable VHDL code for 4-bit binary-to-gray code converter

Note Gray codes are used in the multiple clock domain designs to transfer the control information from one of the clock domains to another clock domain.

Synthesis result for the binary to gray code converter is shown in Fig. 2.24.

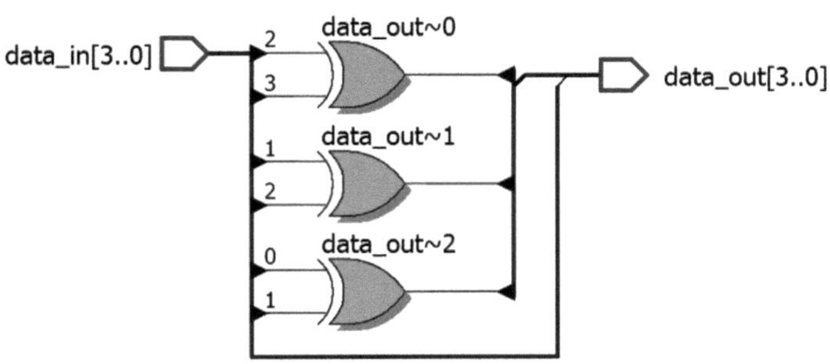

Fig. 2.24 Synthesis result for the 4-bit binary-to-gray converter

2.4.2 Gray-to-Binary Code Converter

Gray-to-binary code converter is reverse of that of binary-to-gray, and the RTL description of 4-bit gray-to-binary code conversion is described in Example 2.12.

```
-- Gray to binary code converter

library ieee;
use ieee.std_logic_1164.all;

entity gray_to_binary is
port( data_in:    in std_logic_vector(3 downto 0);
      data_out: inout std_logic_vector(3 downto 0)
  );
end gray_to_binary;

architecture arch_gray_to_binary of gray_to_binary is
begin

 data_out(3)<=data_in(3);
 data_out(2)<=data_out(3) xor data_in(2);
 data_out(1)<=data_out(2) xor data_in(1);
 data_out(0)<=data_out(1) xor data_in(0);

end arch_gray_to_binary;
```

- The 4 bit gray to binary converter is described by using the XOR logical operator
- 'data_in' is 4 bit Gray input to the code converter
- 'data_out' is the 4 bit Binary output from the code converter

Example 2.12 Synthesizable VHDL code for 4-bit gray-to-binary code converter

Note Gray codes are used in the gray counter implementation and also in the error correcting mechanism.

Synthesis result for the 4-bit gray to binary code converter is shown in Fig. 2.25.

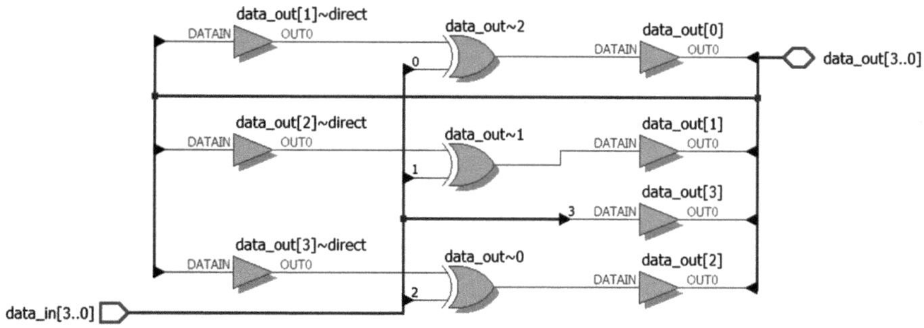

Fig. 2.25 Synthesis result for the 4-bit gray-to-binary converter

2.5 Summary

As discussed already in this chapter, following are important points need to be considered while implementing combinational logic design using VHDL.

1. Use minimum area by using least number of logic gates;
2. NAND and NOR are universal logic gates and used to implement any combinational or sequential logic;
3. Use all the required signals in the sensitivity to avoid simulation and synthesis mismatch;
4. Avoid the usage of tri-state logic and implement the logic required using multiplexers with proper enable circuit.
5. Use less number of adders in design. Adders can be implemented using multiplexers;
6. Subtraction can be implemented using 2's complement addition;
7. MUX can be used as universal logic to realize logic functions;
8. Parity can be checked by using the proper cascading of XOR and XNOR gates;
9. Gray codes are unique cyclic codes and can be used as error correcting codes.

Chapter 3
VHDL and Key Important Constructs

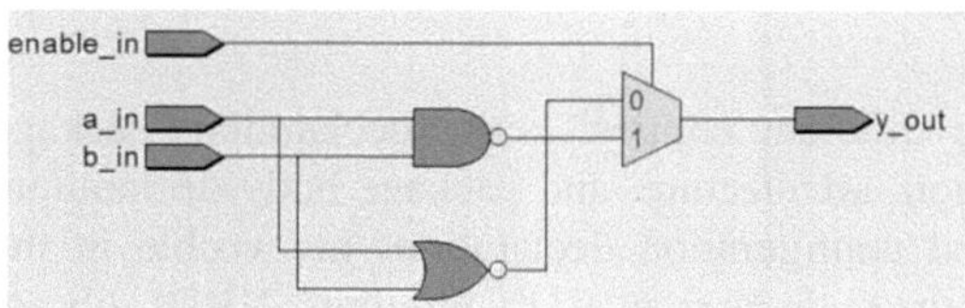

"Logic will get you from A to B. Imagination will take you everywhere." — **Albert Einstein**

To write an efficient RTL using VHDL, it is essential to understand about the VHDL constructs.

VHDL has both concurrent and sequential constructs. So let us understand the VHDL constructs.

Abstract This chapter discusses the key important VHDL constructs. VHDL a is hardware description language and consists of many powerful concurrent and sequential constructs. The key concurrent and sequential constructs are used to describe the design functionality to generate intended hardware. These constructs include process, when else, with select, if then else case, signal and variable declarations and assignments. Even this chapter discusses the important constructs like wait, wait on, wait for, wait until, for loop, and while loop. This chapter is useful for RTL design engineers to understand the VHDL coding styles and synthesizable VHDL. This chapter covers the practical illustrations for every construct. The explanation is given for every synthesizable VHDL code with the synthesis results. This can be useful while working in the FPGA as well as ASIC design domains.

Keywords Concurrent · Sequential · RTL · Assignment · When else · With select · If then else · Case · Process · Nested statements · While · For · Loop · Signal · Variable · Tri-state · MUX · Flip-flop · Latch · Constructs · Architecture · Configuration · Multiple processes · Multiple architectures

As discussed in previous chapters, every VHDL code has one entity and at least one architecture. The major focus of this chapter is to get familiar with the important VHDL constructs. The main objective of design engineer is to write an efficient RTL using the suitable VHDL constructs. Most of the time, due to use of inappropriate

V. Taraate, *PLD Based Design with VHDL*,
DOI 10.1007/978-981-10-3296-7_3

VHDL constructs, it yields in the wrong design. It is essential to use the required constructs to avoid the simulation and synthesis mismatch in the design.

To avoid the simulation and synthesis mismatch, it is essential to use the appropriate VHDL constructs, and it is mandatory to follow the important rules while writing VHDL code. To implement the desired intended design functionality, the following section will play an important role. The following section will discuss about the VHDL programming paradigm, the key statements, and the important rules while coding using VHDL.

3.1 VHDL Design Paradigm

VHDL design paradigm has five different entities: entity declaration, package declaration, configuration declaration, architecture, and package body declaration. Among them, entity, package, and configuration declarations are visible in the VHDL library and hence called as the main or primary design units. VHDL library is the storage area of host environment for compiled design unit.

As the architecture and package body declarations are not visible within library, they are treated as secondary design units. Every design has entity–architecture pair; entity provides port information, and architecture provides the functionality of design.

The packages are used to have the global information. The package and package body contain subprogram and data types which need to be used for other designs. The package body consists of the subprogram declarations, and it should have the same name as that of package.

1. **Entity Declaration**: As discussed earlier, entity provides the port information, that is the interface of the design to any other design or module for the communication is provided by the entity declaration. The entity declaration is used to communicate with the other design units in the same environment. The interface required for communication includes input, output, bidirectional signals and parameterized generic declarations.
2. **Design Architecture**: It is used to describe the functionality of design. Every VHDL code should have at least one architecture. Architecture is concurrent construct, and if VHDL code has multiple architectures, then the configuration can be used to bind entity with the architecture. In most of the practical scenario, it is required to have multiple versions of RTL design, and hence multiple architectures can be used.
3. **Configuration**: It is used to bind entity with one of the architecture. Single configuration statement can be used to define the binding of multiple entity–architecture pairs throughout the design hierarchy.
4. **Package**: If few data types need to be used throughout the multiple design units, then package is used. It consists of the global data types, subprograms, and constants.
5. **Package Body**: Package body is associated with the package declaration, and the name of package body should be same as that of package declaration. Package body consists of functions, procedure, and subprogram.

The description of sequential logic and use of package using VHDL is shown in Example 3.1. The synthesis result for the sequential logic using package is shown in Fig. 3.1.

In the practical scenario, every design should have the list of comments, and these comments are used to identify the functionality of design, design engineer, primary resources, secondary resources, date of creation, version of the design, etc. Every organization has their own methods to maintain the versions of design.

```
package pckg_name is
   subtype nibble is bit_vector ( 3 downto 0);
end;

library ieee;
use ieee.std_logic_1164.all;
use ieee.std_logic_unsigned.all;
use ieee.std_logic_arith.all;
use work.pckg_name.all;

entity design_logic is
   port ( clk, reset_n : in boolean;
           s0_in : in bit;
           s1_in : in bit;
           d3_in : in nibble;
           d2_in : in nibble;
           d1_in : in nibble;
           d0_in : in nibble;
           q_out : out nibble );
end design_logic;

architecture arch_design of design_logic is

 signal reg_sig : nibble;
 signal select_sig : bit_vector ( 1 downto 0);

         begin
            select_sig <= s0_in & s1_in;
        process ( clk, reset_n)
              begin
                   if ( reset_n) then
                      reg_sig <= "0000";
                   elsif ( clk and clk'event ) then
                   case ( select_sig) is
                    when b"00" => reg_sig <= d0_in;
                    when b"01" => reg_sig <= d1_in;
                    when b"10" => reg_sig <= d2_in;
                    when b"11" => reg_sig <= d3_in;
                   end case;
              end if;
    end process;
  q_out <= reg_sig;
end arch_design;
```

Package declaration

Declaration of Library and use of package using 'use' clause.

Entity declaration using the required data types for the input and output.

Functional definition using 'architecture' and internal signal definitions.

Use of concurrent statement process and sequential statement 'if-then-else' and 'case' to define the behavior of design.

Example 3.1 Synthesizable VHDL of sequential design logic

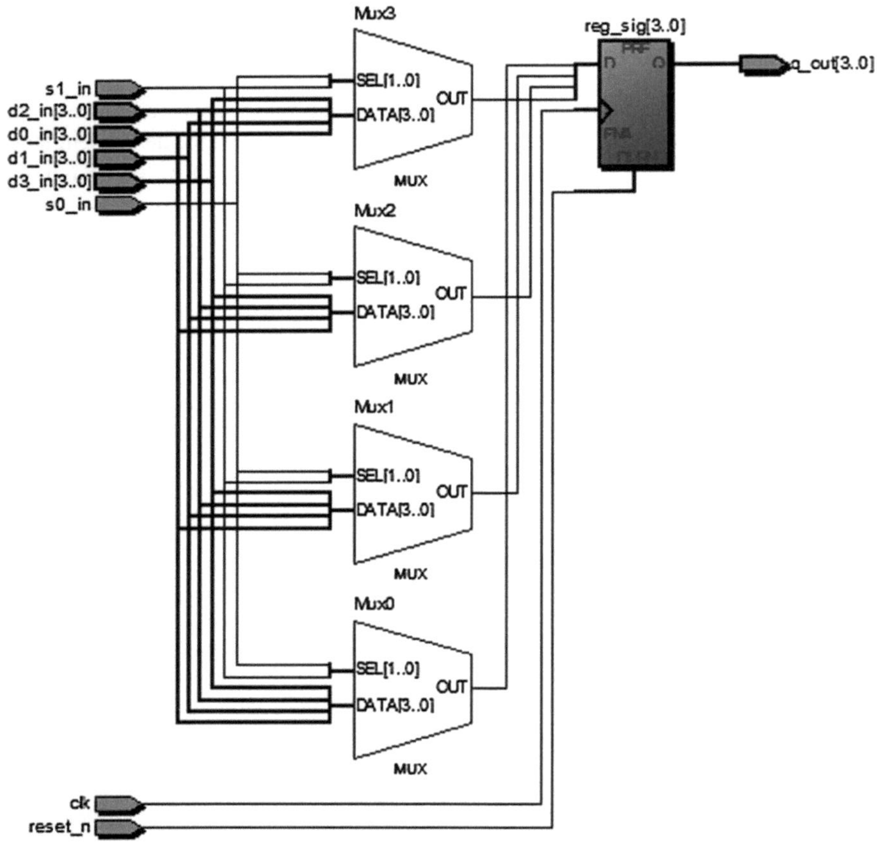

Fig. 3.1 Synthesis result for the sequential design logic

Following are few key comments need to be used at the start of every VHDL design. The comments can be optional but they improves readability.

```
-------------------------------------------------------------------------------
---------------------------------
-- Company Name:
-- Design Engineer:
-- Design Name:
-- Module Name:
-- Project Name:
-- Design Date:
-- Target technology:
-- Design Version:
-- Description:
-- Additional Comments:
-- ----------------------------------------------------------------------------
---------------------------------
```

3.2 Multiple Architectures and Configuration

The description of combinational logic using multiple architecture definitions is shown in Example 3.2. The synthesis result for the combinational logic for multiple architecture design is shown in Fig. 3.2. The last architecture is coupled with entity

```
--multiple architecture definitions
 library ieee;
use ieee.std_logic_1164.all;
use ieee.std_logic_unsigned.all;
use ieee.std_logic_arith.all;

entity design_mult_arch is

    port ( s_in : in std_logic;
           a_in  : in std_logic;
          b_in  : in std_logic;
        y_out : out std_logic);
end design_mult_arch;

architecture arch_design_1 of design_mult_arch is

begin

    y_out <= a_in xor b_in xor s_in;

end arch_design_1;

architecture arch_design of design_mult_arch is

begin

    y_out <= a_in and b_in and s_in;

end arch_design;
```

Architecture defines the functionality of design as three input 'xor' gate.

Architecture is named as 'arch_design_1' and has inputs 'a_in, b_in, s_in'.

Architecture defines the functionality of design as three input 'and' gate.

Architecture is named as 'arch_design' and has inputs 'a_in, b_in, s_in'.

Example 3.2 Synthesizable VHDL using multiple architecture

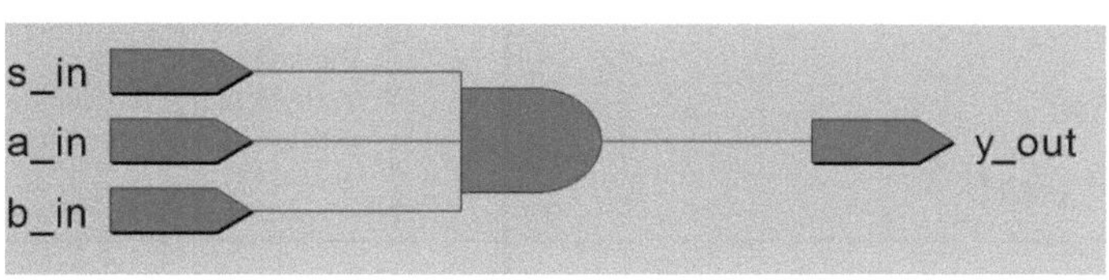

Fig. 3.2 Synthesis result for multiple architecture VHDL

to generate the gate-level netlist. Configuration can be used to bind the required architecture with the entity.

3.2.1 Multiple Architecture and Configuration

The description of combinational logic using multiple architectures and the use of configuration is shown in Example 3.3. The synthesis result for the combinational logic using configuration is shown in Fig. 1.3. Using configuration, the first

```
--multiple architecture definitions and configuration
library ieee;
use ieee.std_logic_1164.all;
use ieee.std_logic_unsigned.all;
use ieee.std_logic_arith.all;

entity conf_mult_arch is

   port ( s_in : in std_logic;
          a_in  : in std_logic;
        b_in  : in std_logic;
        y_out : out std_logic);
end conf_mult_arch;


architecture arch_design_1 of conf_mult_arch is

begin

   y_out <= a_in xor b_in xor s_in;

end arch_design_1;

architecture arch_design of conf_mult_arch is

begin

   y_out <= a_in and b_in and s_in;

end arch_design;

configuration conf_arch1 of conf_mult_arch is

   For arch_design_1

   End For;

end conf_arch1;
```

Architecture defines the functionality of design as three input 'xor' gate.

Architecture is named as 'arch_design_1' and has inputs 'a_in, b_in, s_in'.

Architecture defines the functionality of design as three input 'and' gate.

Architecture is named as 'arch_design' and has inputs 'a_in, b_in, s_in'.

The architecture 'arch_design_1' is coupled with entity using configuration

Example 3.3 Synthesizable VHDL and use of configuration

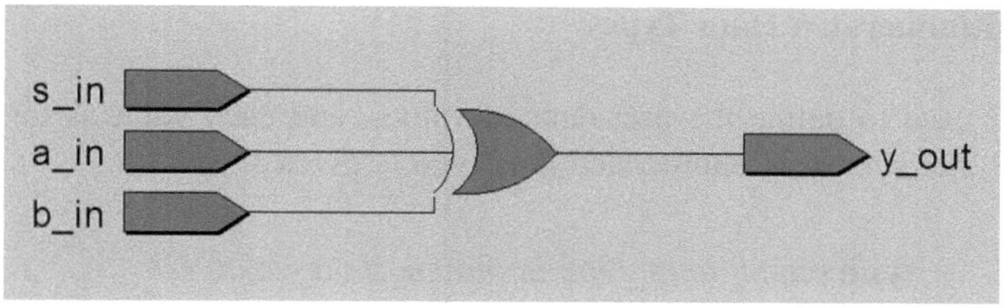

Fig. 3.3 Synthesis result for the configuration

architecture declared is coupled with entity, and hence it generates the gate-level netlist using the assignments declared in the first architecture.

3.3 Objects and Data Types

Data objects are used to pass the information in the design. The information can be passed from one point to another point. Every data object has a collection of value set, and all possible values are defined in the value set. The key VHDL data types and objects are shown in Table 3.1.

Physical, floating point, and access data types are not supported by synthesizer.

3.3.1 Scalar Data Types

The enumeration, integer, physical, and floating point data types are considered the scalar data types.

Table 3.1 Data types and data objects

Data types	Data objects
Scalar types	Constants
Enumerated	Variables
Integer	Signal
Physical	File
Real	
Composite type	
Array	
Record	
Access (pointers)	

3.3.1.1 Enumerated Data Types

These are used to define the user-defined values, and each value is treated as an identifier. The syntax of enumerated data type is given below.

```
type enum_data_type is (enum_data_value {
enum_data_value});
```

In the above syntax, 'enum_data_type' is identifier that is name of the enumerated data type; 'enum_data_value' is identifier, character or literal.

For example,

```
type fsm_state is (s0,s1,s2,s3);
```

The tool assigns the numeric value to each 'fsm_state' according to the order in which the enumerated values are declared. In the above case, for the binary value, s0 = 00, s1 = 01, s2 = 10, s3 = 11.

3.3.1.2 Integer Data Types

These are used to define the range of integer numbers. If the range is not specified, then according to VHDL IEEE standard, the default range of $(2^{-31} + 1)$ to $(2^{31} - 1)$ is used. The syntax is shown below.

```
type type_name is range integer_range
```

In the above syntax, 'type_name' is identifier, that is, the name of the integer data type; integer_range is the range of the integer definition.

For example,

```
type count_value  is range 0 to 7;
type count_value  is range 7 downto 0;
```

3.3.1.3 Physical Data Types

These are used to define the required physical parameters for the design. Time is only physical data type which is predefined in the VHDL standard. Other physical parameters required in the design need to be declared by using physical data type.

```
type type_name is range integer_range
```

In the above syntax, 'type_name' is identifier, that is, name of the physical data type; integer_range is the range of the integer definition.

For example,

```
type data_type is range 0 to 32;

Values

Bit ;
Nibble = 4Bit;
Byte =8 Bit;
Word = 16 Bit;
Double_word = 32Bit;

End Values
```

3.3.1.4 Real Data Type

These data types are used to define the real value for the required variable in the VHDL design. The minimum range according to the VHDL standard is (−1.0E38 to 1.0E38).

Consider the following example,

```
architecture real_data of data_type is
begin
 process ( input1)
 variable tmp : real;
begin
  A_tmp := 2  ;                  -- illegal
 A_tmp := 5ns  ;                 -- illegal
A_tmp := 1.5;                    -- legal
A_tmp := 1.5 E 15;               -- legal

end process;
end architecture;
```

3.3.2 Composite Data Types

These are used while modeling the memory elements. These data types are mainly arrays, records.

3.3.2.1 Arrays

The array declarations as single dimensional or two dimensional are used to group the elements of same type into the design object. The synthesis tool supports the array declaration of single or two dimensional. The range may be unconstrained in the declaration, and then the range can be constrained when array is used in the design. These are generally used in the modeling of memories (RAM or ROM).

The syntax for one-dimensional array is as follows:

```
type name_array is array ( range) of data_type;
```

The example is shown below.

```
type data_bus is array ( 0 to 7) of bit;

architecture arch_array of array_design is

begin

 process ( input1)

 variable tmp1 : bit;
variable  tmp : data_bus;

 begin


   tmp1 := data_bus(7);

end process;

end arch_array;
```

In the above example, the array of 8 entries from '0 to 7' is declared and named as data_bus. Inside the process, the 'tmp1' is declared as of type 'bit' which is the type of array declared. Using the bit select, the tmp value is assigned to 'tmp1'.

3.3.2.2 Records

These are used to group the data elements of different types into the VHDL object. While modeling the data packets, record types can be used. Record is a set of values of same or different types of elements.

The example is shown below.

```
type floating point is ;
record
  sign: std_logic;
  fraction : unsigned ( 7 downto 0);
 exponent : unsigned ( 0 to 7);
end record;
```

3.3.3 Data Objects

The four main types of objects in VHDL are constants, variables, signals, and files. The objects are used to communicate in the design, and every object has its own scope. If object is defined in the package, then it is available to all the designs in the same design environment. If the data object is declared in the entity, then it is available to the architecture associated with that entity only. If the data object is defined in the architecture, then it is available to the architecture only. If data object is declared in the process, then it is local to the process only and available to all the statements inside the process.

3.3.4 Constants

These types of data objects are used to hold the constant value of specified type. The value of data object constant cannot be changed once it is declared.

The syntax to declare the constant is given below.

```
constant constant_name: type_name [:=value];
```

For example,

```
constant data_bus: integer:=8;
```

3.3.4.1 Variable

These types of data objects are used to hold the single value from the values of the specified type. They are mainly used to hold the temporary value within the process and hence local to the process. The variables are updated immediately without any delta delay.

The syntax is shown below.

```
Variable variable_name: type_name [:=value];
```

Following are few examples of variable declaration.

```
Variable opcode: bit_vector(7 down to 0):=00000000;
```

In the above declaration, the variable is named as ‘opcode’ and assigned to value ‘00000000’.

3.3.4.2 Signal

These types of data objects are used to communicate between the VHDL components, and they are used to hold the present and future values. The signals are updated after delta delay at the end of the process.

The syntax is shown below.

```
Signal signal_name: type_name [:=value];
```

Following are few examples of signal declaration:

```
Signal ready: bit ;
ready <= '1' after 10 ns;
```

In the above declaration, the signal is named as ‘ready’ and assigned to value ‘1’ after 10 nano second (ns) time duration.

3.3.4.3 File

These types of data objects are used to communicate with the host environment. Files can be opened for reading and writing. File objects are not supported by synthesis tool. Using the procedures, the read from and write to file is possible. The file can be opened by using file_open() and can be closed by using file_close(). This will be discussed in more detail in the next subsequent chapters.

3.4 Signal Assignments

Signal is used to represent the module interface and is global to the architecture. The interface between the concurrent and sequential statements can be achieved by using signals. Signal assignments can be concurrent or sequential assignments. The concurrent signals assignments can be conditional, selective. The sequential signal assignments are unconditional and hence treated as simple assignments.

```
signal <= expression [after desired delay];
```

While executing the sequential signal assignments inside the process, the right-hand side (RHS) side expression is evaluated, and event is scheduled depending on the delay to change the value of the signal. At the end of the process or at the process suspension, the value of signal is updated.

If the same signal has multiple assignments inside the process, then the synthesizer considers the last assignment as effective assignment. For example,

```
process ( a_in, b_in)
begin

 y_out <= a_in xor b_in;
y_out <= a_in and b_in ;
end process;
```

In the above code as y_out is assigned twice with different functionality, and it infers the hardware as 'AND' of 'a_in, b_in' as the last assignment is effective. So the main important point in the signal assignment is the updating of signal value. All the signals inside the process hold the previous or old value, and all the signal assignments become effective when process suspends, that is, at the end of the process.

3.4.1 Signal Assignments Example

The description of combinational logic using signal assignments is shown in Example 3.4. The synthesis result for the combinational logic using signal assignments is shown in Fig. 3.4. Signals are updated at the end of the process, and hence for the shown example it generates the parallel logic using the assignment statements.

3.5 Variable Assignment

Variables are used to hold the intermediate results within the process. The variables can be declared by using keyword 'variable.' At the initialization phase during the simulation, the initial value is given to variable. Variable assignment statements are used inside the process, and it replaces the current value of variable with the evaluated new value. The 'variable' is declared by using following syntax.

```
variable := expression;
```

```
--signal assignments
 library ieee;
use ieee.std_logic_1164.all;
use ieee.std_logic_arith.all;

entity signal_assignment is

port ( a_in, b_in, c_in, d_in, e_in : in std_logic;
      y1_out , y2_out : out std_logic);

end signal_assignment;

architecture arch_signal of signal_assignment is

signal signal_1, signal_2 : std_logic;

begin

 signal_1 <= a_in xor b_in;

 process ( c_in, signal_1, signal_2)

 begin

    y1_out <= not signal_2;
    signal_2 <= signal_1 xor c_in;

 end process;

  y2_out <= not (e_in xor d_in );

end arch_signal;
```

- Architecture defines the functionality of design.
- Signals are declared to establish the communication.
- Signals are described using 'signal' keyword and of type 'std_logic'
- Signals are updated at the end of the process after delta delay.

Example 3.4 Synthesizable VHDL using signal assignments

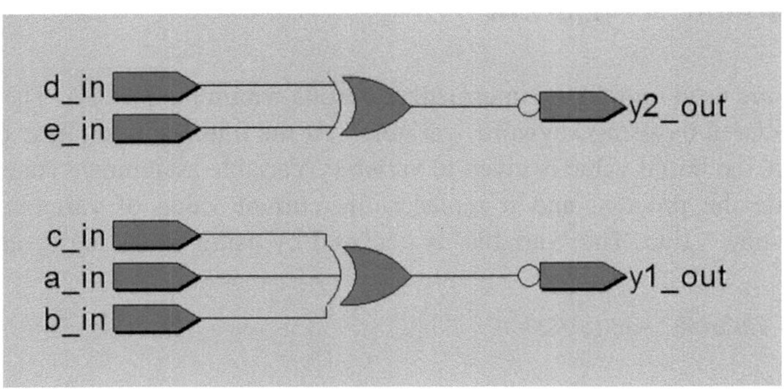

Fig. 3.4 Synthesis result for signal assignment

The expression can contain signals, literals, and variables. Variable assignments are executed immediately in zero simulation time, and hence variable assignments cannot be delayed.

3.5.1 Variable Assignments Example

The description of combinational logic using variable assignments is shown in Example 3.5. The synthesis result for the combinational logic using variable assignments is shown in Fig. 3.5. Variables are updated immediately, and hence for the shown example it generates the cascade logic using the assignment statements.

```
--variable assignments

library ieee;
use ieee.std_logic_1164.all;
use ieee.std_logic_arith.all;

entity variable_assignment is

port ( a_in, b_in, c_in, d_in, e_in : in std_logic;
     y1_out , y2_out : out std_logic);

end variable_assignment;

architecture arch_variable of variable_assignment is

begin

 process ( a_in, b_in, c_in, d_in, e_in)
 variable  variable_1, variable_2, variable_3, variable_4: std_logic;
 begin

           variable_1 := a_in xor b_in;
           variable_2 := variable_1 xor c_in;
           y1_out <= not variable_2;
           variable_3 := not (variable_2 xor e_in );
           variable_4 := not (variable_2 xor d_in );
           y2_out <= variable_4;
 end process;


end arch_variable;
```

- Architecture defines the functionality of design.
- Variables are declared to establish the communication.
- Variables are declared using 'varibale' keyword and of type 'std_logic'
- Variables are updated instant immediately.

Example 3.5 Synthesizable VHDL using variable assignments

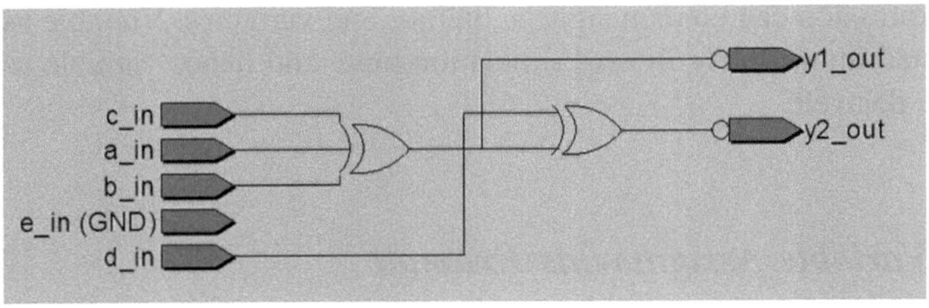

Fig. 3.5 Synthesis result for variable assignment

Table 3.2 Signal versus variable

Signal assignments	Variable assignments
Signals are updated when the process execution suspends, Signals are global to architecture	Variables those are not local to the process are updated immediately, and the event is not scheduled
In the signal assignment, delay can be specified	Variable assignments cannot be delayed
To the same signal inside the process if multiple assignments are used, then the last assignment is effective	Many assignments to the same variable are effective

The difference between the signal and variable assignments is given in Table 3.2.

3.6 Concurrent Constructs

VHDL is one of the powerful HDLs and consists of rich set of statements. VHDL consists of concurrent and sequential statements.

The important and essential concurrent statements are architecture, process, concurrent signal assignments, component instantiation, procedure calls. These statements are executed simultaneously.

3.6.1 When Else

The 'when else' is conditional signal assignment statement. The described functionality using such kind of statement is equivalent to the conditional 'if' statement. The syntax of 'when else' is shown below.

```
output port name or signal <= [expression when condition
else ...] expression;
```

When the Boolean condition is true, then the value of the first expression is assigned to the output or signal. When the condition is false, then the expression corresponding to the 'else' clause is assigned to output or signal. The conditional assignment statement is concurrent statement and hence can be used in the architecture. The major use of this kind of concurrent statement is to assign the values to the signal or to the output port.

The description of combinational logic using concurrent 'when else' construct is shown in Example 3.6. The synthesis result for the combinational logic using 'when else' is shown in Fig. 3.6. When else is a concurrent statement, and it generates the multiplexing logic.

```
--Combinational logic using when else


 library ieee;
use ieee.std_logic_1164.all;
use ieee.std_logic_unsigned.all;
use ieee.std_logic_arith.all;

entity combo_logic is

   port ( enable_in : in std_logic;
                 a_in  : in std_logic;
                b_in  : in std_logic;
                y_out : out std_logic);
end combo_logic;

architecture arch_combo of combo_logic is

begin

  y_out <= ( b_in and a_in)  when (enable_in = '1')
              else ( b_in xor a_in);


end arch_combo;
```

- Architecture defines the functionality of design.
- When-else is concurrent statement.
- When 'enable_in' is true 'y_out' is equal to 'a_in and b_in'.
- When 'enable_in' is false 'y_out' is equal to 'a_in xor b_in'.

Example 3.6 Synthesizable VHDL using when-else

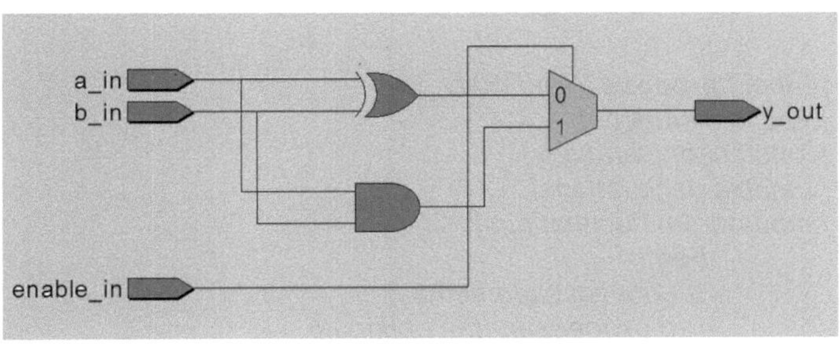

Fig. 3.6 Synthesis result for combo logic using when-else

3.6.2 With Select

The 'with select' is a selected signal assignment statement. It is used to select one of the expressions depending on the condition. This kind of expression uses only single condition to select between multiple options. This kind of statements can be considered as functional equivalent of the 'case' statement. The syntax of 'with select' statement is shown below.

```
with select_condition/expression select
signal <= expression_1 when option_1,
 .
 ;
 ;
expression_n when option_n,
[expression when others];
```

The signal or output is assigned to one of the expressions. All the values specified in the select expression or condition need to be covered. The final option may be using keyword 'others'. While using the 'with select' statement, it is essential to take care that the option values should not overlap each other. If 'others' option is not used, then all the option values should be covered.

The description of combinational logic using 'with select' construct is shown in Example 3.7. The synthesis result for the combinational logic using 'with select' construct is shown in Fig. 3.7. The 'with select' is concurrent construct, and it results into multiplexing logic depending on the specified conditions and expressions.

3.6.3 Process

Process is concurrent statement; multiple processes can be described within the architecture, and all the processes executes concurrently. Process can appear any where inside the architecture body and sequence of statements need to be included within 'begin' and 'end process.' Process name or label is optional while writing VHDL code. The structure of process declaration is shown below.

```
[Label:] process [(sensitivity_list)]
[type declarations]
[constant declarations]
[variable declarations]
[subprogram declarations]
          begin
          sequential statements
          end process [name optional];
```

```
--Combinational logic using with select


library ieee;
use ieee.std_logic_1164.all;
use ieee.std_logic_unsigned.all;
use ieee.std_logic_arith.all;

entity combo_logic_select is

   port ( enable_in : in std_logic;
                   a_in  : in std_logic;
                   b_in  : in std_logic;
                   y_out : out std_logic);
end combo_logic_select;

architecture arch_combo of combo_logic_select is

begin
  with ( enable_in ) select
  y_out <= ( b_in nand a_in)  when  '1',
                ( b_in nor  a_in)  when  '0';

end arch_combo;
```

- Architecture defines the functionality of design.
- With-select is concurrent statement.
- When 'enable_in' is true 'y_out' is equal to 'a_in nand b_in'.
- When 'enable_in' is false 'y_out' is equal to 'a_in nor b_in'.

Example 3.7 Synthesizable VHDL using with-select

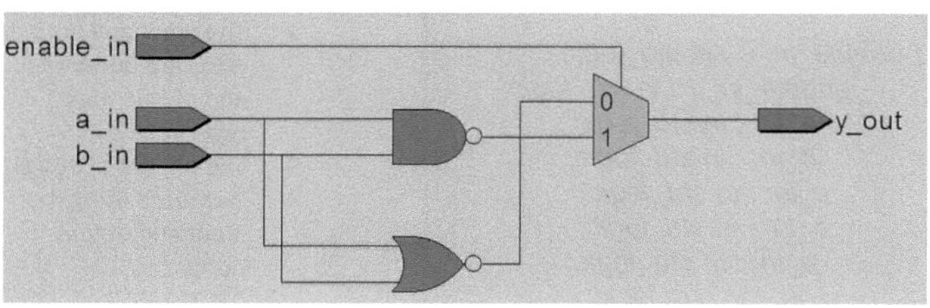

Fig. 3.7 Synthesis result for combo logic using with-select

Following are important points need to be considered while describing functionality using process statement.

1. The process is declared using keyword 'process' and ends with 'end process.'
2. The name or label to any process is optional.
3. Every process is sensitive to list of inputs or signals. The list of inputs or signals is called as sensitivity list.

4. The sequence of statements inside the process is executed sequentially and starts with the keyword 'begin.' Between the 'process' and 'begin' keyword, the types, constants, variables, functions, and procedures which are local to the process can be declared.
5. Inside process, the signal declaration is not allowed and even concurrent statements are not allowed.
6. Process can be invoked by event on any signal specified in the sensitivity list.
7. In VHDL, 'end process' does not mean the end of process execution; the process is executed in the indefinite loop.
8. For missing sensitivity list, the process must have the wait statement to suspend and to activate the process depending on the event or true condition.
9. To avoid the simulation and synthesis mismatch, it is recommended to specify all the required signals in the sensitivity list of process.

The description of combinational logic using multiple concurrent processes is shown in Example 3.8. The synthesis result for the combinational logic using multiple processes is shown in Fig. 3.8. Multiple process constructs executes concurrently. The statements inside the process execute sequentially.

```
--Combinational logic using multiple processes.

library ieee;
use ieee.std_logic_1164.all;
use ieee.std_logic_unsigned.all;
use ieee.std_logic_arith.all;

entity multi_process is

   port ( enable_in_0 : in std_logic;
              enable_in_1 : in std_logic;
                a_in  : in std_logic;
                b_in  : in std_logic;
               c_in  : in std_logic;
               d_in  : in std_logic;
                e_in  : in std_logic;
               f_in  : in std_logic;
               y_out : out std_logic;
               y1_out: out std_logic);
end multi_process;
```

- ➤ Entity defines the input and output ports.
- ➤ The input and output ports are declared and of 'std_logic' type.
- ➤ The pin out of design is created using the input and output ports.

Example 3.8 Synthesizable VHDL using multiple processes

```
--functional definition using multiple processes.

architecture arch_multi_process of multi_process is

begin

   P1: process ( enable_in_0, a_in, b_in, c_in, d_in)

               begin

                     if ( enable_in_0 ='1') then

                        y_out <= b_in xor a_in;

                     else

                        y_out <= d_in xor c_in;

                     end if;

               end process;

               P2: process ( enable_in_1, a_in, b_in, e_in, f_in)

               begin

                     if ( enable_in_1 ='1') then

                        y1_out <= e_in xor f_in;

                     else

                        y1_out <= b_in xor a_in;

                     end if;

               end process;

end arch_multi_process;
```

- Process named as 'P1' and is sensitive to the inputs 'enable_in_0, a_in, b_in, c_in, d_in'.
- For true value on 'enable_in_0' , the 'y_out is equal to xor of 'a_in, b_in'
- For false value on 'enable_in_0' , the 'y_out is equal to xor of 'c_in, d_in'

- Process named as 'P2' and is sensitive to the inputs 'enable_in_1, a_in, b_in, e_in, f_in'.
- For true value on 'enable_in_1' , the 'y1_out is equal to xor of 'e_in, f_in'
- For false value on 'enable_in_0' , the 'y1_out is equal to xor of 'a_in, b_in'

Example 3.8 (continued)

3.7 Sequential Constructs

The key important sequential statements are 'if then else,' 'case,' 'block,' 'next,' and 'loop,' and these are executed sequentially as they appear within the subprogram or process. The following section focuses on the key sequential constructs and

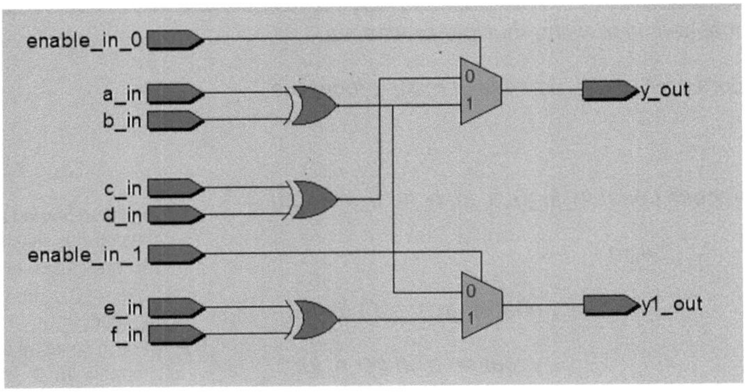

Fig. 3.8 Synthesis result for combinational design using concurrent processes

other constructs like 'NEXT,' 'BLOCK,' and 'Assert' will be discussed in the next subsequent chapters.

3.7.1 If Then Else

It is a sequential statement and is used inside the process. It is used to select one or more statements for execution depending on the specified condition. Condition specified in the Boolean expression is evaluated as true or false. For the true condition, the statements specified after 'if' keyword are executed. For the false condition, the sequence of statements after 'else' clause is executed. The syntax of 'if then else' is shown below.

```
if condition then
Sequence of statements
else
Sequence of statements
end if;
```

The description of tri-state logic using 'if then else' construct is shown in Example 3.9. The synthesis result for the tri-state logic is shown in Fig. 3.9. 'If then else' is sequential construct and is used inside the process statement.

```
--tri state buffer using if-then-else construct.


 library ieee;
use ieee.std_logic_1164.all;
use ieee.std_logic_unsigned.all;
use ieee.std_logic_arith.all;

entity tri_state_logic is

    port ( enable_in : in std_logic;
                  a_in  : in std_logic;
                 b_in  : in std_logic;
                  y_out : out std_logic);
end tri_state_logic;

architecture arch_tri_state of tri_state_logic is

begin

     process ( enable_in, a_in, b_in)

                begin

                        if ( enable_in ='1') then

                            y_out <= b_in and a_in;

                        else

                            y_out <= 'Z';        <----

                        end if;

                end process;


end arch_tri_state;
```

- Process is sensitive to the inputs 'enable_in, a_in, b_in'.
- For true value on 'enable_in' , the 'y_out is equal to and of 'a_in, b_in'
- For false value on 'enable_in', the 'y_out is equal to high impedance.

Example 3.9 Synthesizable VHDL of tri-state logic

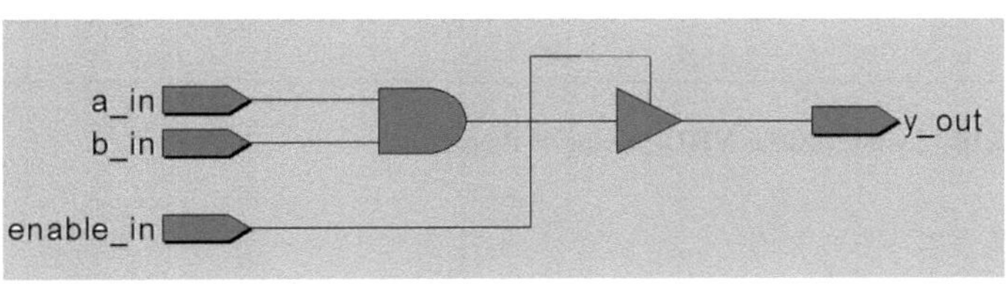

Fig. 3.9 Synthesis result for tri-state logic using if-then-else construct

The description of two-to-one multiplexer using 'if then else' construct is shown in Example 3.10. The synthesis result for the two-to-one multiplexer is shown in Fig. 3.10. 'If then else' construct generates multiplexer.

```
--Two to one mux using if-then-else construct.


library ieee;
use ieee.std_logic_1164.all;
use ieee.std_logic_unsigned.all;
use ieee.std_logic_arith.all;

entity mux_logic is

   port ( enable_in : in std_logic;
          a_in  : in std_logic;
          b_in  : in std_logic;
          y_out : out std_logic);
end mux_logic;

architecture arch_mux of mux_logic is

begin

   process ( enable_in, a_in, b_in)

          begin

                if ( enable_in ='1') then

                   y_out <= b_in;

                else

                   y_out <= a_in;   <----

                end if;

          end process;


end arch_mux;
```

- Process is sensitive to the inputs 'enable_in, a_in, b_in'.
- For true value on 'enable_in' , the 'y_out is equal to 'b_in'
- For false value on 'enable_in', the 'y_out is equal to 'a_in'.

Example 3.10 Synthesizable VHDL using if-then-else

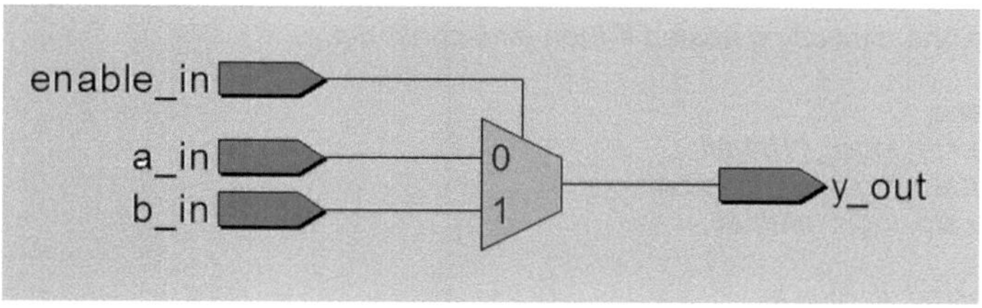

Fig. 3.10 Synthesis result for MUX using if-then-else construct

3.7.2 Nested If Then Else

It is a sequential statement and is used inside the process. It is used to select one or more statements for execution depending on the specified condition. Condition specified in the Boolean expression is evaluated as true or false. For the true condition, the statements specified after 'if' keyword are executed. For the false condition, the sequence of statements after 'else if' clause is executed, and if none of the condition is matched then the sequence of statements after 'else' clause will be executed. It infers the priority logic; the syntax of nested 'if-then-else' is shown below.

```
if condition then
Sequence of statements
[elsif condition then
Sequence of statements...]
[else
Sequence of statements]
end if;
```

The description of four-to-one multiplexer using nested 'if then else' construct is shown in Example 3.11. The synthesis result for the four-to-one multiplexer is shown in Fig. 3.11. Nested 'if then else' construct generates priority logic in case of four-to-one multiplexer.

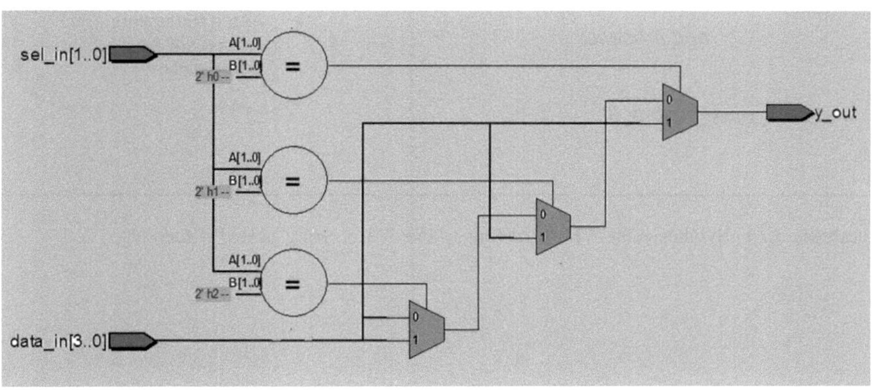

Fig. 3.11 Synthesis result for nested if-then-else construct

```
--Four to one mux using nested if-then-else construct.

library ieee;
use ieee.std_logic_1164.all;
use ieee.std_logic_unsigned.all;
use ieee.std_logic_arith.all;

entity nested_if_mux is

   port (   sel_in : in std_logic_vector ( 1 downto 0);
                 data_in  : in std_logic_vector (3 downto 0);
                 y_out : out std_logic);
end nested_if_mux;

architecture arch_mux_nested_if of nested_if_mux is

begin

    process ( sel_in, data_in)

                begin

                        if ( sel_in ="00") then

                            y_out <= data_in(0);

                        elsif ( sel_in ="01") then

                            y_out <= data_in(1);

                        elsif ( sel_in ="10") then

                            y_out <= data_in(2);

                        else

                            y_out <= data_in (3);

                        end if;

                end process;

end arch_mux_nested_if;
```

- Process is sensitive to the inputs 'sel_in, data_in'.
- The functionality is defined to generate four to one multiplexer using nested if-then-else construct.
- Depending on the status of select lines, one of the input is assigned to output.
- All the statements inside process are executed sequentially.

Example 3.11 Synthesizable VHDL of four-to-one MUX using nested if-then-else

3.7.3 Case

It is a sequential statement and used inside the process. It is used to select one of the statements specified based on the value of expression. The expression may or may not be Boolean and can be character array, variables, and signals. When there is large number of alternatives to generate the required output, the 'case' statement is useful. It generates the parallel logic. The case statement syntax is shown below and consists of several 'when' clauses with one or more options. The expression value is compared with the option. If the expression value is equal to the option specified, then the sequence of statements specified after => symbol are executed. The 'when others' must be the last option.

```
case conditional expression is
when condition_1 =>
Sequence of statements
when condition_n =>
Sequence of statement
[when others =>
Sequence of statements]
end case;
```

The description of two-to-one multiplexer using 'case' construct is shown in Example 3.12. The synthesis result for the two-to-one multiplexer is shown in Fig. 3.12. The 'case' construct generates parallel logic in the case of two-to-one multiplexer.

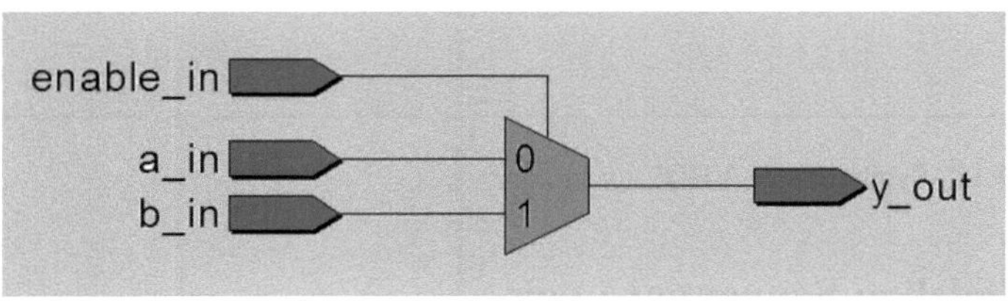

Fig. 3.12 Synthesis result for MUX logic using case construct

```
--Two to one mux using 'case' construct.

library ieee;
use ieee.std_logic_1164.all;
use ieee.std_logic_unsigned.all;
use ieee.std_logic_arith.all;

entity mux_logic_case is

   port ( enable_in : in std_logic;
                a_in  : in std_logic;
               b_in  : in std_logic;
                y_out : out std_logic);
end mux_logic_case;

architecture arch_mux of mux_logic_case is

begin

    process ( enable_in, a_in, b_in)

              begin

                        case (enable_in) is

                        when '0' => y_out <= a_in;
                        when '1' => y_out <= b_in;

                        end case;

              end process;


end arch_mux;
```

- Process is sensitive to the inputs 'enable_in, a_in, b_in'.
- The functionality is defined to generate two to one multiplexer using 'case'.
- Depending on the status of enable_in, one of the inputs is assigned to output.
- All the statements inside process are executed sequentially.

Example 3.12 Synthesizable VHDL using case

3.8 Modeling Sequential Logic

In the sequential logic, an output is a function of the present input and the past output, and hence it has memory or storage capacity. VHDL constructs like process, 'if then else,' and 'case' can be efficiently used to write synthesizable RTL for sequential logic elements. The key elements are register or flip-flop (edge triggered) and latch (level sensitive). These can be efficiently modeled for the intended design functionality using the VHDL constructs. The following section discusses the key concepts for modeling sequential logic. The details of sequential logic design will be discussed in few subsequent chapters.

3.8.1 Four-Bit Register

The description of four-bit register using 'if then else' construct is shown in Example 3.13. The synthesis result for the four-bit register having asynchronous reset and positive edge-triggered clock is shown in Fig. 3.13. 'If-then-else' construct generates multiplexer, but as the else clause is missing in the nested 'if then else' construct it generates the sequential logic which is triggered on the rising edge of clock due to use of 'clk='1'' and clk'event'.

```
--Four bit register using 'if-then-else'


library ieee;
use ieee.std_logic_1164.all;
use ieee.std_logic_unsigned.all;
use ieee.std_logic_arith.all;

entity register_4bit is

    port (        clk, reset : in std_logic;
                  data_in  : in std_logic_vector (3 downto 0);
                  y_out : out std_logic_vector (3 downto 0));
end register_4bit;

architecture arch_register_4bit of register_4bit is

begin

     process ( clk, reset)

               begin

                if ( reset = '1') then          <----

                     y_out <= "0000";

                elsif ( clk='1' and clk'event) then
                     y_out <= data_in;

               end if;

               end process;


end arch_register_4bit;
```

- Process is sensitive to the inputs 'clk, reset'
- The functionality is defined to generate four bit register using 'if'.-then-else
- For 'reset=1' output 'y_out' is equal to "0000".
- For rising edge of clk signal 'data_in' is assigned to 'y_out'.

Example 3.13 Synthesizable VHDL of four-bit register

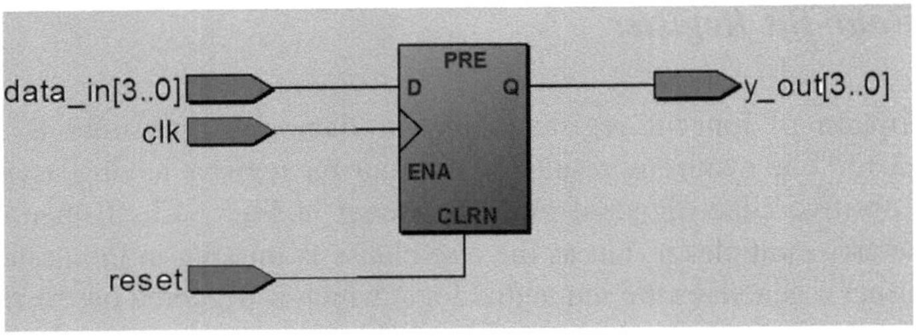

Fig. 3.13 Synthesis result for edge triggered logic

3.8.2 *Four-Bit Latch*

The description of four-bit latch using 'if then else' construct is shown in Example 3.14. The synthesis result for the four-bit positive level sensitive latch is shown in Fig. 3.14. 'If then else' construct generates multiplexer but as the else clause is missing in the nested 'if then else' construct, it generates the sequential logic which is positive level sensitive.

3.9 Wait Statements

For the process declaration without any sensitivity list, the process body must contain at least one 'wait' statement. 'Wait' statement inside process body is used to suspend the process execution. Even the 'wait' statement inside process body is used to activate the suspended process depending on the specified condition. When the condition specified in the 'wait' statement is met, the sequence of statements are executed until it encounters another 'wait' statement. One or more than one 'wait' statement can be used inside the process. Few important wait statement syntax are shown below.

```
wait on sensitivity list;
wait for time expression;
wait until conditional expression;
```

3.9.1 *Wait On*

The 'wait on' statement needs to be used inside the process having empty sensitivity list. The syntax of 'wait on' is shown below.

```
wait on sensitivity list;
```

```
--Four bit level sensitive latch using 'if-then-else'


library ieee;
use ieee.std_logic_1164.all;
use ieee.std_logic_unsigned.all;
use ieee.std_logic_arith.all;

entity latch_4bit is

   port (    enable_in, reset_in : in std_logic;
                 data_in  : in std_logic_vector (3 downto 0);
                 y_out : out std_logic_vector (3 downto 0));
end latch_4bit;

architecture arch_latch_4bit of latch_4bit is

begin

    process ( enable_in, reset_in, data_in)

              begin

                      if ( reset_in = '1') then

                        y_out <= "0000";

                      elsif ( enable_in ='1') then
                                y_out <= data_in;

                      end if;

              end process;


end arch_latch_4bit;
```

- Process is sensitive to the inputs 'enable_in, reset_in, data_in'
- The functionality is defined to generate four bit latch using 'if'.-then-else
- For 'reset_in=1' output 'y_out' is equal to "0000".
- For active high value on 'enable_in' input 'data_in' is assigned to 'y_out'.

Example 3.14 Synthesizable VHDL of four-bit latch

Now consider the following VHDL code,

```
process
begin
y_out <= a_in xor b_in xor c_in;
wait on a_in, b_in, c_in;
end
```

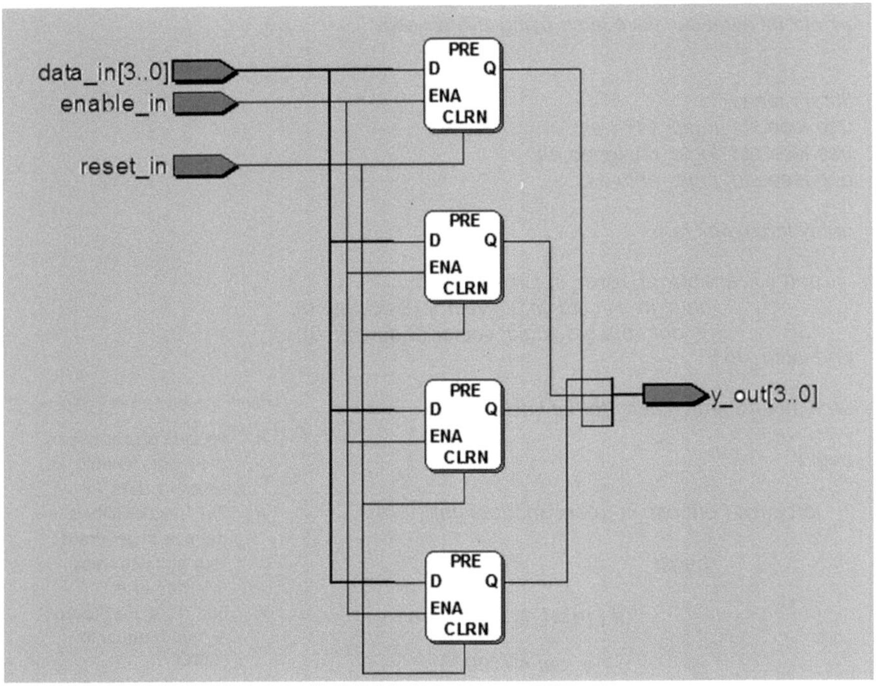

Fig. 3.14 Synthesis result for sequential logic

As shown in the above VHDL code, for any event on either 'a_in, b_in, c_in', the process is executed and generates the combinational logic. To generate the combinational logic, only one 'wait' statement should be present inside the 'process.'

3.9.2 Wait For

It is used to suspend the specified process execution for the given time duration. The syntax of 'wait for' is shown below.

```
wait for time expression;
```

If we use the statement in the process, then the process is suspended for the 5 nano second (ns) time duration. The syntax to wait for 5 nano second is shown below.

```
wait for 5ns;
```

3.9.3 *Wait Until*

This statement is used inside the 'process' having empty sensitivity list. The process is suspended until the specified condition is true. The process is activated when the conditional expression is true. The 'wait until' is used to infer the synchronous sequential logic. Most of the time, the 'wait until clk='1'' is used, and it should be the first statement inside the process. By using the 'wait until,' sequential logic with asynchronous reset cannot be inferred.

The description of sequential logic using 'wait until' construct is shown in Example 3.15. The synthesis result for the sequential logic having synchronous

```
--sequential logic using 'wait-until'


library ieee;
use ieee.std_logic_1164.all;
use ieee.std_logic_unsigned.all;
use ieee.std_logic_arith.all;

entity sequential_logic is

   port (   clk, reset_in : in std_logic;
            data_in  : in std_logic_vector (3 downto 0);
            y_out : out std_logic_vector (3 downto 0));
end sequential_logic;

architecture arch_sequential_logic of sequential_logic is

begin

    process

             begin

                   wait until ( clk = '1');

                   if ( reset_in ='1') then

                      y_out <= "0000";

                   else
                      y_out <= data_in;

                   end if;

             end process;


end arch_sequential_logic;
```

- Process doesn't have any sensitivity list.
- The functionality is defined to generate four bit register using 'if'.-then-else
- For active high value on 'clk' input executes 'if-then-else' statement.
- For 'reset_in=1' output 'y_out' is equal to "0000".
- For 'reset_in=0' output 'y_out' is equal to data_in.

Example 3.15 Synthesizable VHDL using wait until construct

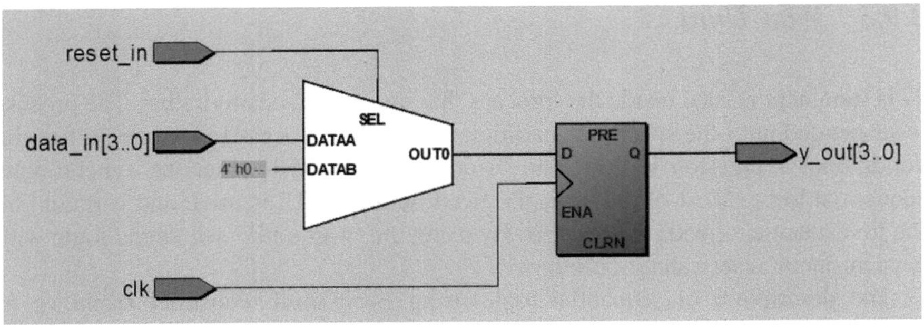

Fig. 3.15 Synthesis result for sequential logic using wait until construct

reset and positive edge-triggered clock is shown in Fig. 3.15. 'If-then-else' construct generates multiplexer and is used inside the 'wait until (clk='1')'. So it infers the sequential logic which is positive edge-triggered.

3.10 Loops

Loops are used when the repeated execution of sequence of statements is required. Simple 'loop' statement is indefinite execution of sequence of statements. In the 'for loop,' the sequence of statements executed depends on the count value. In the 'while' loop, the sequence of statement is executed until specified condition is false. While writing synthesizable code, only 'for loop' is used as the number of loop iterations is fixed.

3.10.1 Loop

The syntax of 'loop' statement is shown below.

```
[label:] loop
Sequence of Statements
end loop [label];
```

The loop label is optional, and the statements within the 'loop' body are repeatedly executed unlimited times. It is essential to use the 'exit' statement to end the execution.

3.10.2 While Loop

It is a conditional loop statement, and the syntax is shown below.

```
[label:] while condition loop
Sequence of Statements
end loop [label];
```

Before the execution of the loop, the condition is evaluated. For the true condition, the sequence of statements inside loop body is executed and control is transferred to beginning of the loop. When the condition evaluated becomes false, the loop execution terminates. Under such circumstances, the statements that follow the 'end loop' clause are executed.

3.10.3 For Loop

For the fixed number of times for the repeated execution of statements, the 'for loop' is used. The syntax is shown below.

```
[label:] for cout in range loop
Sequence of Statements
end loop [label];
```

The label is optional, and the loop consists of the count value. The sequence of statements inside loop body is executed when the count is in the specified range. After the completion of every iteration, the count value is assigned to the next value specified in the range. The ascending range is specified by keyword 'to' and the descending range is specified by keyword 'downto.'

The description of parity generator using 'for' loop is shown in Example 3.16. The synthesis result for the parity generator is shown in Fig. 3.16. The inferred gate level netlist is a combinational logic and consists of four-input XOR library cell.

```
--Parity Generator using For loop


library ieee;
use ieee.std_logic_1164.all;
use ieee.std_logic_unsigned.all;
use ieee.std_logic_arith.all;

entity parity_generator is

    port (     data_in  : in std_logic_vector (3 downto 0);
                                     y_out : out std_logic);
end parity_generator;

architecture arch_parity_generator of parity_generator is

begin

     process (data_in)

          variable temp_q_out : std_logic;

    begin

          temp_q_out := '0';

          for count in 0 to 3 loop

          temp_q_out := temp_q_out xor data_in (count);

          end loop;

          y_out <= temp_q_out;

    end process;

end arch_parity_generator;
```

- Process is sensitive to data_in.
- The for loop executes four times.
- The functionality to generate the parity is declared by using xor'.

Example 3.16 Synthesizable VHDL of parity generator

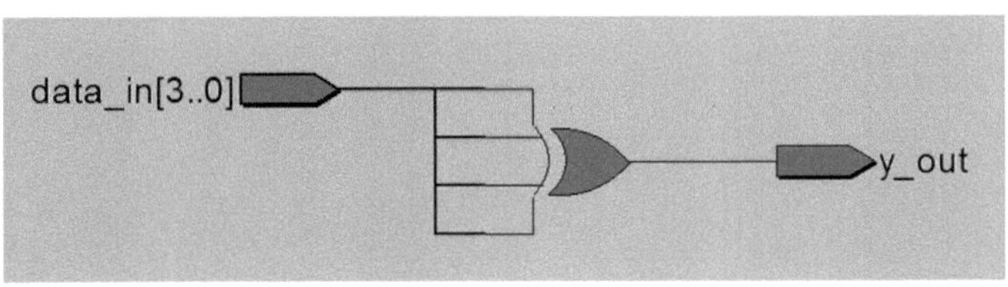

Fig. 3.16 Synthesis result for parity generator using loop

3.11 Summary

As discussed in this chapter, following are key important points to summarize the chapter

1. VHDL supports concurrent and sequential statements, and VHDL is case-insensitive language.
2. VHDL code should have one entity and at least one architecture.
3. In the multiple architecture code, the last architecture is coupled with the entity to generate the synthesis result.
4. Architecture is concurrent statement and used to define the functionality of design.
5. Concurrent statements like 'when else' and 'with select' are used inside the architecture. These statements can not be used inside 'process'.
6. Process is concurrent statement, and VHDL architecture consists of one or more than one processes.
7. All the statements inside process are executed sequentially.
8. It is essential to use all the required signals in the sensitivity list of process.
9. If the sensitivity list is missing, then to suspend and activate the process, wait statement need to be used.
10. Sequential statements like 'if then else', 'case' are used inside the process. These are used to define the combinational design or sequential design functionality.
11. Sequential logic element as register can be inferred using 'wait until' or 'clk='1' and clk'event'.
12. If 'else' clause is eliminated in the if-then-else statement, then it infers storage element either latch or flip-flop depending on the use of the construct.
13. Loops are used for repetitive statement execution. Only the 'for loop' generates synthesizable result.
14. Signals are updated after delta delay at the end of the process.
15. Variables are updated immediately and local to the process.

References

1. Altera Quartus II Evaluation Licence (Quartus II 32-bit web edition).
2. Xilinx ISE suite 14.6 evaluation Licence.

Chapter 4
Combinational Logic Design Using VHDL Constructs

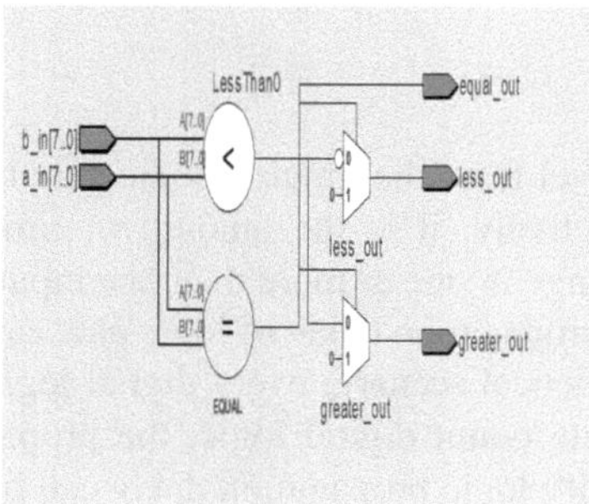

"It is the supreme art of the teacher to awaken joy in creative expression and knowledge." --- Albert Einstein

Try to use the VHDL sequential and concurrent construct to design the combinational logic.

Abstract This chapter discusses the RTL coding and synthesis using VHDL for the key combinational arithmetic resources such as adders, subtractors, multipliers, and comparators. This chapter is useful for the beginners to understand about the use of the concurrent and sequential VHDL constructs such as process, if then else, case, and their use in the design of combinational logic. Even this chapter discusses the code converters, data selectors as multiplexers, decoders, and encoders. This chapter is organized in such a way that it covers simple logic design and gate delay concepts to the priority logic design. This chapter concludes with the summary.

Keywords Propagation delay · Glitch · Cascade logic · Priority logic · Parallel logic · Adders · Subtractors · Multipliers · Code converter · BCD · Excess-3 · Seven_segment · MUX · Decoder · Encoder · Priority_logic · Case · If-then-else · Process · Sensitivity · Memory · RTL · Area · Speed · Synthesis

As discussed in the previous chapter, the VHDL has rich set of concurrent and sequential statements. The HDL can be used to design combinational and sequential logic. The designer can choose the constructs efficiently to generate the intended

---

The original version of this chapter was revised: The incorrect information have been corrected. The erratum to this chapter is available at 10.1007/978-981-10-3296-7_12

V. Taraate, *PLD Based Design with VHDL*,
DOI 10.1007/978-981-10-3296-7_4

design functionality. In the combinational design, the output of digital circuit is dependent on the present input and the logic does not have the storage. This chapter focuses on the key combinational design elements such as adders, subtractors, multiplier, comparator, and code converters. Even this chapter discusses the data selectors as multiplexers, decoders, and encoders. The discussion in this chapter is important for the design of complex logic and even for the synthesis of the complex design. The VHDL synthesizable RTL is described and covered with the synthesis results and the description about the functionality of the design.

4.1 Combinational Logic and Delays

The amount of time required for the signal to travel from the input of logic gate to the output is called as propagation delay. Effectively, it is the amount of time required for output to reflect the changes after change in one or more than one input. The propagation delay is represented by $t_{pd.}$ The propagation delay of logic gate can have maximum or minimum value, and in the practical scenario every digital logic has the propagation delay. For the minimum gate count digital logic, the propagation delay is shorter, but for the high gate density logic propagation delay can be higher. For the complement logic (inverter), the propagation delay is shown in Fig. 4.1.

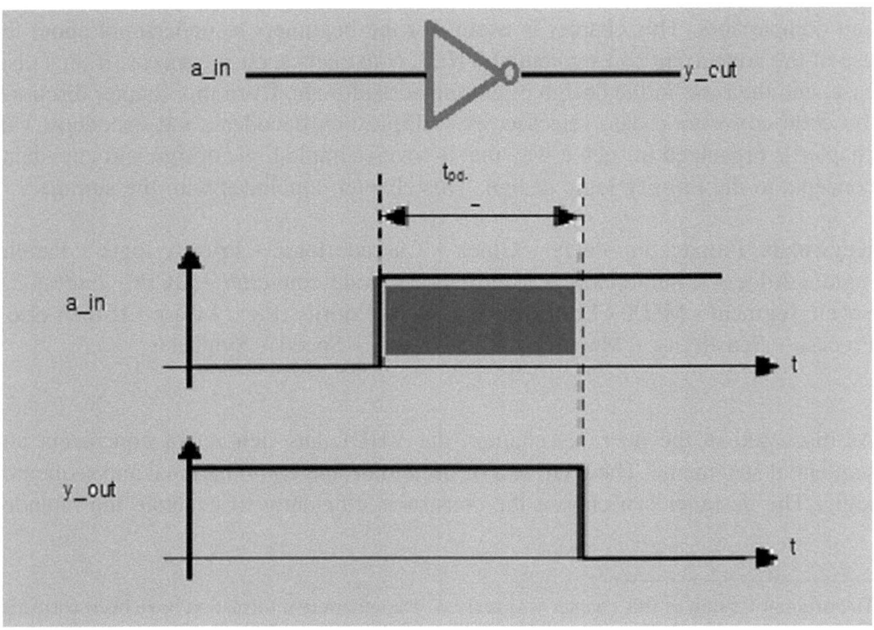

Fig. 4.1 Example of propagation delay

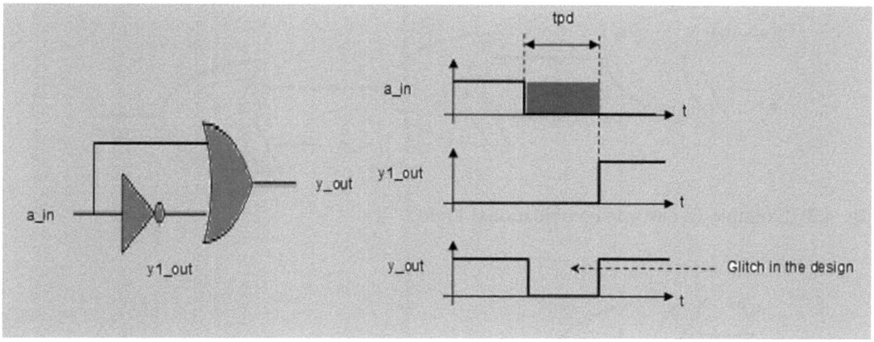

Fig. 4.2 Example of glitch in the digital circuit

The propagation delay for the various paths can be different, and it is essential to consider the longest delay path in the design while computing the delay. Due to different path delays, the design can be prone to glitches.

Glitch in the design results into unwanted output in the digital circuit. Even the glitch propagation can result in the wrong output and it affects on the output of subsequent stage. The unpredicted output in the design results into the unintended design behavior.

Figure 4.2 shows information about the glitch in the design.

As shown in the above example, when ‘a_in=1,’ the output ‘y_out’ is logic ‘1’. But when an input ‘a_in’ transits from logic ‘1’ to logic ‘0’, then due to the propagation delay of the OR gate, the output ‘y_out’ stays in the logic ‘0’ level and both inputs of OR gate are treated as logic ‘0’ for the duration of the propagation delay time. In this example, the delay of OR gate is considered as zero. Glitches can be avoided by using the latches or flip-flops as timed circuit elements. The speed of the design is dependent on the delay of the logic gates, and hence, propagation delay is treated as one of the most important parameter in the design of the logic circuits.

4.1.1 Cascade Combinational Logic

When multiple numbers of logic elements are cascaded, then the overall propagation delay is the addition of the individual gate propagation delay. This generates cumulative effect and slows down the speed of digital logic. If the circuit has all the combinational elements, then the timing path is called as the combinational path. Considering Fig. 4.3, in this example, three XOR logic gates are cascaded, and hence, the overall propagation delay of combinational path is $3{*}t_{pd}$. The propagation delay of each XOR gate is ‘$t_{pd.}$’ If we consider every logic gate has propagation delay of ‘1 ns,’ then the overall propagation delay is 3 ns.

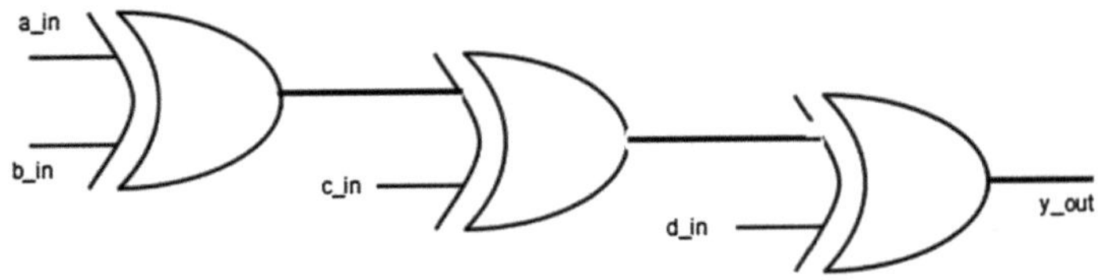

Fig. 4.3 Example of cascade combinational logic

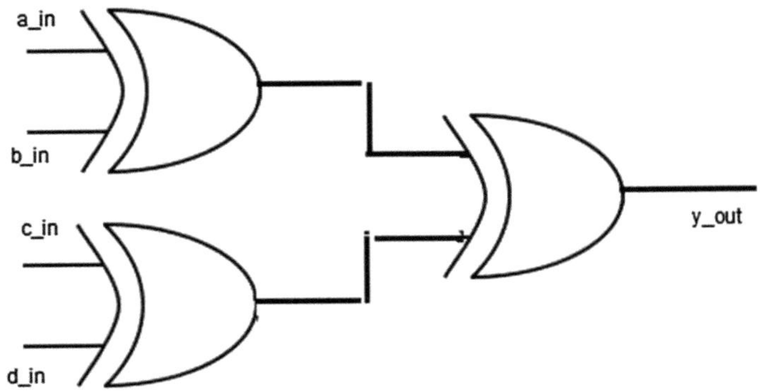

Fig. 4.4 Example of the parallel logic

4.1.2 Parallel Combinational Logic

If we consider Fig. 4.4, then the overall propagation delay is $2^*t_{pd.}$ If every XOR gate has delay of 1 ns, then the overall propagation delay of this logic is 2 ns. At the inputs, the two XOR gates are parallel and perform the operation concurrently. Hence, parallel logic has shorter delay as compare to cascade logic.

4.2 Arithmetic Circuits

The key arithmetic logic circuit elements are adders, subtractors, multipliers, and dividers. These elements can be described efficiently using the concurrent and sequential constructs. While prototyping care needs to be taken that, the synthesis result should have lesser area and least data path delay. If target technology is ASIC, then during synthesis, these elements are implemented using the standard cells, and if the target technology is programmable ASIC (FPGA), then these elements can be implemented using LUTs or the dedicated arithmetic resources. In most of the practical design scenarios, it is observed that adder consumes more area as compare to the multiplexers. This section discusses the efficient RTL using VHDL for multibit adders, subtractors, and multipliers.

Consider the design to perform the two operations: 'a_in+b_in' and 'a_in-b_in.' The addition operation is performed when op_code is equal to logic '0', and subtraction operation is performed when op_code is logic '1'. This can be represented by the following logic diagram. As shown in the logic diagram, the subtraction is performed by using 2's complement of b_in. Hence subtraction (a_in-b_in = a_in + b_in +1) operation uses the same resources and this technique is called as resource sharing.

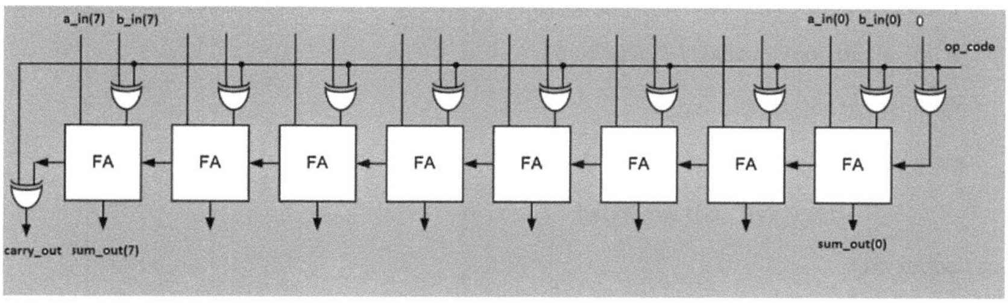

4.2.1 *Multibit Adder*

If we consider any processor, then the adders are used to perform the addition or subtraction operations. For 8-bit processor, the ALU can consist of 8-bit adder. The subtraction operation can be implemented as 2's complement addition. There are many efficient techniques to reduce the area, and these techniques are resource sharing, optimization, and grouping and will be discussed in the next subsequent chapters. This section describes the multibit adder–subtractor RTL using VHDL.

The RTL description of 8-bit synthesizable adder using VHDL is shown in Example 4.1, and the synthesis result is shown in Fig. 4.5.

As shown in the above figure, the 8-bit adder is implemented using two 8-bit half adders, and the overall propagation delay of combinational logic is $2{*}t_{pd.}$ If the standard cell is available as full adder, then 8-bit adder can be implemented using the 8 full adders.

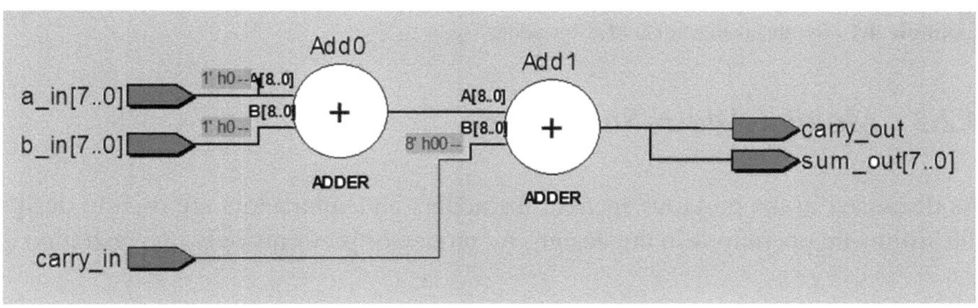

Fig. 4.5 Synthesis result of 8-bit adder

```
--8-Bit adder
library ieee;
use ieee.std_logic_1164.all;
use ieee.std_logic_arith.all;
use ieee.std_logic_unsigned.all;
use ieee.numeric_bit.all;
entity adder_8bit is
port ( a_in : in std_logic_vector ( 7 downto 0);
     b_in : in std_logic_vector ( 7 downto 0);
    carry_in : in std_logic;
    sum_out : out std_logic_vector ( 7 downto 0);
                carry_out : out std_logic);
end adder_8bit;
architecture arch_adder_8bit of adder_8bit is
signal temp_result : std_logic_vector ( 8 downto 0);
signal temp_sig1, temp_sig2, temp_sig3 : std_logic_vector (8 downto 0 );
begin
  temp_sig1 <= '0' &a_in;
  temp_sig2 <= '0' &b_in;
  temp_sig3 <= "00000000" &carry_in;
  temp_result <= (temp_sig1) + (temp_sig2) + (temp_sig3);
  sum_out <= temp_result (7 downto 0);
  carry_out <= temp_result (8);
end arch_adder_8bit;
```

- Architecture defines the functionality of design.
- The code generates parallel logic using signal assignment statements.
- The 'temp_result' holds the intermediate result. The size of 'temp_result' is declared as 9-bit and it is of std_logic type.
- The 'sum_out' is 8-bit output and is assigned from temp_result(7 downto 0).
- The 'carry_out' is single bit and is assigned from temp_result(8).

Example 4.1 Synthesizable RTL of 8-bit adder

4.2.2 Multibit Adder–Subtractor

As discussed in the previous section, the adders and subtractors are used to design the arithmetic operations in the design. As processor performs only one operation at a time, the synthesizer uses the adders to perform the subtraction. The subtraction is performed using 2's complement addition. The 8-bit adder–subtractor RTL using VHDL is shown in Example 4.2, and the synthesis result is shown in Fig. 4.6

```
--8-Bit adder subtractor
library ieee;
use ieee.std_logic_1164.all;
use ieee.std_logic_arith.all;
use ieee.std_logic_unsigned.all;
use ieee.numeric_bit.all;
entity adder_sub_8bit is
port ( a_in : in std_logic_vector ( 7 downto 0);
      b_in : in std_logic_vector ( 7 downto 0);
     carry_in : in std_logic;
                  op_code : in std_logic;
     sum_out : out std_logic_vector ( 7 downto 0);
                  carry_out : out std_logic);
end adder_sub_8bit;
architecture arch_adder_sub_8bit of adder_sub_8bit is
signal temp_result : std_logic_vector ( 8 downto 0);
signal temp_sig1, temp_sig2, temp_sig3 : std_logic_vector (8 downto 0 );
begin
 temp_sig1 <= '0' &a_in;
 temp_sig2 <= '0' &b_in;
 temp_sig3 <= "00000000" & carry_in;
 temp_result <= ((temp_sig1) + (temp_sig2) + (temp_sig3))
                  when (op_code ='1') else
                   ((temp_sig1) - (temp_sig2) - (temp_sig3));
 sum_out <= temp_result (7 downto 0);
 carry_out <= temp_result (8);
end arch_adder_sub_8bit;
```

- Architecture defines the functionality of design.
- The code generates parallel logic using signal assignment statements.
- The 'temp_result' holds the intermediate result. The size of 'temp_result' is declared as 9-bit and it is of std_logic type.
- The 'sum_out' is 8-bit output and is assigned from temp_result(7 downto 0).
- The 'carry_out' is single bit and is assigned from temp_result(8).
- For 'op_code=1' it performs the addition operation and for 'op_code=0' it performs the subtraction.

Example 4.2 Synthesizable RTL of 8-bit adder–subtractor

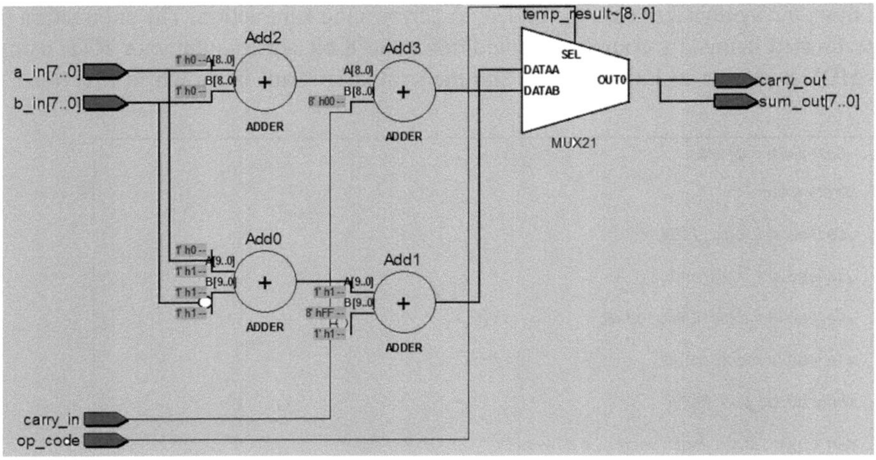

Fig. 4.6 Synthesis result of 8-bit adder–subtractor

As shown in the synthesis result, it uses four half adders of 8 bit and multiplexing logic to select the output from adder or subtractor depending on the status of opcode. The logic inferred can be minimized using the resource sharing and is discussed in Chap. 8.

4.2.3 Multiplier

Multipliers are used in the digital signal processing applications. It is the requirement that multiplier should have lesser area and higher speed. This will reduce the overall combinational delay. The multiplication is perfumed using operator '*'. There are different types of multiplication algorithms can be used in the design of digital circuit, and one of the best algorithms is Booth multiplier. This section discusses the basic 8-bit multiplier using VHDL operator '*'. The RTL is described in Example 4.3, and the synthesis result is shown in Fig. 4.7.

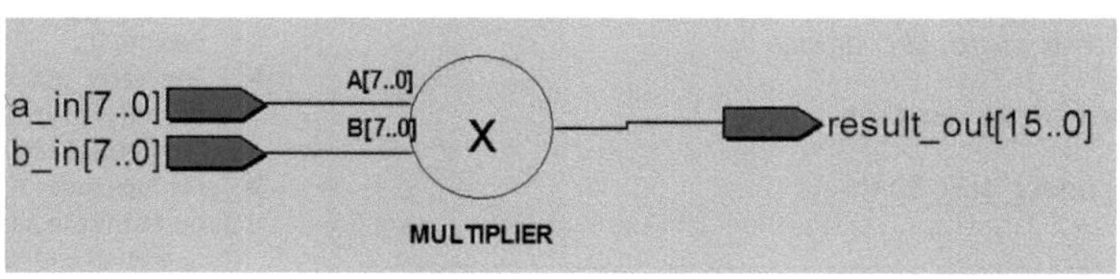

Fig. 4.7 Synthesis result of 8-bit multiplier

```
--8 bit multiplier
library ieee;
use ieee.std_logic_1164.all;
use ieee.std_logic_arith.all;
use ieee.std_logic_unsigned.all;
use ieee.numeric_bit.all;
entity multiplier_8bit is
port ( a_in : in std_logic_vector ( 7 downto 0);
        b_in : in std_logic_vector ( 7 downto 0);
       result_out : out std_logic_vector ( 15 downto 0));
end multiplier_8bit;
architecture arch_multiplier_8bit of multiplier_8bit is
begin
 process ( a_in , b_in )
 begin
        result_out <= a_in * b_in;
 end process;
end arch_multiplier_8bit;
```

- Architecture defines the functionality of design.
- Process is sensitive to 'a_in', and 'b_in'. Any event on one of the signal executes the process.
- The 'result_out' is 16 bit and the multiplication of 'a_in' and 'b_in' is assigned to 'result_out'.

Example 4.3 Synthesizable RTL of 8-bit multiplier

4.2.4 Comparators

The comparators are used to compare the magnitude of two numbers. If both numbers are having same magnitude, then it generates an output 'equal_out' to logic '1'; if a_in is less as compare to b_in, then it generates an output 'greater_out' equal to logic '1'; and if a_in is less as compare to b_in, then it generates an output 'less_out' equal to logic '1'. The RTL using VHDL is shown in Example 4.4, and the synthesis result is shown in Fig. 4.8.

```
--8 bit comparator
library ieee;
use ieee.std_logic_1164.all;
use ieee.std_logic_unsigned.all;
use ieee.std_logic_arith.all;
entity comparator_8bit is
 port ( a_in : in std_logic_vector (7 downto 0);
        b_in : in std_logic_vector (7 downto 0);
       equal_out : out std_logic;
       less_out : out std_logic;
       greater_out : out std_logic);
end comparator_8bit;
architecture arch_comparator of comparator_8bit is
begin
 equal_out <= '1' when (a_in = b_in) else '0';
 less_out <= '1' when (a_in < b_in) else '0';
 greater_out <= '1' when (a_in > b_in) else '0';
 end arch_comparator ;
```

- Architecture defines the functionality of design.
- The code generates parallel logic using signal assignments.
- Signal assignments are continuous in nature and for (a_in = b_in) it generates binary '1' at 'equal_out'.
- For (a_in > b_in) it generates binary '1' at 'greater_out'.
- For (a_in < b_in) it generates binary '1' at 'lessl_out'.

Example 4.4 Synthesizable VHDL code of 8-bit comparator

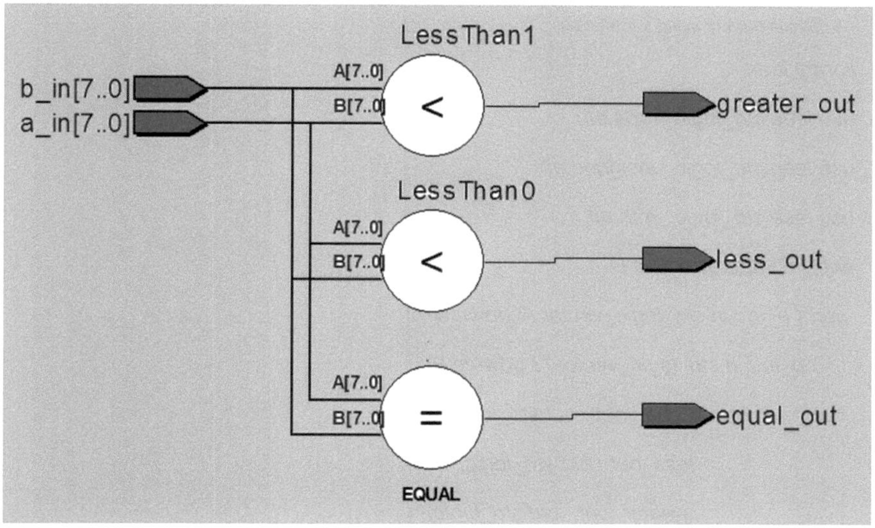

Fig. 4.8 Synthesis result of 8-bit comparator

As shown in the above synthesis result, it infers the parallel logic using the comparator elements, for this design it uses three 8-bit comparators. The RTL code can be modified by using the 'if then else' construct to reduce the area of the design.

The RTL using if then else construct VHDL is shown in Example 4.5, and the equivalent synthesis result is shown in Fig. 4.9.

As shown in the synthesis result, it uses more number of multipliers and less number of comparators to generate the comparison results; hence, there is area reduction. In the PLD based designs the multiplexing logic occupies the lesser area.

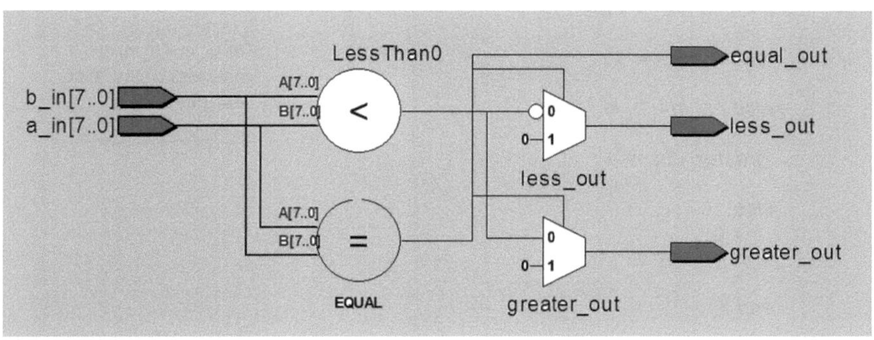

Fig. 4.9 Synthesis result of 8-bit comparator using if-then-else

```vhdl
--8 bit comparator using if-then-else
library ieee;
use ieee.std_logic_1164.all;
use ieee.std_logic_unsigned.all;
use ieee.std_logic_arith.all;
entity comparator_8bit is
 port ( a_in : in std_logic_vector (7 downto 0);
        b_in : in std_logic_vector (7 downto 0);
                    equal_out : out std_logic;
                    less_out : out std_logic;
                    greater_out : out std_logic);
end comparator_8bit;
architecture arch_comparator of comparator_8bit is
begin
  process ( a_in, b_in )
  begin
   equal_out <= '0';
   less_out <= '0';
   greater_out <= '0';
         if ( a_in = b_in ) then
          equal_out <= '1';
         elsif ( a_in > b_in ) then
          greater_out <='1';
         else
          less_out <='1';
         end if;
 end process;
 end arch_comparator ;
```

- Architecture defines the functionality of design.
- The code generates parallel logic using signal assignments.
- Nested If-then-else is used inside the process. Process is sensitive to input 'a_in' and 'b_in'.
- For (a_in = b_in) it generates binary '1' at 'equal_out'.
- For (a_in > b_in) it generates binary '1' at 'greater_out'.
- For (a_in < b_in) it generates binary '1' at 'less_out'.

Example 4.5 Synthesizable VHDL for the 8-bit comparator

4.3 Code Converter

In many applications, the code converters are used. The code converters are used to convert one form of code into another form. For example, binary-to-gray code converter converts the binary number into the gray, and BCD-to-Excess-3 code converter converts the binary number into the Excess-3. In the similar way, the BCD-to-seven-segment decoder is used to convert the BCD code into the equivalent seven-segment representation.

The gray codes are used in the multiple clock domain designs as only one bit changes in the two successive gray codes. Seven-segment code representations are used to display the BCD number on the seven-segment display.

4.3.1 Binary-to-Excess-3 Code Converter

As the name indicates, the Excess-3 code can be generated by adding binary '0011' in the binary number. The RTL using VHDL is shown in Example 4.6, and the synthesis result is shown in Fig. 4.10.

As shown in the synthesis result, the hardware inferred is parallel and combinational in nature. The BCD-to-Excess-3 code can be implemented by using the addition operator by adding the value '0011' in the respective input. The modified architecture is shown below, and the synthesis result for the modified architecture is shown in Fig. 4.11.

```
architecture arch1_bin_to_excess3 of binary_to_excess3 is

begin

 excesss3_out <= bin_in + "0011"

;end arch1_bin_to_excess3;
```

```
--Binary to Excess 3 code converter
library ieee;
use ieee.std_logic_1164.all;
use ieee.std_logic_unsigned.all;
use ieee.std_logic_arith.all;
entity binary_to_excess3 is
 port ( bin_in : in std_logic_vector (3 downto 0);
      excesss3_out : out std_logic_vector (3 downto 0));
end binary_to_excess3;
architecture arch_bin_to_excess3 of binary_to_excess3 is
begin
 process ( bin_in )
 begin
  case ( bin_in) is
      when "0000" => excesss3_out <= "0011";
      when "0001" => excesss3_out <= "0100";
      when "0010" => excesss3_out <= "0101";
      when "0011" => excesss3_out <= "0110";
      when "0100" => excesss3_out <= "0111";
      when "0101" => excesss3_out <= "1000";
      when "0110" => excesss3_out <= "1001";
      when "0111" => excesss3_out <= "1010";
      when "1000" => excesss3_out <= "1011";
      when "1001" => excesss3_out <= "1100";
      when "1010" => excesss3_out <= "1101";
      when "1011" => excesss3_out <= "1110";
      when "1100" => excesss3_out <= "1111";
      when others => excesss3_out <= "zzzz";
      end case;
end process;
end arch_bin_to_excess3;
```

- Architecture defines the functionality of design.
- The code generates parallel logic using case construct.
- Process is sensitive to input 'bin_in'.
- Depending on the binary code at 'bin_in' it generates the equivalent 'Excess 3' code at output ' excess3_out'.

- For the binary inputs (case conditions) "1101, 1110 and 1111", during synthesis use the **when others => excess3_out<="ZZZZ";**
- For the BCD to Excess-3 conversion use the case conditions from "0000" to "1001" and for the remaining conditions use **when others => excess3_out<="ZZZZ";**
- Synthesis result will be the combinational logic using multiplexers.

Example 4.6 Synthesizable RTL for binary-to-excess-3 code converter

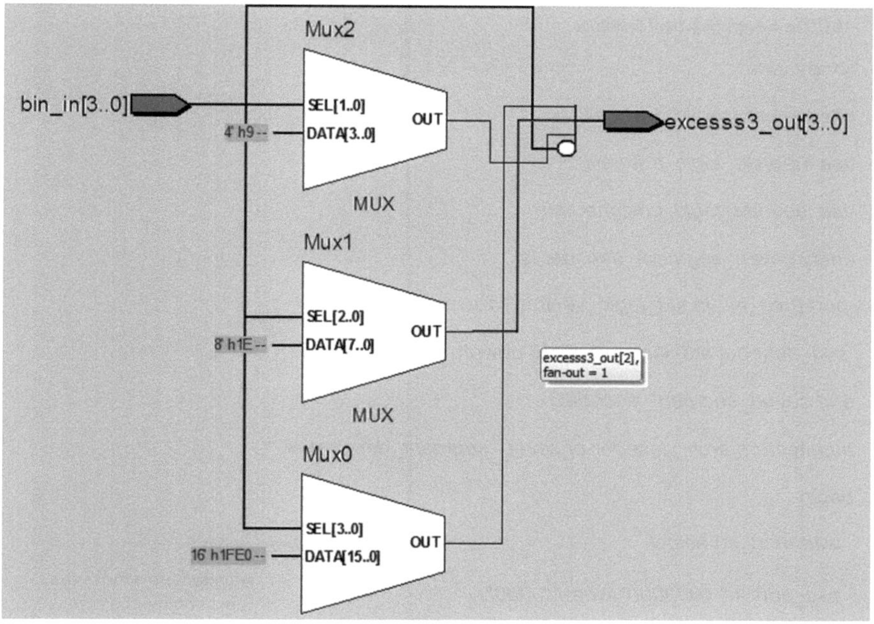

Fig. 4.10 Synthesis result for binary-to-excess-3 code converter

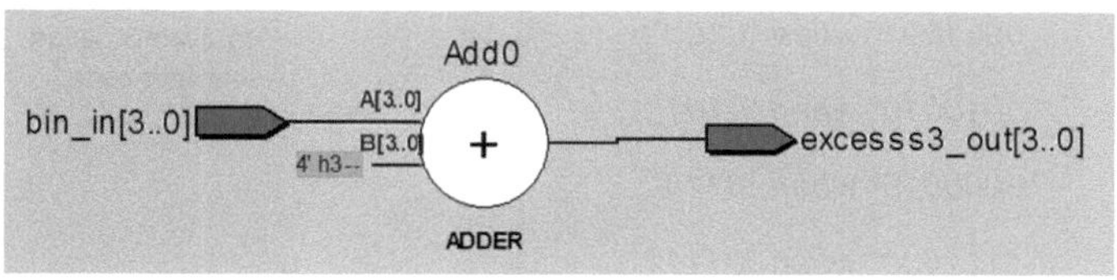

Fig. 4.11 Synthesis result for binary-to-excess-3 code converter using addition operator

4.3.2 BCD-to-Seven-Segment Decoder

The given BCD number can be converted using the BCD-to-seven-segment decoder and can be used in the system design to display the result. The RTL for the BCD-to-seven-segment decoder is described in Example 4.7, and the synthesis result is shown in Fig. 4.12. It is assumed that zero at the 'seg_out' enables the segment.

```
--BCD to Seven Segment Decoder
library ieee;
use ieee.std_logic_1164.all;
use ieee.std_logic_arith.all;
use ieee.std_logic_unsigned.all;
entity seven_segment_decoder is
 port (bcd_in : in std_logic_vector(3 downto 0);
 seg_out : out std_logic_vector(6 downto 0));
end seven_segment_decoder;
architecture arch_decoder of seven_segment_decoder is
begin
  with (bcd_in) select
  seg_out<= " 1000000" when "0000";
              " 1111001" when "0001";
              " 0100100" when "0010";
              " 1110000" when "0011";
              " 0011001" when "0100";
              " 0010010" when "0101";
              " 0000010" when "0110";
              " 1111000" when "0111";
              " 0000000" when "1000";
              " 0010000" when "1001";
              "-------" when others;
end  arch_decoder ;
```

- Architecture defines the functionality of design.
- The code generates parallel logic using concurrent assignment statement.
- For the 4-bit 'bcd_in' input it generates the equivalent seven segment code.

Example 4.7 Synthesizable VHDL RTL of BCD-to-seven-segment decoder

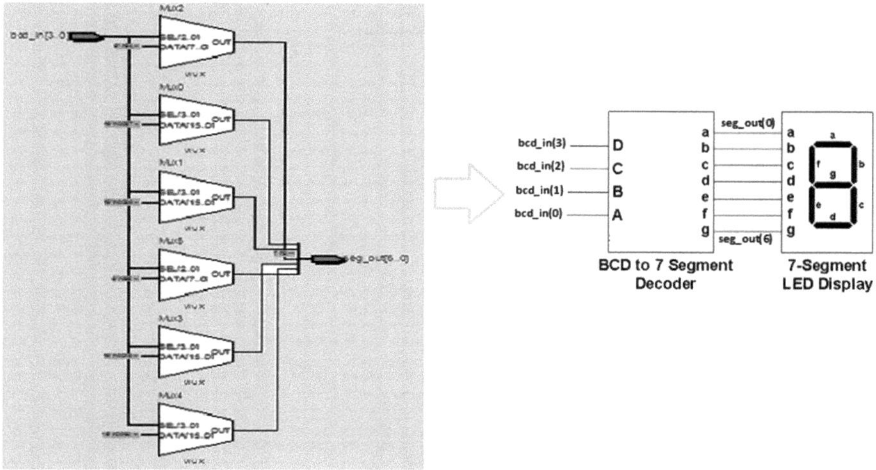

Fig. 4.12 Synthesis result of BCD-to-seven-segment decoder

4.4 Multiplexers

Multiplexers are used to select one of the inputs from many. Multiplexers are also called as universal logic, and terminology used in the practical world is MUX. By using the suitable multiplexers, any of the combinational logic function can be realized. Multiplexers are used as selection logic in ASIC and FPGA-based designs. Multiplexer consumes lesser area as compare to adders, and most of the time, multiplexers are used to implement arithmetic components such as adders and subtractors.

The block diagram of n:1 MUX is shown in Fig. 4.13, and it consists of ‘n’ input lines, ‘m’ select lines, and one output line. Input lines are denoted by ‘i(0), i(1), …, i(n − 1)’; select lines by ‘s(0), s(1), …, s(m − 1)’; and output line by ‘y’.

As shown in Fig. 4.13, multiplexer has ‘n’ input lines, ‘m’ select lines, and single output line. Relation between the input lines and select lines is given by $n = 2^m$. For example, for 4:1 MUX, input lines are four so $m = \log_2 n$, that is select lines are equal to two.

Let us consider 4:1 MUX having four input lines ‘a_in(0) to a_in(3),’ two select lines ‘sel_in(0) to sel_in(1),’ and single output line ‘y_out,’ at a time instance the information on one of the input line is available on the output and is shown in Fig. 4.14.

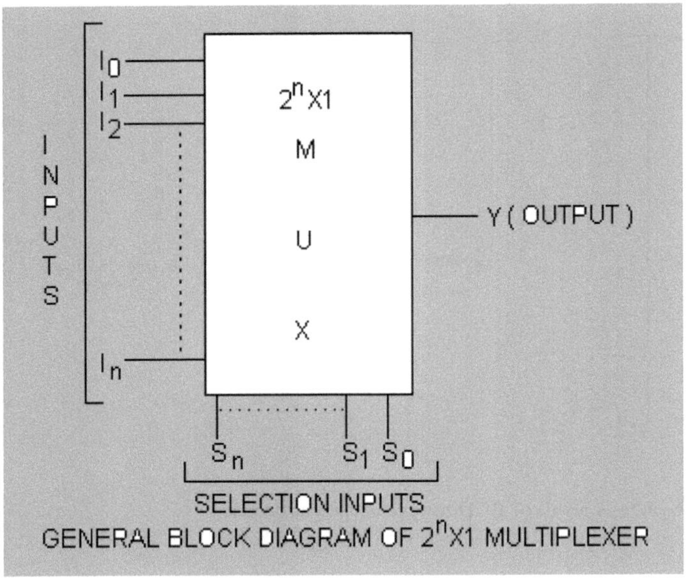

Fig. 4.13 Block diagram of n:1 MUX

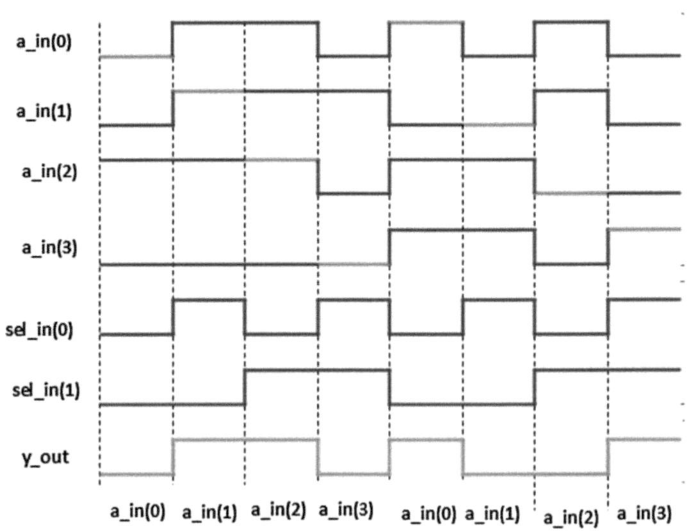

Fig. 4.14 Timing sequence of 4:1 MUX

4.4.1 Multiplexer as Universal Logic

As discussed earlier, multiplexer is treated as universal logic as all types of combinational logic functions can be realized using MUX.

The logic realization of NOT gate using single 2:1 MUX is shown in Fig. 4.15.

As shown in Fig. 4.15, the a_in is used as select input, and when it is logic '0', the output y_out is logic '1'. When a_in is logic '1', the output y_out is logic '0'.

Figure 4.16 shows the realization of two-input XOR logic using the 2:1 MUX.

As shown in Fig. 4.16, a_in is used as select line of 2:1 MUX, the output y_out is equal to b_in for a_in is equal to logic '0'. For a_in is equal to logic '1', the output y_out of 2:1 MUX is complement of b_in. In this, it is assumed that NOT gate is realized using 2:1 MUX. So to implement XOR logic, two tow to one multiplexers are required. The concept of realizing logic using MUX is used in the design of configurable or programmable logic and will be discussed in the subsequent chapters.

The implementation of 2-input OR gate is shown in Fig. 4.17, and as shown, it uses the single MUX to realize the OR logic.

As shown in Fig. 4.17, a_in is used as select line of 2:1 MUX, the output y_out is equal to b_in for the a_in is equal to logic '0'. For a_in is equal to logic '1', the output y_out of 2:1 MUX is logic '1'. Readers can implement the AND, XNOR, NOR, and NAND logics using minimum number of multiplexers.

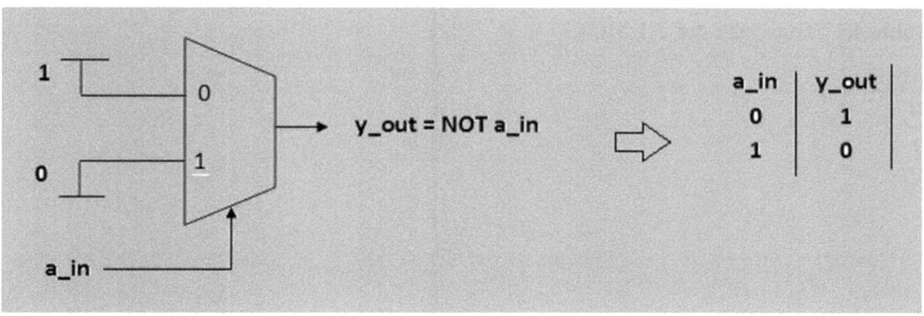

Fig. 4.15 NOT logic realization using 2:1 MUX

a_in	b_in	y_out
0	0	0
0	1	1
1	0	1
1	1	0

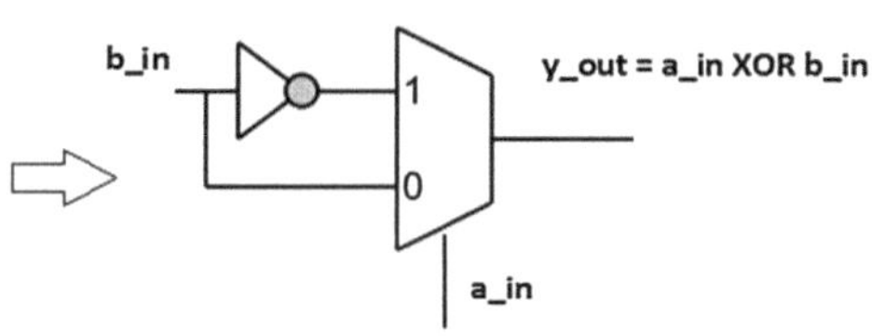

Fig. 4.16 XOR realization using 2:1 MUX

Fig. 4.17 OR logic realization using 2:1 MUX

4.4.1.1 2:1 MUX

A 2:1 MUX has two input lines, one select line, and one output line. When 'sel_in' input is logical '0', output 'y_out' is assigned to 'a_in' and input 'b_in' is assigned to 'y_out' for 'sel_in' equal to logical '1'. Table 4.1 describes the truth table of 2:1 MUX, and implementation using logic gates is represented in Fig. 4.18.

The RTL for the 2:1 MUX using 'if then else' construct is shown in Example 4.8, and the synthesis result is shown in Fig. 4.19.

Note If then else is used to infer the multiplexer. If-else clause is eliminated, then it infers latches.

Table 4.1 Truth table for 2:1 MUX

sel_in	y_out
0	a_in
1	b_in

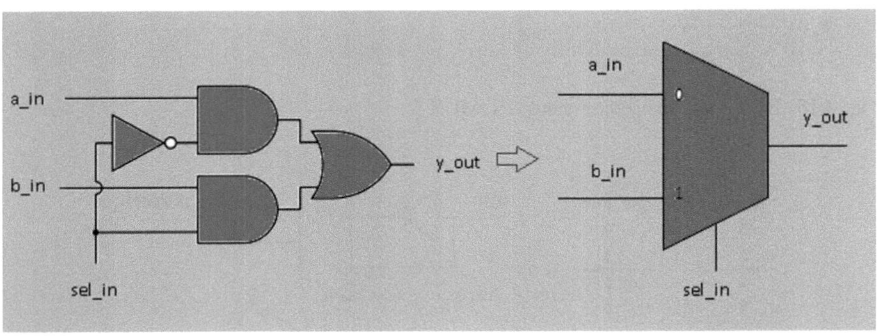

Fig. 4.18 2:1 MUX as universal logic cell

```
--mux 2 to1 using if then else
library ieee;

use ieee.std_logic_1164.all;

use ieee.std_logic_arith.all;

use ieee.std_logic_unsigned.all;

entity mux_2to1 is

port ( a_in : in std_logic;

        b_in : in std_logic;

       sel_in : in std_logic;

      y_out : out std_logic);

end mux_2to1;

architecture arch_mux_2to1 of mux_2to1 is

begin

 P1:  process ( a_in, b_in, sel_in)

    begin

                    if ( sel_in ='1') then

                       y_out <= b_in;

                   else

                       y_out <= a_in;

                  end if;

      end process;

end arch_mux_2to1;
```

- Architecture defines the functionality of design.
- Process is sensitive to 'a_in', 'b_in' and 'sel_in'. Any event on one of the signal invokes the process.
- If-then-else is sequential statement and used inside the process.
- For true 'sel_in' condition the input 'b_in' is assigned to 'y_out'.
- For false 'sel_in' condition the input 'a_in' is assigned to 'y_out'

Example 4.8 Synthesizable VHDL RTL for 2:1 MUX

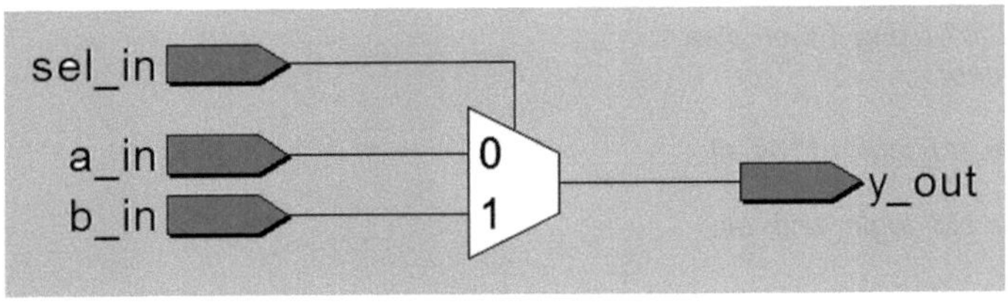

Fig. 4.19 Synthesized 2:1 MUX. *Note* A 2:1 multiplexer symbolic representation is used to describe the implementation of higher complexity multiplexers. Multiplexer is treated as universal logic. Using multiplexers, all possible combinational logic can be realized

Fig. 4.20 Two-input XOR logic using 2:1 MUX

a	b	y
0	0	0
0	1	1
1	0	1
1	1	0

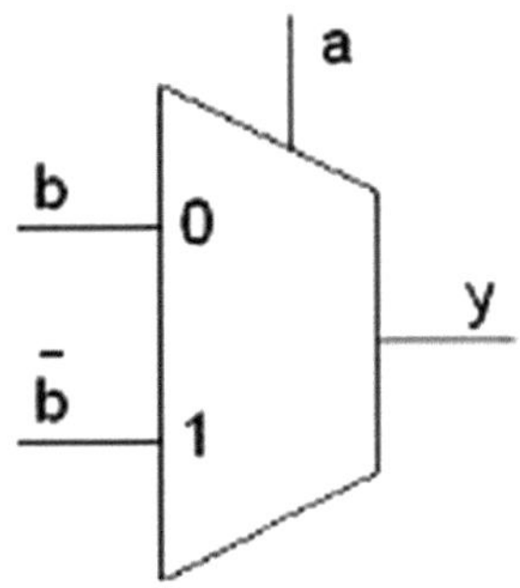

The reason for using MUX as universal logic is because it is easy to understand and is simple to implement. Figure 4.20 describes how 2:1 MUX is used to implement the two-input XOR logic gate. Consider XOR logic gate has two inputs 'a', 'b' and output 'y'. The implementation of two-input XOR logic gate using 2:1 MUX is shown in Fig. 4.20.

Let us discuss the other ways to describe the 2:1 MUX. There are different ways in which 2:1 MUX can be described. It can be described by using 'if then else' or by using 'case' construct. The VHDL RTL of 2:1 MUX using 'case' construct is shown in Example 4.9, and the synthesis result is shown in Fig. 4.21.

4.4.1.2 4:1 MUX Using Nested 'If Then Else'

The 4:1 MUX has four input lines and single output line. The 4:1 MUX has two select line and is used to select one of the inputs at a time. The truth table of 4:1 MUX is shown in Table 4.2, and Example 4.10 describes the synthesizable RTL for 4:1 MUX.

```
--mux 2 to1 using case
library ieee;
use ieee.std_logic_1164.all;
use ieee.std_logic_arith.all;
use ieee.std_logic_unsigned.all;
entity mux_2to1_case is
port ( a_in : in std_logic;
        b_in : in std_logic;
       sel_in : in std_logic;
        y_out : out std_logic);
end mux_2to1_case;
architecture arch_mux_2to1 of mux_2to1_case is
begin
 P1:  process ( a_in, b_in, sel_in)
     begin
          case (sel_in) is
               when '0'     => y_out <= a_in;
              when others => y_out <= b_in;
        end case;
        end process;
end arch_mux_2to1;
```

- Architecture defines the functionality of design.
- Process is sensitive to 'a_in', 'b_in' and 'sel_in'. Any event on one of the signal invokes the process.
- Case is sequential statement and used inside the process.
- For true 'sel_in' condition the input 'b_in' is assigned to 'y_out'.
- For false 'sel_in' condition the input 'a_in' is assigned to 'y_out'

Example 4.9 Synthesizable VHDL RTL for 2:1 MUX using 'case'

An equivalent synthesis result for the 4:1 MUX described in the above example is shown in Fig. 4.22. As shown in Fig. 4.22, input 'a_in(0)' has highest priority as compare to other inputs. Input 'a_in(3)' has least priority.

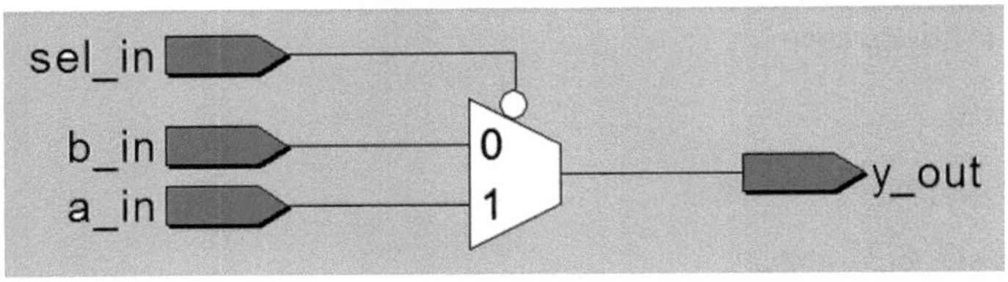

Fig. 4.21 Synthesis result of 2:1 MUX using 'case' construct. *Note* 'if then else' generates priority logic, and 'case' generates parallel logic. It is recommended to use 'case' statement to describe MUX and decoders. It is recommended to use 'if then else' to describe priority logic

Table 4.2 Truth table of 4:1 MUX

sel_in(1)	sel_in(0)	y_out
0	0	a_in(0)
0	1	a_in(1)
1	0	a_in(2)
1	1	a_in(3)

4.4.1.3 4:1 MUX Using 'Case' Construct

The 4:1 MUX is described by using the 'case' sequential construct, and it is described in Example 4.11. The synthesis result is shown in Fig. 4.23. As shown in the figure, 'case' construct generates the parallel logic (Example 4.11).

4.5 Decoders

Decoder has 'n' select lines or input lines and 'm' output lines and is used to generate either active high output or active low output. The relation between select lines and output lines is given by $m = 2^n$. Depending on the logic status on 'n' input lines, at a time one of the output lines goes high or low.

If we consider the decoder having two select lines 'sel_in(0) and sel_in(1)' and four output lines 'y_out(0) to y_out(3),' then depending on the status of select inputs, one of the output lines goes high and is shown in Fig. 4.24.

4.5.1 3 Line to 8 Decoder with Enable Using 'Case'

Figure 4.25 shows 3:8 decoder; X_2, X_1, and X_0 are select inputs, and Y_0 to Y_7 are active high output lines.

The truth table of 3:8 decoder is shown in Table 4.3. For the decoder having active high output, at a time one of the output lines is active high.

```
--mux 4 to1 using nested if-then-else
library ieee;
use ieee.std_logic_1164.all;
use ieee.std_logic_arith.all;
use ieee.std_logic_unsigned.all;
entity mux_4to1 is
port ( a_in : in std_logic_vector ( 3 downto 0);
       sel_in : in std_logic_vector (1 downto 0);
       y_out : out std_logic);
end mux_4to1;
architecture arch_mux_4to1 of mux_4to1 is
begin
 P1:  process ( a_in,sel_in)
     begin
         if ( sel_in ="00") then
             y_out <= a_in (0);
         elsif ( sel_in ="01") then
             y_out <= a_in (1);
         elsif ( sel_in ="10") then
             y_out <= a_in (2);
         else
             y_out <= a_in (3);
         end if;
  end process;
end arch_mux_4to1;
```

- Architecture defines the functionality of design.
- Process is sensitive to 'a_in', and 'sel_in'. Any event on one of the signal invokes the process.
- Nested if-then-else is used inside the process.
- Depending on the 'sel_in' codition one of the input is assigned to output 'y_out'.
- Nested if-the-else infers the priority logic.
- In this 'a_in(0)' has highest priority and input 'a_in(3) has the least priority among the inputs.

Example 4.10 Synthesizable VHDL RTL of 4:1 MUX using nested 'if-then-else'

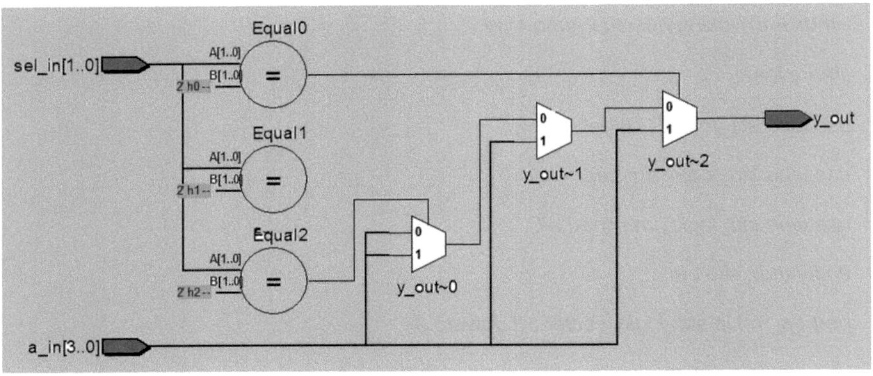

Fig. 4.22 Synthesized 4:1 MUX priority logic

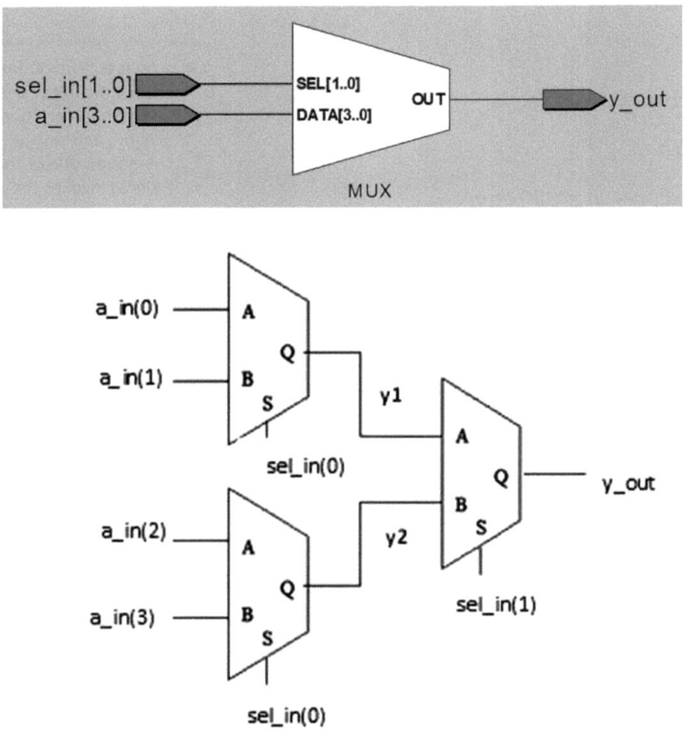

Fig. 4.23 Synthesis result for 4:1 MUX using 'case'

```
--mux 4 to1 using case
library ieee;
use ieee.std_logic_1164.all;
use ieee.std_logic_arith.all;
use ieee.std_logic_unsigned.all;
entity mux_4to1_case is
port ( a_in : in std_logic_vector ( 3 downto 0);
       sel_in : in std_logic_vector (1 downto 0);
        y_out : out std_logic);
end mux_4to1_case;
architecture arch_mux_4to1_case of mux_4to1_case is
begin
 P1: process ( a_in,sel_in)
     begin
     case (sel_in) is
      when "00" => y_out <= a_in (0);
      when "01" => y_out <= a_in (1);
      when "10" => y_out <= a_in (2);
     when others => y_out <= a_in (3);
     end case;
    end process;
end arch_mux_4to1_case;
```

- Architecture defines the functionality of design.
- Process is sensitive to ‘a_in’, and ‘sel_in’. Any event on one of the signal invokes the process.
- Case is used inside the process.
- Depending on the ‘sel_in’ condition one of the input is assigned to output ‘y_out’.
- Case generates the parallel logic.
- The last condition in the case is defined by using when others keyword.

Example 4.11 Synthesizable VHDL RTL using ‘case’

```
--Decoder 3 to 8 using case
library ieee;

use ieee.std_logic_1164.all;

use ieee.std_logic_unsigned.all;

use ieee.std_logic_arith.all;

entity decoder_3to8 is

port ( sel_in : in std_logic_vector ( 2 downto 0);

        enable_in : in std_logic;

        y_out : out std_logic_vector ( 7 downto 0));

end decoder_3to8;

architecture arch_decoder_3to8 of decoder_3to8 is

begin

    process ( sel_in, enable_in)

            begin

            if ( enable_in='1') then

                    y_out <= "11111111";

            else

             case ( sel_in) is

                    when "000" => y_out <= "11111110";

                    when "001" => y_out <= "11111101";

                    when "010" => y_out <= "11111011";

                    when "011" => y_out <= "11110111";

                    when "100" => y_out <= "11101111";

                    when "101" => y_out <= "11011111";

                    when "110" => y_out <= "10111111";

                    when "111" => y_out <= "01111111";

                    when others => null;

                    end case;

                    end if;

            end process;

end arch_decoder_3to8;
```

- Architecture defines the functionality of design.
- Process is sensitive to ‘sel_in’, and ‘enable_in’. Any event on one of the signal invokes the process.
- If-then-else is sequential statement and used inside the process.
- For true ‘enable_in’ condition all output lines assigned to logic ‘1’.
- For ‘enable_in’ active low input decoder is enabled and one of the output line is active low.
- The described code generates parallel logic.

Example 4.12 Synthesizable RTL of 3:8 decoder using ‘case’

Table 4.3 is the truth table of 3:8 decoder without the enable input. The truth table described aboveholds good for the decoder with active high enable ‘en=1.’ When ‘en=0,’ decoder is disabled and itgenerates an output ‘Y=00000000.’ For decoder having active high enable input the gate levelrepresentation (Fig. 4.25) can be modified by using four input AND gates.

The RTL description by using synthesizable VHDL constructs for 3:8 decoder having active low enable input and active low output lines is shown in Example 4.12.

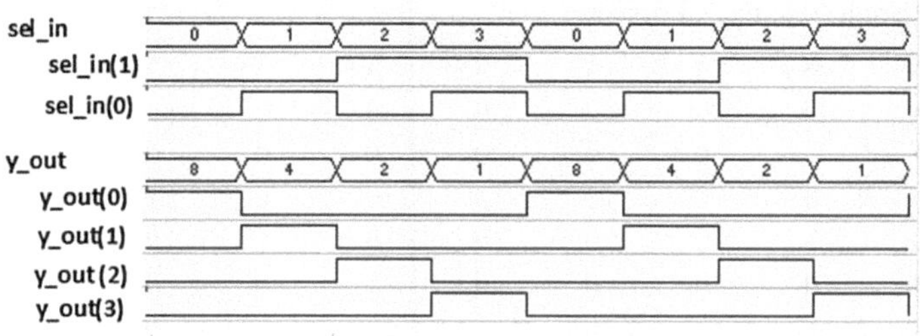

Fig. 4.24 Timing sequence of 2:4 decoder

Fig. 4.25 Gate-level representation of 3:8 decoder

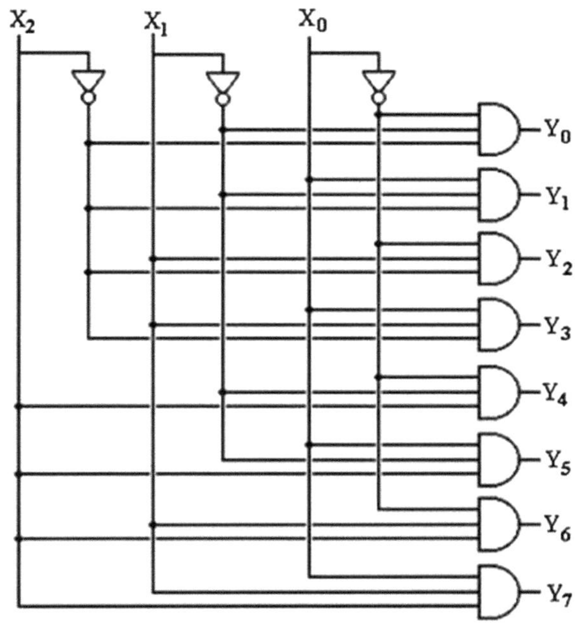

4.5.2 2 Line to 4 Decoder with Enable Using 'Case'

The 2 line to 4 or (2:4) decoder has two select inputs 'sel_in (1), sel_in(0),' enable input 'enable_in,' and four output lines 'y_out(0) to y_out(3).' The truth table and equivalent representation are shown in Table 4.4

The synthesizable VHDL RTL is described in Example 4.13, and the equivalent hardware inferred is shown in Fig. 4.26 (Examples 4.13).

Table 4.3 Truth table for 3:8 decoder

X_2	X_1	X_0	Y_7	Y_6	Y_5	Y_4	Y_3	Y_2	Y_1	Y_0
0	0	0	0	0	0	0	0	0	0	1
0	0	1	0	0	0	0	0	0	1	0
0	1	0	0	0	0	0	0	1	0	0
0	1	1	0	0	0	0	1	0	0	0
1	0	0	0	0	0	1	0	0	0	0
1	0	1	0	0	1	0	0	0	0	0
1	1	0	0	1	0	0	0	0	0	0
1	1	1	1	0	0	0	0	0	0	0

Note In the practical applications, decoders are used to select one of the memories or input–output device at a time. To enable the expansion of decoder, decoder always has either active high enable or active low enable

Table 4.4 Truth table for 2:4 decoder

enable_in	sel_in(1)	sel_in(0)	y_out(3)	y_out(2)	y_out(1)	y_out(0)
0	0	0	0	0	0	1
0	0	1	0	0	1	0
0	1	0	0	1	0	0
0	1	1	1	0	0	0
1	X	X	0	0	0	0

4.6 Encoders

The function of an encoder is the reverse of the decoder. Encoder has ‘n’ input lines and ‘m’ output lines, and the relation between input lines and output lines is given by $n = 2^m$. For example, consider 4:2 encoder. The number of input lines is n = 4 and output lines is m = 2.

If we consider the encoder having two output lines ‘y_out(1) and y_out(0)’ and four input lines ‘sel_in(0) to sel_in(3),’ then depending on the status of select inputs, output is generated the timing sequence and is shown in Fig. 4.27.

The truth table is described in Table 4.5.

The VHDL RTL description for 4:2 encoder is described in Example 4.14. The VHDL RTL infers the hardware as shown in Fig. 4.28 (Example 4.14).

4.6.1 Priority Encoders

Priority encoders are used in the practical applications and have ‘n’ input lines and ‘m’ output lines, and the relation between input lines and output lines is given by $n = 2^m$. For example, consider 4:2 priority encoder. The number of input lines is

```
--Decoder 2 to 4
library ieee;
use ieee.std_logic_1164.all;
use ieee.std_logic_arith.all;
use ieee.std_logic_unsigned.all;
entity decoder_2to4 is
port ( sel_in : in std_logic_vector (1 downto 0);
       enable_in : in std_logic;
        y_out : out std_logic_vector ( 3 downto 0));
end decoder_2to4;
architecture arch_decoder_2to4 of decoder_2to4 is
begin
  P1:  process ( enable_in, sel_in)
     begin
       if (enable_in ='1') then
         y_out <= "0000" ;
       else
         case (sel_in) is
           when "00" => y_out <= "0001" ;
           when "01" => y_out <= "0010" ;
           when "10" => y_out <= "0100" ;
           when "11" => y_out <= "1000" ;
           when others => null;
         end case;
     end if;
end process;
end arch_decoder_2to4;
```

- Architecture defines the functionality of design.
- Process is sensitive to 'enable_in', and 'sel_in'. Any event on one of the signal invokes the process.
- Case is used inside the process. For active low value on 'enable_in' input the case statement is executed.
- Depending on the 'sel_in' condition one of the output line goes high at a time.
- Case generates the parallel logic.
- The last condition in the case is defined by using when others keyword.

Example 4.13 Synthesizable VHDL RTL for 2:4 decoder

```
--Encoder 4to2 using if-then-else
library ieee;

use ieee.std_logic_1164.all;

use ieee.std_logic_unsigned.all;

use ieee.std_logic_arith.all;

entity encoder_4to2 is

port ( sel_in : in std_logic_vector ( 3 downto 0);

     enable_in : in std_logic;

     y_out : out std_logic_vector ( 1 downto 0));

end encoder_4to2;

architecture arch_encoder_4to2 of encoder_4to2 is

begin

   process ( sel_in, enable_in)

         begin

          if ( enable_in='1') then

                 y_out <= "00";

         else

           if ( sel_in ="1000") then

                 y_out <= "11";

                elsif ( sel_in ="0100") then

                 y_out <= "10";

                elsif ( sel_in ="0010") then

                 y_out <= "01";

                else

                  y_out <= "00";

                end if;

        end if;

        end process;

end arch_encoder_4to2;
```

- Architecture defines the functionality of design.
- Process is sensitive to 'sel_in', and 'enable_in'. Any event on one of the signal invokes the process.
- If-then-else is sequential statement and used inside the process.
- For true 'enable_in' condition all output lines assigned to logic '0'.
- For 'enable_in' active low , encoder is enabled and depending on the priority of signal the two bit output is generated at 'y_out'.
- The described code generates priority logic, Input 'sel_in(3)' has highest priority and 'sel_in(0)' has the least priority.
- If more than one input line is active this logic will not be able to assign the priority output.

Example 4.14 Synthesizable VHDL RTL for 4:2 encoder

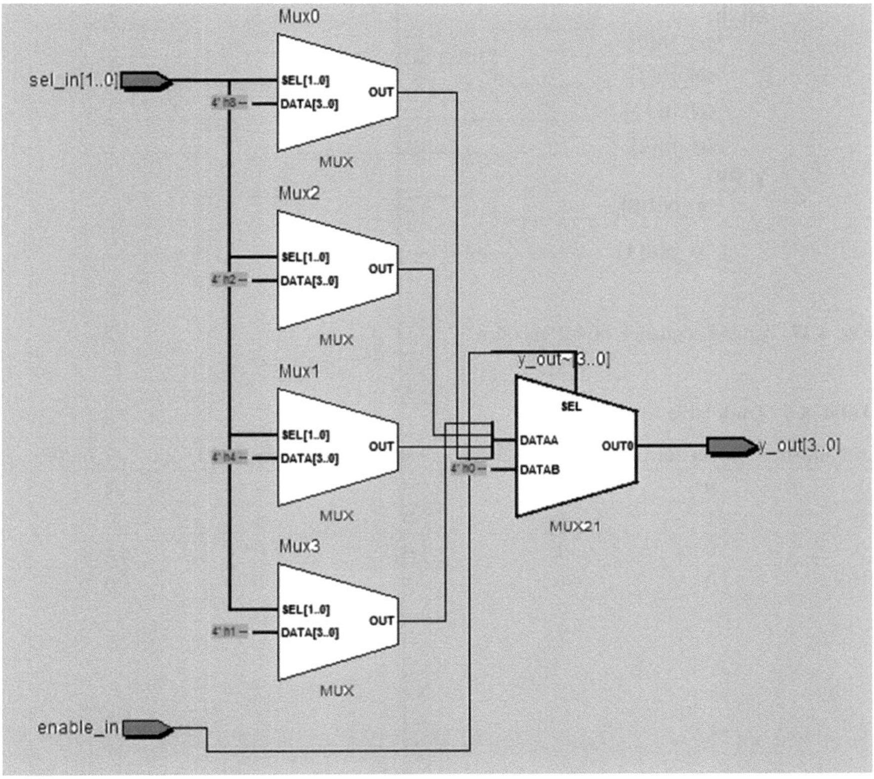

Fig. 4.26 2:4 decoder with active low enable input

n = 4 and output lines is m = 2. The truth table is described in Table 4.6. The input sel_in(3) has highest priority, and the sel_ in[0] has lowest priority, where 'X' indicates the don't care.

The VHDL RTL description for 4:2 priority encoder is described in Example 4.15. The VHDL RTL infers the hardware as shown in Fig. 4.29.

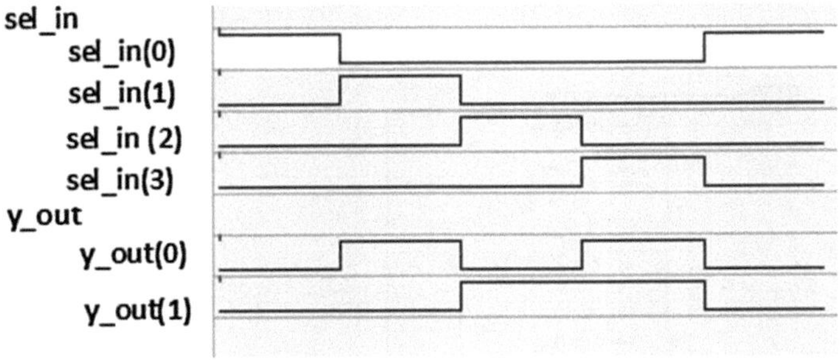

Fig. 4.27 Timing sequence of 4:2 encoder

Table 4.5 Truth table for 4:2 encoder

sel_in(3)	sel_in(2)	sel_in(1)	sel_in(0)	y_out(1)	y_out(0)
1	0	0	0	1	1
0	1	0	0	1	0
0	0	1	0	0	1
0	0	0	1	0	0

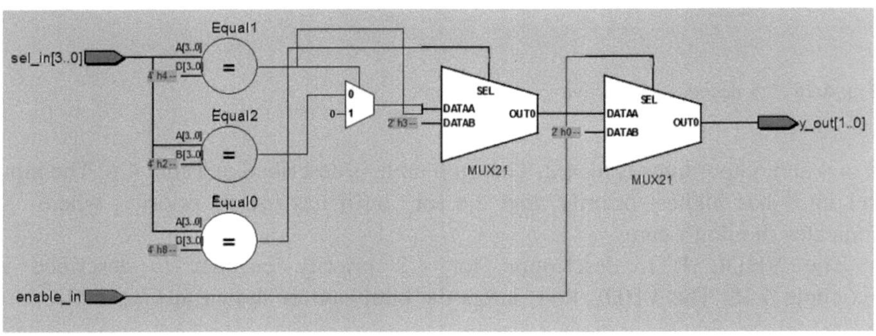

Fig. 4.28 Synthesis result of 4:2 encoder

Table 4.6 Truth table for 4:2 priority encoder

sel_in(3)	sel_in(2)	sel_in(1)	sel_in(0)	y_out(1)	y_out(0)
1	X	X	X	1	1
0	1	X	X	1	0
0	0	1	X	0	1
0	0	0	1	0	0

```
--Encoder 4to2 using if-then-else
library ieee;
use ieee.std_logic_1164.all;
use ieee.std_logic_unsigned.all;
use ieee.std_logic_arith.all;
entity encoder_4to2 is
port ( sel_in : in std_logic_vector ( 3 downto 0);
      enable_in : in std_logic;
    y_out : out std_logic_vector ( 1 downto 0));
end encoder_4to2;
architecture arch_encoder_4to2 of encoder_4to2 is
begin
    process ( sel_in, enable_in)
          begin
           if ( enable_in='1') then
                  y_out <= "00";
          else
            if ( sel_in(3) ='1') then
                   y_out <= "11";
                  elsif ( sel_in(2) ='1') then
                   y_out <= "10";
                  elsif ( sel_in(1) ='1') then
                   y_out <= "01";
                  else
                    y_out <= "00";
                 end if;
        end if;
        end process;
end arch_encoder_4to2;
```

- Architecture defines the functionality of design.
- Process is sensitive to 'sel_in', and 'enable_in'. Any event on one of the signal invokes the process.
- If-then-else is sequential statement and used inside the process.
- For true 'enable_in' condition all output lines assigned to logic '0'.
- For 'enable_in' active low , encoder is enabled and depending on the priority of signal the two bit output is generated at 'y_out'.
- The described code generates priority logic, Input 'sel_in(3)' has highest priority and 'sel_in(0)' has the least priority.

Example 4.15 Synthesizable VHDL RTL for 4:2 priority encoder

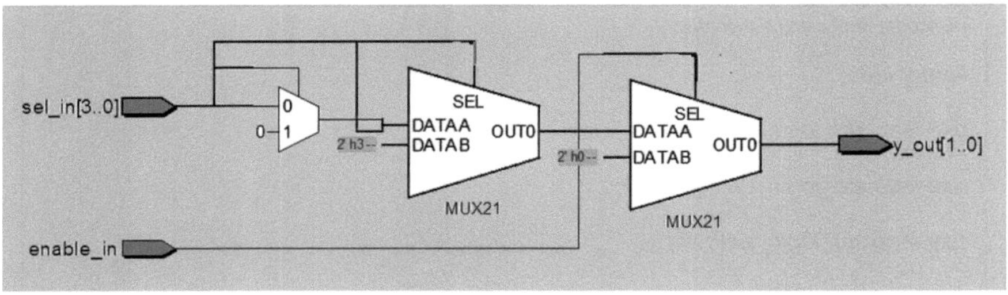

Fig. 4.29 Synthesized 4:2 priority encoder logic. *Note* In the practical applications, encoders are used to design the control logic. As 'case' generates the parallel logic and 'if then else' generates the priority logic, 'case' is used to describe the behavior of encoder. 'If else' is used to describe the behavior of priority encoder. Priority encoders can be used to sense the level sensitive interrupts

4.7 Summary

As discussed in this chapter, the combinational logic using VHDL can be efficiently implemented using the concurrent and sequential VHDL constructs and following are key points to summarize.

1. Multiplexer is universal logic and used to design any combinational functionality.
2. The propagation delay of cascade logic is more as compare to parallel logic.
3. Signal assignments execute concurrently. Adder consumes more area as compare to multiplexers.
4. The process is concurrent statement, and all the processes inside the architecture execute in parallel.
5. 'If then else' generates the 2:1 MUX, and 'nested if' generates the priority logic.
6. 'case' is used to model the parallel logic and used inside the process.
7. 'when others' condition in the 'case' is used to describe the non-specified conditions in the design functionality.
8. The synthesis tool ignores the sensitivity list specified in the process blocks.
9. Decoders are used to select one of the memories or input–output device at a time.
10. Priority encoders are used in the design of interrupt control logic, and logic can be described by using nested 'if else then.'

Chapter 5
Sequential Logic Design

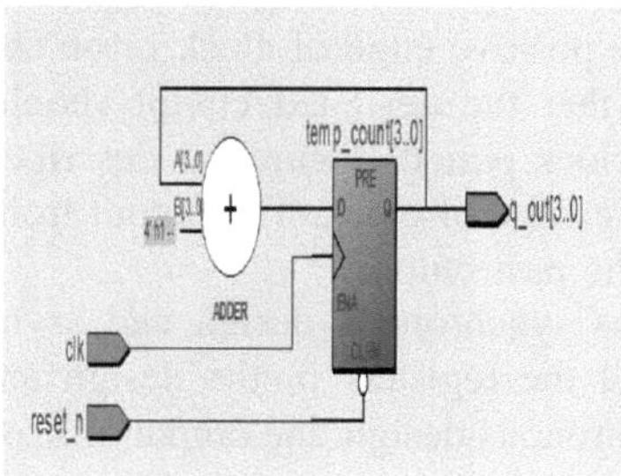

" Learn from yesterday, live for today, hope for tomorrow. The important thing is not to stop questioning." --- Albert Einstein

Try to revise the fundamentals and design the efficient sequential design using the VHDL Constructs.

Abstract This chapter describes the practical understanding about the sequential logic designs. RTL coding using VHDL is described in detail with the practical scenarios and concepts. The VHDL RTL for the flip-flops, latches, various counters, and shift registers is covered with the synthesis results and explanations. Even this chapter describes the timing parameters for the sequential logic and the maximum frequency calculation for the design. The practical do's and don'ts are explained with the meaningful diagrams and timing sequences. This chapter is useful for the ASIC and FPGA designers while coding for the sequential logic. This chapter also covers the asynchronous sequential circuits and issues like metastability in the design. How to overcome the metastability is explained with meaningful example and design scenarios.

Keywords Latch · Flip-flop · D flip-flop · Toggle flip-flop · Edge-triggered · Level sensitive · Asynchronous · Synchronous · Toggle · Cumulative delay · Up–down · Shift register · Ripple · Johnson · Ring · Metastability · Synchronous clear · Asynchronous clear · Synchronous preset · Asynchronous preset · Maximum frequency · Setup time · Hold time · Clock to q delay · Level synchronizer · BCD counters · Gray counters · Timing paths · Register-to-register path · Combinational path · Input-to-register path · Register-to-output path · Multiple clock domain designs

V. Taraate, *PLD Based Design with VHDL*,
DOI 10.1007/978-981-10-3296-7_5

5.1 Sequential Logic

Sequential logic is described as the digital logic whose output is the function of present input and past output. So the sequential logic holds the binary data. Sequential logic elements are latches and flip-flops and used as logic elements to design the sequential logic for the given design functionality. For the RTL design engineer, it is essential to understand the efficient RTL design for clock-based logic circuits. The sequential logic is used to hold the larger amount of data in the complex designs. The logic is triggered on the active edge of the clock. The chapter discusses the efficient VHDL RTL to describe the required functionality of the sequential logic. In the practical applications, it is always essential to describe the logic circuit which can be triggered either on the positive edge of clock or on the negative edge of clock. It is always expected that the designed circuit should generate the stable output for finite duration of clock period. Figure 5.1 describes the basic sequential logic triggered on the positive edge of clock. The output from the logic is the function of a present input and the past output.

Even the sequential logic can be classified as synchronous design and asynchronous design. In the synchronous design, all the registers in the design are triggered by the same clock sources. In the asynchronous design, the output of least significant bit (LSB) register is used as clock input to the next register. Even the design that uses the different clock sources of the same or different frequency is called asynchronous designs or multiple clock domain designs.

Figure 5.2 shows the synchronous design where all the registers in the design are triggered by the same clock source. Hence, the overall propagation delay to update the output is 'tpd.' If every flip-flop has delay of 'tpd,' then the overall frequency is dependent on the 'tpd,' combinational delay 'tcombo,' and setup time 'tsu' of the register.

For the synchronous design, the 'clk_1' and 'clk_2' are triggered at the same time instant and there is no phase difference between the 'clk_1' and 'clk_2.' So the clock skew is zero between 'clk_1' and 'clk_2,' and hence, both clocks will arrive at the same time instance at the 'clk' input of register. It is assumed that wire delays are zero.

The 'clk_1' and 'clk_2' waveforms are shown in Fig. 5.3 and generated from the same clock source 'clk.' Here assumption is the wire or net delay is zero.

The asynchronous sequential design is shown in the following Fig. 5.4. As shown in the figure, the output of LSB flip-flop is used to drive the clock of the next subsequent flip-flop; hence, the overall propagation delay is the cumulative effect.

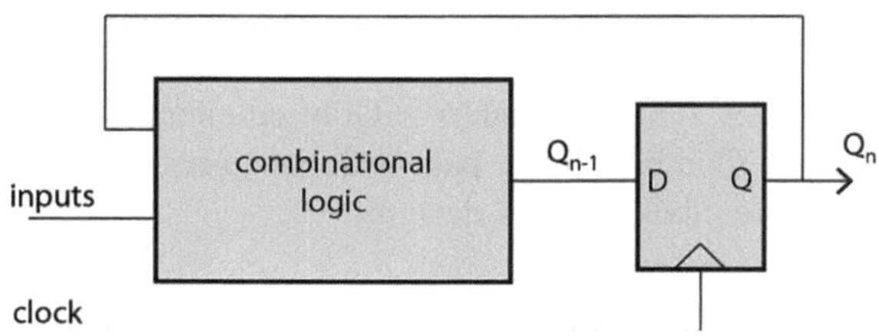

Fig. 5.1 Basic sequential logic

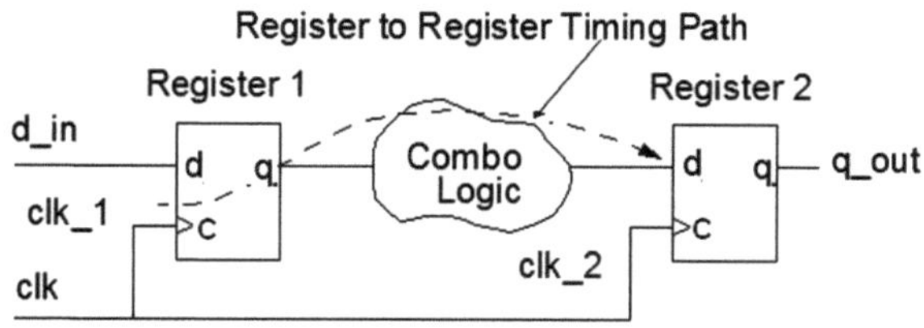

Fig. 5.2 Synchronous sequential logic

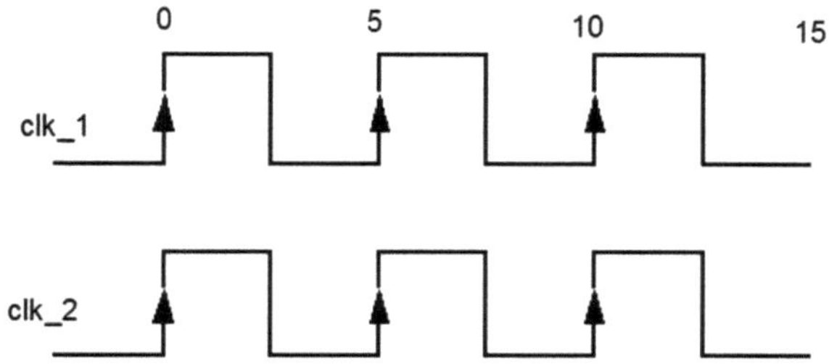

Fig. 5.3 Timing sequence for synchronous clock generation

For four-stage counter, the overall propagation delay is four times the propagation delay of flip-flop. If every flip-flop has the propagation delay of 1 ns, then the overall propagation delay is 4 ns.

These kinds of logic circuits are called asynchronous logic. Figure 5.4 shows the asynchronous ripple counter using JK flip-flop, where every JK flip-flop acts as toggle flip-flop. In the practical ASIC design scenarios, the D flip-flops are used to design the sequential logic. The sequential logic in this chapter is described by using the D flip-flops. The timing sequence for the 4-bit ripple counter is shown in the Fig. 5.5.

The multiple clock domain design is also treated as asynchronous design and shown in Fig. 5.6. As shown in Fig. 5.6, the two different modules are triggered by the clock sources 'clk_1' and 'clk_2.' respectively. If clock frequency is same or

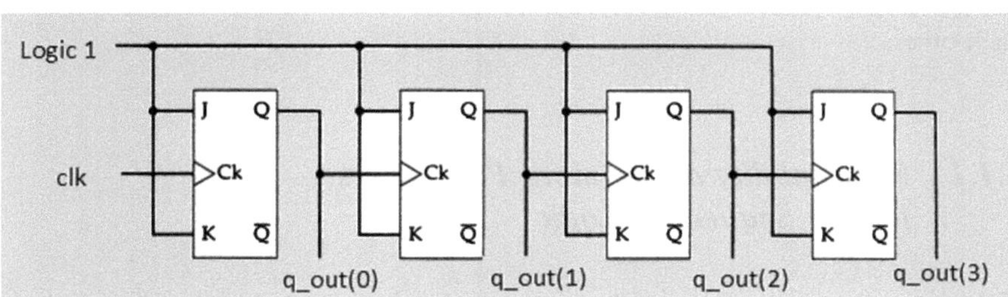

Fig. 5.4 Asynchronous four-bit counter

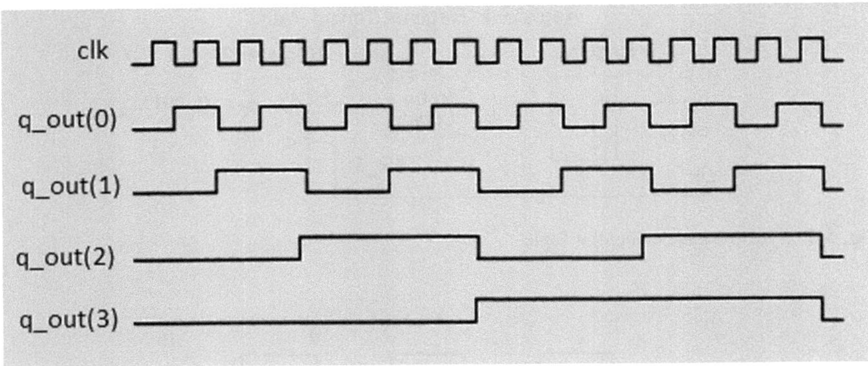

Fig. 5.5 Timing sequence for asynchronous counter

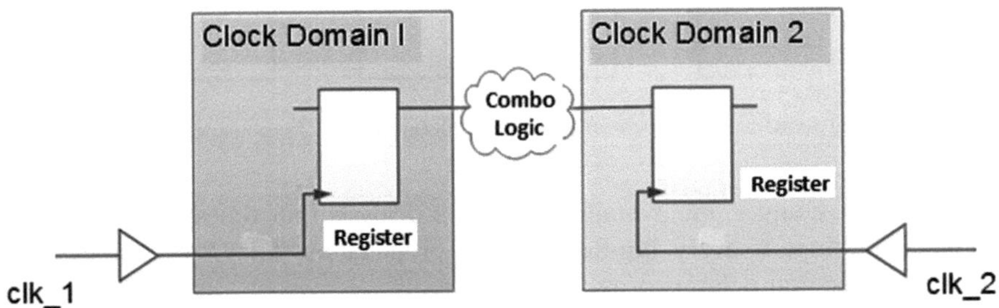

Fig. 5.6 Multiple clock domain design

different, the 'clk_1' and 'clk_2' might have the phase difference while triggering the register. Due to the phase difference between the 'clk_1' and 'clk_2,' both clock domain logics are not triggered at the same time. Hence, it is recommended to use the multiple clock domain design concepts while establishing the communication between clock domain 1 and clock domain 2. Few techniques are discussed in the next subsequent chapter.

The clock generation using two different clock sources with the phase difference or clock skew is shown in Fig. 5.7. As shown the clocks are skewed with respect to each other.

5.1.1 Metastability and Timing Parameters for the Sequential Logic

If the timing parameters in the design are violated, then the flip-flop goes into the metastable state. The main timing parameters in the design are flip-flop propagation

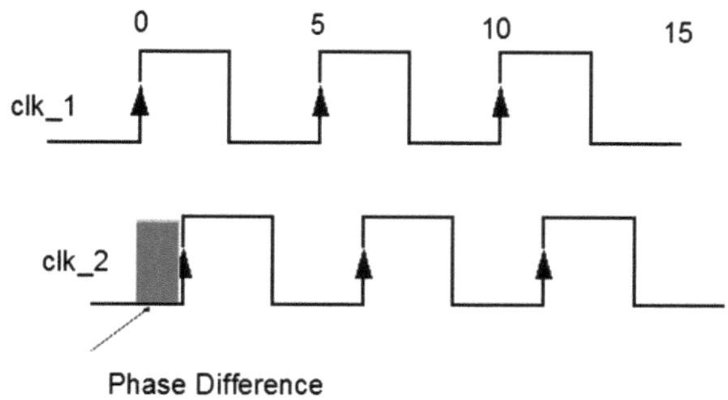

Fig. 5.7 The clocks with the phase difference

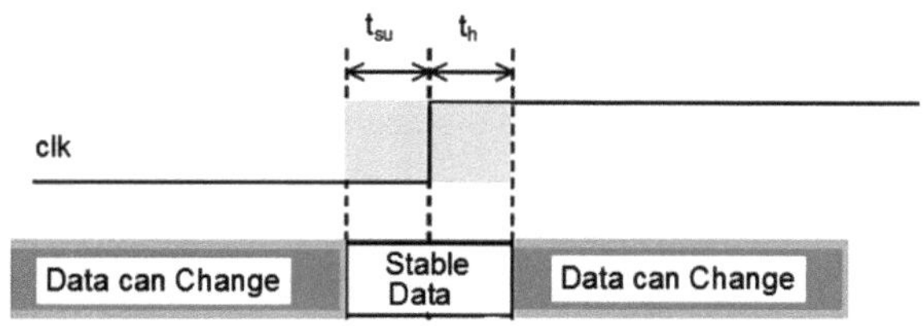

Fig. 5.8 Timing parameters of flip-flop

delay (t_{pd}), setup time (t_{su}), and hold time (t_h). The timing parameters of the D flip-flop are shown in Fig. 5.8.

As shown in the figure, the data at the 'D' input should be stable for the duration of setup and hold time. Data can change outside the widow of the setup and hold time. If the data is not stable during the setup and hold time window, then the flip-flop goes into the metastable state.

5.1.1.1 Setup Time

The amount of time for which the data at the flip-flop 'D' input should be stable before arrival of the active edge of clock is called setup time.

5.1.1.2 Hold Time

The amount of time for which the data at the flip-flop 'D' input should be stable after arrival of the active clock edge is called hold time.

5.1.1.3 Propagation Delay of Flip-Flop

The amount of time required for the flip-flop to generate the valid output after arrival of the active clock edge is called propagation delay of flip-flop. This is also named as clock to output (q) delay.

As stated earlier if any of the timing parameter is violated then the flip-flop goes into the metastable state. Consider the scenario described in Fig. 5.9.

As shown in Fig. 5.9, the register 0 is triggered by clock source 'clk_1' and the register 1 is triggered by another clock source 'clk_2,' so due to the different arrival time of the 'clk_1' and 'clk_2,' the register 1 goes into the metastable state. The timing sequence is shown in Fig. 5.10. It is assumed that D input of register 0 is logic '1'.

As shown in Fig. 5.10, the d_in input of register 1 has changed during the rising edge of the clk_2 and hence has the timing violation. Under such circumstances, the output of register 1 goes into the metastable state.

To avoid the metastability, the two-stage level synchronizer can be used. Figure 5.11 describes the use of the two-stage level synchronizer in the design to solve the metastable issue.

As shown in Fig. 5.11, although register 1 goes into the metastable state on the next rising edge of the clock 'clk_2,' the output 'q_out' is forced into the valid state.

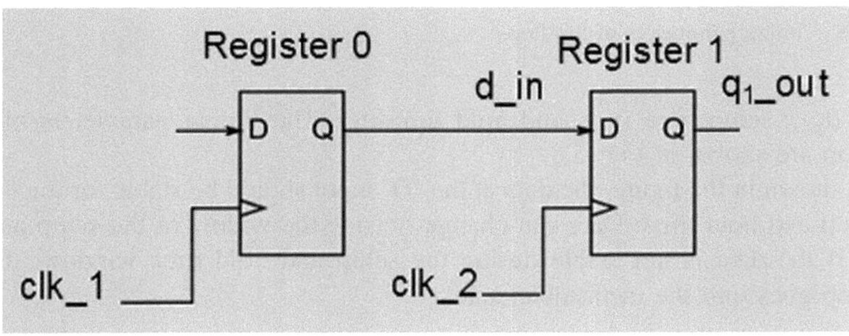

Fig. 5.9 The design with metastable state

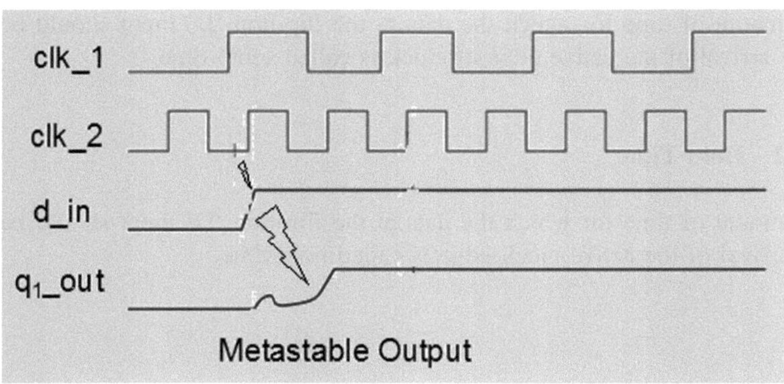

Fig. 5.10 Timing sequence with metastable output

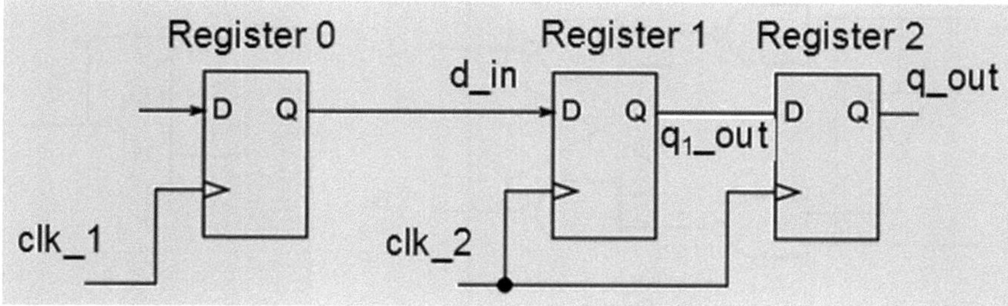

Fig. 5.11 Sampling d_in using two-stage level synchronizer

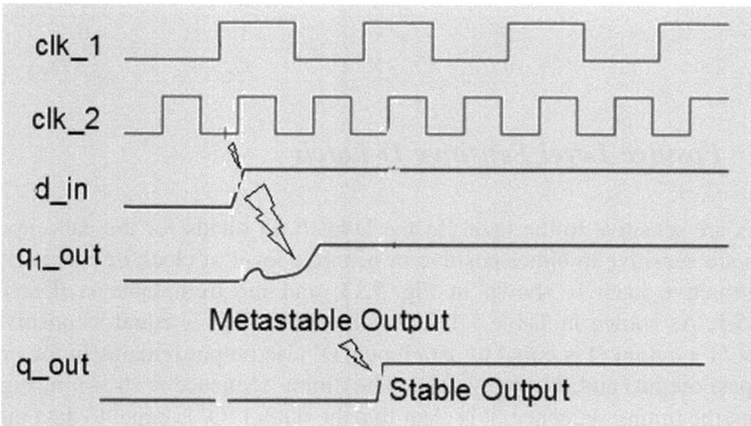

Fig. 5.12 The timing sequence using two-stage level synchronizer

So by adding one more register in the output path, the metastability issue is eliminated. Always, setup and hold parameters of the register 1 are violated. So during synthesis, it is essential to disable the timing from 'clk_1' to register 1's output 'q_1_out.'

The timing sequence for sampling of the 'd_in' using two-stage level synchronizer is shown in Fig. 5.12.

5.2 D-Latches in the Design

Most of the time, the designer is confused while using the sequential elements during the RTL design. The main sequential design elements are latch and flip-flop. Latch is level sensitive, and flip-flop is edge-triggered. The following section gives the information about the efficient RTL using VHDL for the positive and negative level sensitive latch.

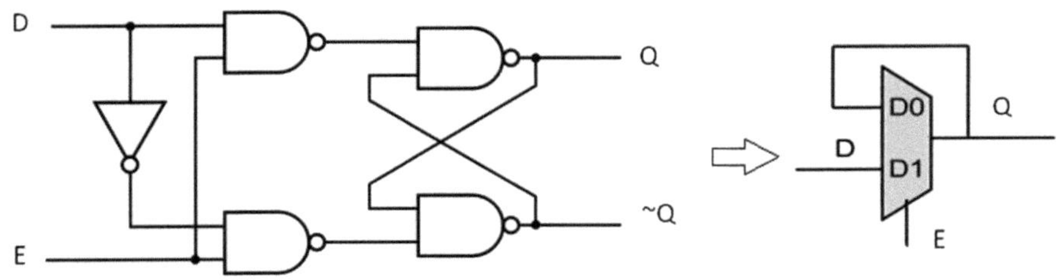

Fig. 5.13 Positive level sensitive D-latch

Table 5.1 Truth table for positive level sensitive D-latch

E	D	Q	~Q
1	0	0	1
1	1	1	0
0	X	Q_{n-1}	$\sim Q_{n-1}$

5.2.1 *Positive Level Sensitive D-Latch*

Latches are sensitive to the level. In the D-latch, D stands for the data input. The latches are sensitive to either positive or negative level of clock or enable. Positive level sensitive latch is shown in Fig. 5.13, and the truth table is described in Table 5.1. As shown in Table 5.1 for latch enable ('E') is equal to positive level (logical '1') output Q is equal to data input 'D' else output remains in the previous state (past output) and shown by Q_{n-1}. The timing sequence is shown in Fig. 5.14.

From the timing sequence, it is clear that the output 'Q' is equal to data input 'D' during the time period for which enable input 'E' is equal to positive level. So D-latch acts transparently during this period. During negative level (logical '0') of enable 'E', D-latch holds the previous value.

Now, the important point in your mind is how to describe the positive level sensitive D-latch using VHDL. It is very simple to visualize and to describe. Example 5.1 describes the RTL using VHDL for the positive level sensitive D-Latch, and the synthesis result is shown in Fig. 5.15.

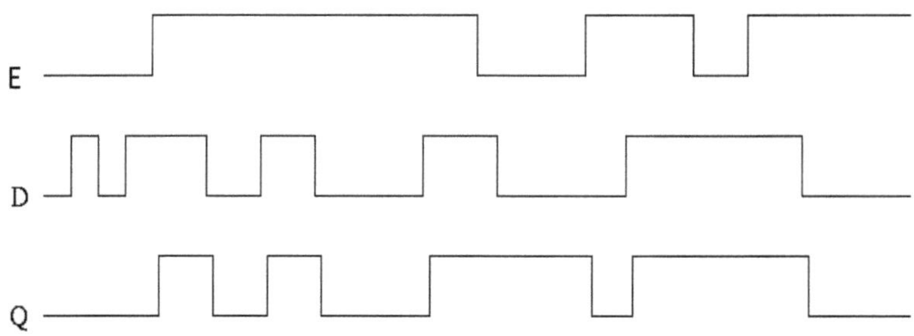

Fig. 5.14 Timing sequence for positive level sensitive D-latch

```
--positive level sensitive D latch

library ieee;
use ieee.std_logic_1164.all;
use ieee.std_logic_arith.all;
use ieee.std_logic_unsigned.all;

entity d_latch is

port ( d_in : in std_logic;
     latch_enable : in std_logic;
     q_out : out std_logic );

end d_latch;

architecture arch_d_latch of d_latch is
begin

  process ( d_in, latch_enable)

        begin

         if (latch_enable ='1') then

         q_out <= d_in;

         end if;

        end process;

end arch_d_latch ;
```

- Architecture defines the functionality of design.
- Process is sensitive to 'd_in', and 'latch_enable'. Any event on one of the signal invokes the process.
- If-then-else is sequential statement and used inside the process.
- For true value of 'latch_enable' condition the input 'd_in' is assigned to 'q_out'.
- For false 'latch_enable' condition the previous output value is hold. As else clause is eliminated it infers latch.

Example 5.1 Synthesizable RTL for positive level sensitive D-latch

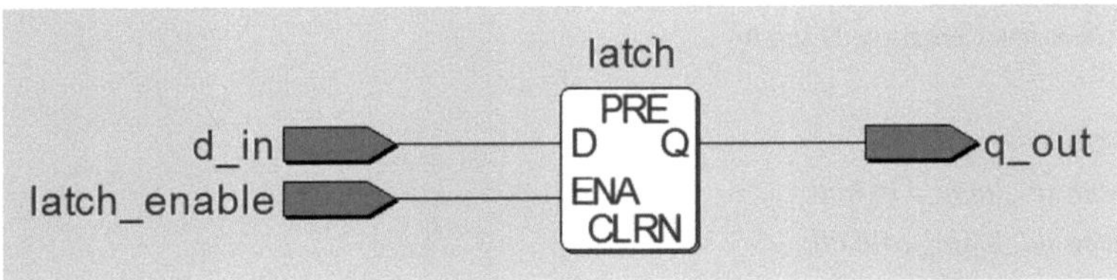

Fig. 5.15 Positive level sensitive D-latch

5.2.2 Negative Level Sensitive D-Latch

The truth table of the negative level sensitive D-Latch is described in Table 5.2, and it has active low or negative level sensitive latch enable ('E'): data input 'D' and output 'Q.'

The equivalent gate-level representation is shown in Fig. 5.16. The latch acts transparently on the negative level of 'E' and holds the data during the positive level of 'E'. The timing sequence is shown in Fig. 5.17.

The RTL using VHDL is shown in Example 5.2, and the synthesis result is shown in Fig. 5.18.

Table 5.2 Truth table for negative level sensitive D-latch

E	D	Q	~Q
0	0	0	1
0	1	1	0
1	X	Q_{n-1}	$\sim Q_{n-1}$

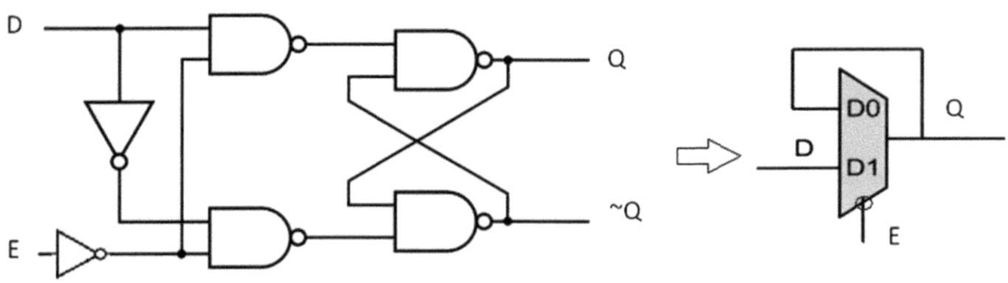

Fig. 5.16 Negative level sensitive D-latch

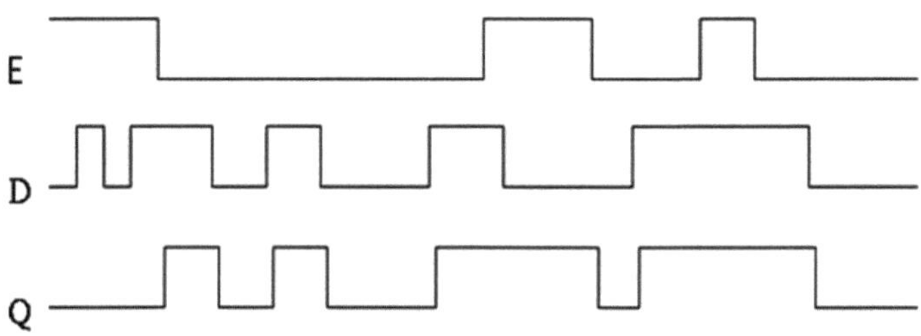

Fig. 5.17 Timing sequence for negative level sensitive latch

```
--Negative level sensitive D latch
library ieee;
use ieee.std_logic_1164.all;
use ieee.std_logic_arith.all;
use ieee.std_logic_unsigned.all;

entity d_latch is

port ( d_in : in std_logic;
       latch_enable : in std_logic;
        q_out : out std_logic );

end d_latch;

architecture arch_d_latch of d_latch is
begin

  process ( d_in, latch_enable)

          begin

            if (latch_enable ='0') then

            q_out <= d_in;

            end if;

          end process;

end arch_d_latch ;
```

- Architecture defines the functionality of design.
- Process is sensitive to 'd_in', and 'latch_enable'. Any event on one of the signal invokes the process.
- If-then-else is sequential statement and used inside the process
- For false (logic 0) 'latch_enable' condition the input 'd_in' is assigned to 'q_out'.
- For true (logic 1) 'latch_enable' condition the previous value is hold at the output. As else clause is missing it infers latch.

Example 5.2 Synthesizable VHDLRTL for negative level sensitive D-latch

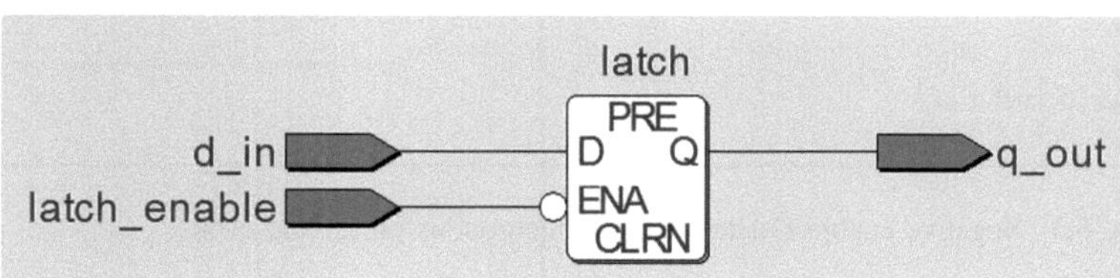

Fig. 5.18 Synthesis result for negative level sensitive latch

```
--Negative enable D latch with asynchronous preset and clear.
library ieee;
use ieee.std_logic_1164.all;
use ieee.std_logic_arith.all;
use ieee.std_logic_unsigned.all;

entity d_latch_pre_clr is
port ( d_in : in std_logic;
     latch_enable : in std_logic;
                    preset_in : in std_logic;
                    clear_in : in std_logic;
                    q_out : out std_logic );
end d_latch_pre_clr;

architecture arch_d_latch of d_latch_pre_clr is
begin

  process ( d_in, latch_enable, preset_in, clear_in)

          begin

            if ( clear_in ='0') then

             q_out <= '0';

          elsif ( preset_in ='0') then

             q_out <= '1';

            elsif (latch_enable ='0') then

            q_out <= d_in;

            end if;

          end process;

end arch_d_latch ;
```

- Architecture defines the functionality of design.
- Process is sensitive to 'd_in', 'latch_enable', 'preset_in' and 'clear_in'. Any event on one of the signal invokes the process.
- If-then-else is sequential statement and used inside the process.
- The 'clear_in' signal has the highest priority
- The 'preset_in has the second priority.
- The 'latch_enable' has the last priority. For 'latch_enable' is equal to logic '0' the output q_out is equl to d_in.
- As else clause is eliminated it infers latch.

Example 5.3 Negative enable D-latch using asynchronous preset and clear

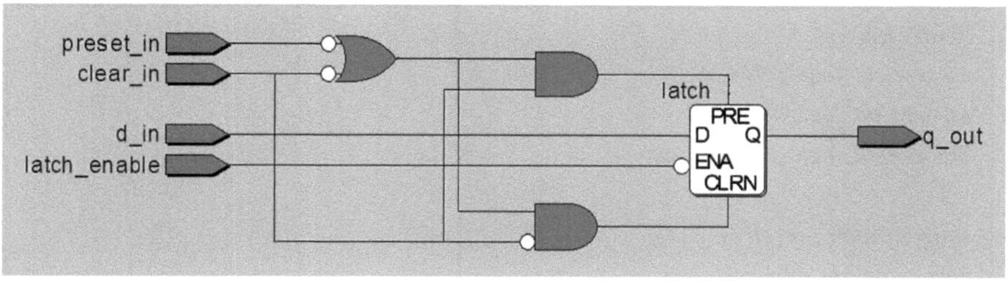

Fig. 5.19 Synthesis result for the negative level sensitive D-latch with asynchronous inputs

5.2.3 Negative Level Sensitive D-Latch with Preset and Clear

The sequential elements can be described by incorporating the asynchronous or synchronous preset and clear (reset) input signals. Depending on the design requirements, the asynchronous preset, reset or synchronous preset or reset can be used in the design. Asynchronous preset and clear inputs have no logic in data path, whereas the synchronous preset or clear inputs have the combinational logic in the data path. Asynchronous inputs can arrive irrespective of the active clock edge to change the output of the sequential cell. But the synchronous inputs are sampled on the active clock edge to make the changes in the output.

The RTL using VHDL is shown in Example 5.3. As shown described in the example, the input 'clear_in' has the highest priority over the 'preset_in' and the output assignment is irrespective of the 'latch_enable.' These kind of asynchronous inputs gives the clean data path. The synthesis result is shown in Fig. 5.19, and it infers the negative level sensitive D-latch with the asynchronous logic circuit at the clear and preset inputs.

5.2.4 Positive Level Sensitive D-Latch with Asynchronous Preset and Clear

The positive Level Sensitive D-latch with asynchronous preset and clear is described in Example 5.4.

The synthesis result is shown in Fig. 5.20.

```
library ieee;
use ieee.std_logic_1164.all;
use ieee.std_logic_arith.all;
use ieee.std_logic_unsigned.all;

entity d_latch_pre_clr is
port ( d_in : in std_logic;
     latch_enable : in std_logic;
                  preset_in : in std_logic;
                  clear_in : in std_logic;
                  q_out : out std_logic );
end d_latch_pre_clr;

architecture arch_d_latch of d_latch_pre_clr is
begin

  process ( d_in, latch_enable, preset_in, clear_in)

        begin

         if ( clear_in ='0') then

          q_out <= '0';

        elsif ( preset_in ='0') then

          q_out <= '1';

         elsif (latch_enable ='1') then

         q_out <= d_in;

         end if;

        end process;

end arch_d_latch ;
```

- Architecture defines the functionality of design.
- Process is sensitive to 'd_in', 'latch_enable', 'pre-set_in' and 'clear_in'. Any event on one of the signal invokes the process.
- If-then-else is sequential statement and used inside the process.
- The 'clear_in' signal has the highest priority.
- The 'preset_in has the second priority.
- The 'latch_enable' has the last priority. For 'latch_enable' is equal to logic '1' the output q_out is equl to d_in.
- As else clause is eliminated it infers latch.

Example 5.4 Synthesizable VHDL RTL for the positive level sensitive D-latch with asynchronous inputs

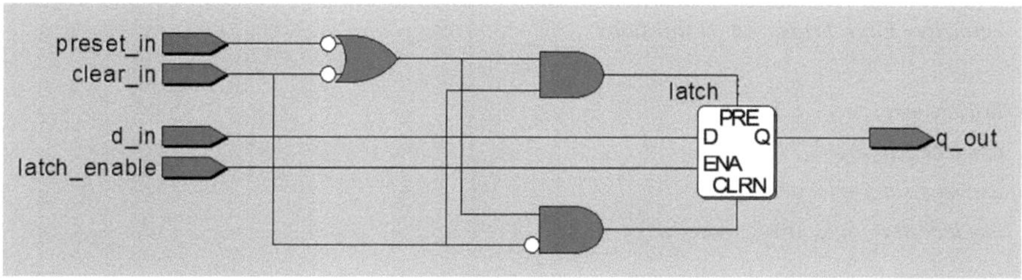

Fig. 5.20 Synthesis result for positive level sensitive D-latch with asynchronous inputs

5.3 Flip-Flop

Flip-flop is an edge-triggered logic circuit. It can be triggered either on positive edge of clock or on negative edge of clock. Flip-flop can be realized by using positive and negative level sensitive latches in cascade. Flip-flop is used as a memory storage element. Flip-flops are set–reset (SR), JK, D, and toggle. In an ASIC or FPGA design, the D flip-flop is used as a memory storage element, where D stands for the data input. The subsequent section discusses on the positive and negative edge-triggered flip-flop.

5.3.1 Positive Edge-Triggered D Flip-Flop

Positive edge-triggered D flip-flop is triggered on positive edge of clock. Practically, there is no logic gate which can be triggered on edge! Positive edge flip-flop is realized by using negative level sensitive latch followed by positive level sensitive latch. The logic circuit for positive edge-triggered D flip-flop is shown in Fig. 5.21.

The synthesizable RTL using VHDL is shown in the following Example 5.5. The synthesis result is shown in Fig. 5.22.

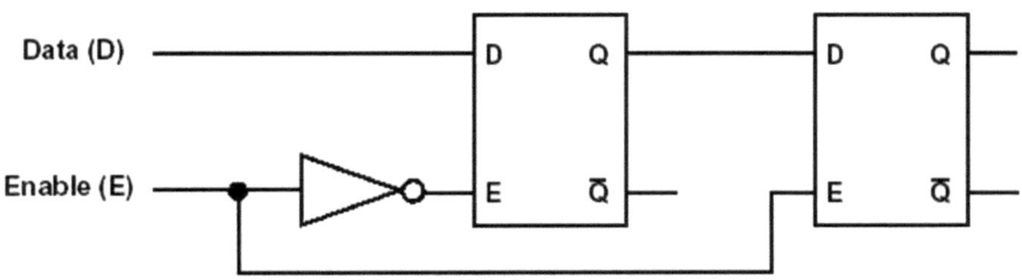

Fig. 5.21 Positive edge-triggered D flip-flop

```
--positive Edge Triggered  D flip-flop

library ieee;
use ieee.std_logic_1164.all;
use ieee.std_logic_arith.all;
use ieee.std_logic_unsigned.all;

entity d_flipflop is

port ( d_in : in std_logic;
      clk : in std_logic;
      q_out : out std_logic );

end d_flipflop;

architecture arch_d_flipflop of d_flipflop is
begin

  process ( d_in, clk)

         begin

           if (clk='1' and clk'event) then

           q_out <= d_in;

           end if;

         end process;

end arch_d_flipflop ;
```

- Architecture defines the functionality of design.
- Process is sensitive to 'd_in, 'clk'. Any event on one of the signal invokes the process.
- If-then-else is sequential statement and used inside the process.
- For rising edge of clock the data input 'd_in' is assigned to 'q_out'.
- Due to missing else clause it generated D flip-flop which is triggered on positive edge of clock.

Example 5.5 Synthesizable VHDL RTL for positive edge-triggered D flip-flop

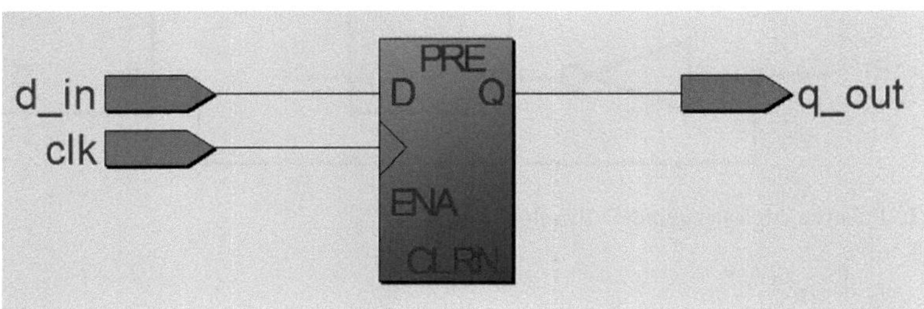

Fig. 5.22 Synthesis result for the positive edge-triggered flip-flop

```
--Negative Edge Triggered D flip-flop

library ieee;
use ieee.std_logic_1164.all;
use ieee.std_logic_arith.all;
use ieee.std_logic_unsigned.all;

entity d_flipflop is

port ( d_in : in std_logic;
     clk : in std_logic;
     q_out : out std_logic );

end d_flipflop;

architecture arch_d_flipflop of d_flipflop is
begin

  process ( d_in, clk)

       begin

        if (clk='0' and clk'event) then

        q_out <= d_in;

        end if;

       end process;

end arch_d_flipflop ;
```

- Architecture defines the functionality of design.
- Process is sensitive to ‘d_in, ‘clk’. Any event on one of the signal invokes the process.
- If-then-else is sequential statement and used inside the process.
- For falling edge of clock the data input ‘d_in’ is assigned to ‘q_out’.
- Due to missing else clause it generated D flip-flop which is triggered on negative edge of clock.

Example 5.6 Synthesizable VHDL RTL for negative edge-triggered flip-flop

5.3.2 Negative Edge-Triggered D Flip-Flop

Negative edge-triggered D flip-flop is triggered on negative edge of clock. Negative edge flip-flop is realized by using positive level sensitive latch followed by the negative level sensitive latch. The logic circuit for negative edge-triggered D flip-flop is shown in Fig. 5.23.

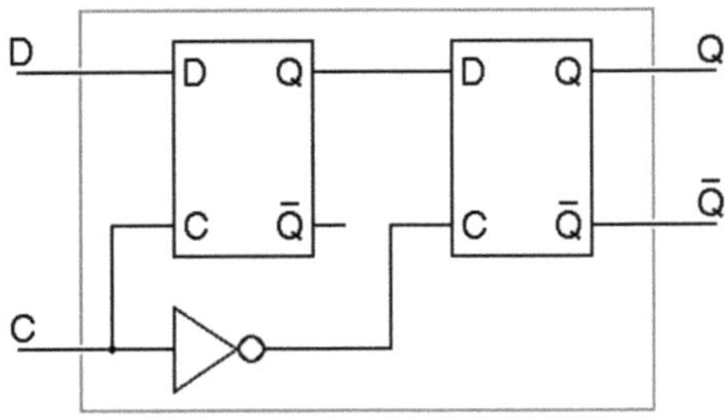

Fig. 5.23 Negative edge-triggered D flip-flop

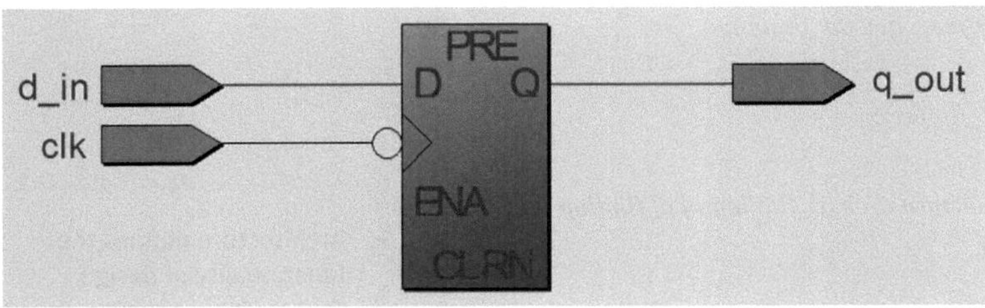

Fig. 5.24 Synthesis result for the negative edge-triggered D flip-flop

The synthesizable RTL using VHDL for the negative edge-triggered D flip-flop is described in the following Example 5.6. The synthesis result is shown in Fig. 5.24.

5.4 Synchronous and Asynchronous Reset

There is always confusion while using asynchronous or synchronous reset for the ASIC or FPGA designs. Synchronous reset signal is sampled on active clock edge and the reset logic is part of the data path, whereas asynchronous signal is sampled irrespective of active clock edge and logic is not a part of the data path or data input logic. This section describes about the RTL using VHDL for D flip-flop using asynchronous and synchronous resets.

5.4.1 D Flip-Flop with Asynchronous Reset

Asynchronous reset logic is not a part of data path and used to initialize flip-flop irrespective of active clock edge and hence named as asynchronous reset. This technique to initialize flip-flop is not recommended for internal reset signal generation as it is prone to glitches. Care needs to be taken by designer to synchronize

```
library ieee;
use ieee.std_logic_1164.all;
use ieee.std_logic_arith.all;
use ieee.std_logic_unsigned.all;

entity d_flip_flop is
port ( d_in : in std_logic;
     clk : in std_logic;
                    preset_in : in std_logic;
                    clear_in : in std_logic;
                    q_out : out std_logic );
end d_flip_flop;

architecture arch_d_flip_flop of d_flip_flop is
begin

  process ( d_in, clk, preset_in, clear_in)
        begin

         if ( clear_in ='0') then

          q_out <= '0';

        elsif ( preset_in ='0') then

          q_out <= '1';

         elsif (clk ='1' and clk'event) then

         q_out <= d_in;

         end if;

        end process;
end arch_d_flip_flop ;
```

- Architecture defines the functionality of design.
- Process is sensitive to 'd_in, 'clk', 'preset_in' and 'clear_in'. Any event on one of the signal invokes the process.
- If-then-else is sequential statement and used inside the process.
- The 'clear_in', 'preset_in' are asynchronous inputs and they are active low. The input 'clear_in' has highest priority as compare to 'preset_in'.
- For rising edge of clock the data input 'd_in' is assigned to 'q_out'.
- Due to missing else clause it generated D flip-flop which is triggered on positive edge of clock.

Example 5.7 D flip-flop with asynchronous active low clear input

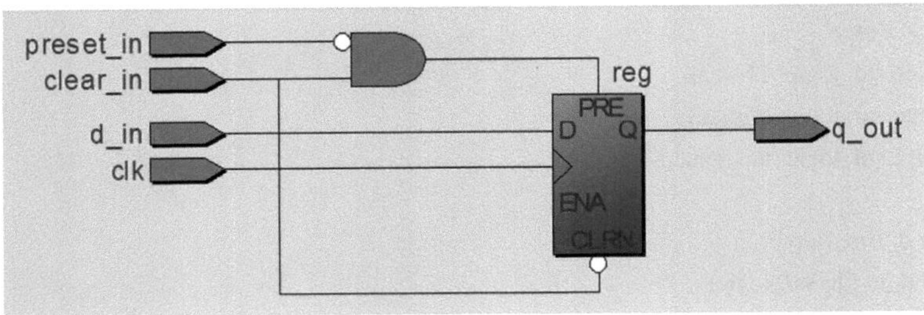

Fig. 5.25 Synthesis result of D flip-flop with asynchronous active low reset input

this reset signal internally to avoid the glitches. The internally synchronized reset signal is applied to the storage elements. The reset deassertion is the main problem in the asynchronous reset signals, and this problem can be overcome by using two-stage level synchronizer. Level synchronizer avoids the race-around conditions during reset deassertion.

Synthesizable RTL using VHDL is shown in Example 5.7 and uses active low asynchronous reset signal 'clear_in' and preset signal 'preset_in.' The synthesis result is shown in Fig. 5.25.

5.4.2 D Flip-Flop with Synchronous Reset

In synchronous reset, the reset logic is part of data input that is data path and reset signal is sampled on the active clock edge. The synchronous reset does not have issues of glitches or hazards, so this approach is best suited for the design. This mechanism does not require the additional synchronization circuit.

The RTL using VHDL is described in Example 5.8 and uses active low synchronous reset signal 'clear_in' and active low preset input 'preset_in.'

In most of the practical applications, multiple asynchronous inputs are required. Consider an application where it is required to load the input data when enable input is active and it is essential to initialize register when reset signal is active and valid. If both asynchronous inputs arrive at a time, then the output assignment should be dependent on the priority of these signals.

The synthesis result for positive edge-triggered D flip-flop with synchronous reset input is shown in Fig. 5.26. As shown in the figure, the 'preset_in' and 'clear_in' logic circuit is part of the data path. In this 'clear_in' has highest priority as compare to 'preset_in.'

```
library ieee;
use ieee.std_logic_1164.all;
use ieee.std_logic_arith.all;
use ieee.std_logic_unsigned.all;

entity d_flip_flop is
port ( d_in : in std_logic;
     clk : in std_logic;
                    preset_in : in std_logic;
                    clear_in : in std_logic;
                    q_out : out std_logic );
end d_flip_flop;

architecture arch_d_flip_flop of d_flip_flop is
begin
  process ( d_in, clk, preset_in, clear_in)
        begin

         if (clk ='1' and clk'event) then

          if ( clear_in ='0') then

           q_out <= '0';

         elsif ( preset_in ='0') then
           q_out <= '1';
          else
          q_out <= d_in;

          end if;

          end if;

         end process;

end arch_d_flip_flop ;
```

- Architecture defines the functionality of design.
- Process is sensitive to 'd_in, 'clk', 'preset_in' and 'clear_in'. Any event on one of the signal invokes the process.
- If-then-else is sequential statement and used inside the process.
- The 'clear_in', 'preset_in' are synchronous inputs and they are active low. The input 'clear_in' has highest priority as compare to 'preset_in'.
- For rising edge of clock the data input 'd_in' is assigned to 'q_out'.
- Due to missing else clause it generated D flip-flop which is triggered on positive edge of clock.

Example 5.8 D flip-flop with active low synchronous reset input

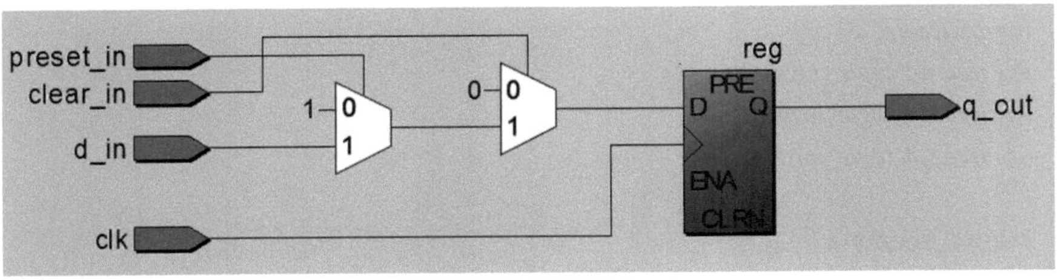

Fig. 5.26 Synthesis result for D flip-flop with synchronous reset

5.5 Sequential Circuit Timing

As discussed earlier, the sequential circuit has the timing parameters and they are flip-flop propagation delay, setup time, and hold time. While designing sequential logic, it is essential to take care that there should not be any timing violation. Consider the simple scenario shown in Fig. 5.27.

As shown in the figure, the synchronous circuit has the timing path from register 1 to register 2. Practically, there can be timing paths. The path from the d_in to d input of register 1 and called input-to-register path. The path from clock input 'clk_2' of register 2 to q_out is called register-to-output path, and the path from 'clk_1' of register 1 to the d input of register 2 is called register-to-register path. The design can have input-to-output path also, and it is purely combinational. In the above figure, there is no combinational path.

The maximum operating frequency for the design is dependent upon the register-to-register path, and in the above figure, the data required time is $T_{clk} - t_{su.}$ That is, data should arrive at the d input of the register 2 before the setup time of the register 2, where T_{clk} is the timing period of the clock. If we consider the data arrival time, then data at the d input of the register 2 is arriving at the time duration '$t_{pd} + t_{combo}$.' So for positive slack, the required time minus arrival time should have either zero or positive value.

Under such circumstances, there is no violation in the design and the clock time period is given by $T_{clk} - t_{su} = t_{pd} + t_{combo}$, that is, $T_{clk} = t_{su} + tpd + t_{combo}$. If the setup time of flip-flop is 1 ns, combinational delay is 2 ns, and the propagation

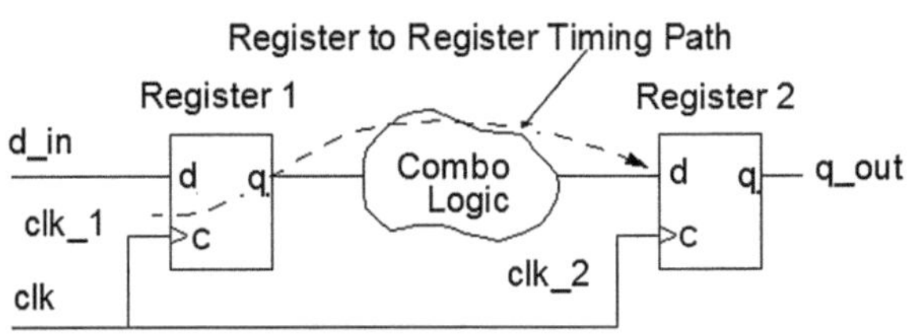

Fig. 5.27 Synchronous circuit timing path

delay of flip-flop is 2 ns, then the clock period is 5 ns. So the design operates at the maximum frequency of 200 MHz.

The static timing analysis (STA) tool are used to find out the timing violation and to perform the timing analysis for the design.

5.6 Synchronous Counters

If all the storage elements are triggered by the same source clock signal, then the design is said to be synchronous. The advantage of synchronous design is that, the overall propagation delay for the design is equal to the propagation delay of flip-flop or storage element. STA is very easy for the synchronous logic, and even the performance improvement is possible by using the pipelining. Most of the ASIC or FPGA implementation uses the synchronous logic. This section describes the synchronous counter design.

Four-bit binary counter is used to count from '0000' to '1111,' and the four-bit BCD counter is used to count from the '0000' to '1001.' Figure 5.28 shows the four-bit binary counter where every sequential logic stage is divided by two counters.

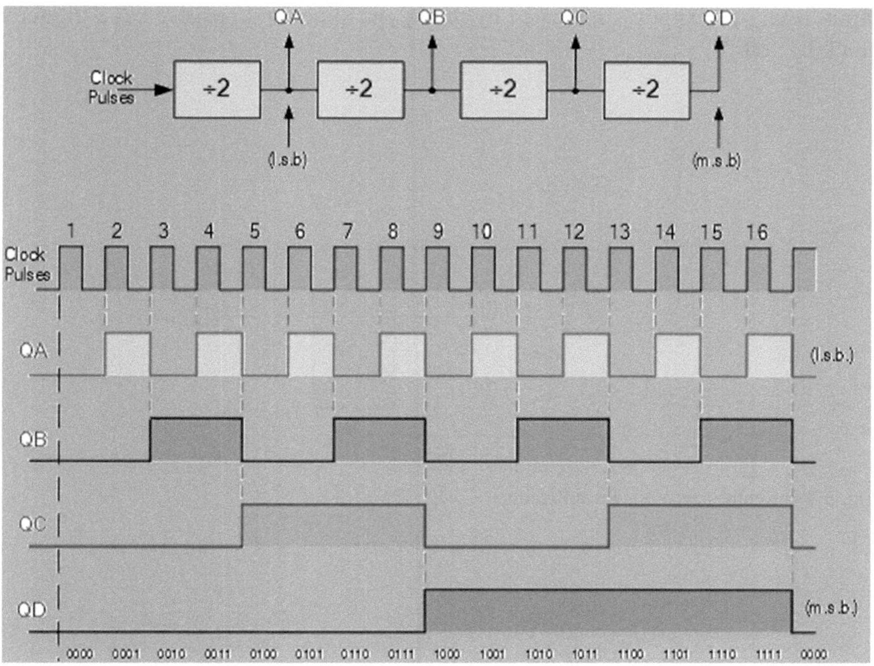

Fig. 5.28 Four-bit binary counter

As shown in Fig. 5.28, the counter has four output lines 'QA, QB, QC, QD' where 'QA' is LSB and 'QD' is MSB. The output at 'QA' toggles on every clock pulse and hence divided by two. The output at 'QB' toggles for every two clock cycles, and hence, it is divide by four, at 'QC' output toggles for every four clock cycles and hence the output is divided by eight. Similarly, the output at 'QD' toggles for every eight clock cycles, and hence, output at 'QD' is divided by sixteen of the input clock frequency. In the practical applications, counters are used as clock divider network. Even counters are used in the frequency synthesizers to generate variable frequency outputs.

5.6.1 Four-Bit Up Counter

Counters are used to generate the predefined or required count sequence on the active edge of clock. In an ASIC or FPGA design, it is essential to write an efficient RTL code for the clock divider network by using the synthesizable constructs. Four-bit up counter is described by using synthesizable VHDL constructs. Counter counts from '0000' to '1111' on the positive edge of the clock and wraps around to '0000' on the next positive edge of the count. The counter described in Example 5.9 has active low asynchronous 'reset_n' input and when it is active low the status on output line 'q_out' is '0000.' During normal operation, 'reset_n' is active high.

The synthesis result is shown in Fig. 5.29 and as shown it has active low reset input 'reset_n.' Output is indicated by the 'q_out' lines and positive edge-triggered clock by 'clk.'

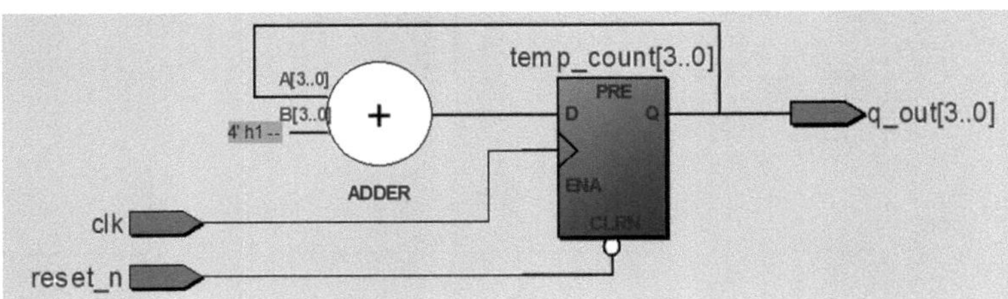

Fig. 5.29 Synthesized four-bit up counter

```
library ieee;
use ieee.std_logic_1164.all;
use ieee.std_logic_arith.all;
use ieee.std_logic_unsigned.all;

entity binary_up_counter is

port ( clk : in std_logic;
                    reset_n : in std_logic;
                    q_out : out std_logic_vector (3 downto 0) );

end binary_up_counter;

architecture arch_counter of binary_up_counter is

signal temp_count : std_logic_vector ( 3 downto 0);

begin

  process ( clk, reset_n )

        begin

          if ( reset_n ='0') then

           temp_count <= "0000";

          elsif (clk ='1' and clk'event) then

          temp_count <= temp_count + "0001";

          end if;

        end process;

        q_out <= temp_count;

end arch_counter ;
```

- Architecture defines the functionality of design.
- Process is sensitive to 'clk', 'and 'reset_n'. Any event on one of the signal invokes the process.
- If-then-else is sequential statement and used inside the process.
- For logic '0' 'reset_n' condition the output 'q_out' is assigned to zero. During normal operation 'reset_n' is active high and counter increments.
- Counter increments on positive edge of clock.

Example 5.9 VHDLRTL for four-bit up counter

5.6.2 *Four-Bit Down Counter*

Four-bit down counter is described by using synthesizable VHDL constructs. Counter counts from '1111' to '0000' and triggered on the positive edge of the clock and wraps around to '1111' on the next positive edge of the count after reaching to count value '0000.' The counter is described in Example 5.10. Counter

```
library ieee;
use ieee.std_logic_1164.all;
use ieee.std_logic_arith.all;
use ieee.std_logic_unsigned.all;

entity binary_down_counter is

port ( clk : in std_logic;
                    reset_n : in std_logic;
                    q_out : out std_logic_vector (3 downto 0) );

end binary_down_counter;

architecture arch_counter of binary_down_counter is

signal temp_count : std_logic_vector ( 3 downto 0);

begin

  process ( clk, reset_n )

         begin

          if ( reset_n ='0') then

           temp_count <= "0000";

          elsif (clk ='1' and clk'event) then

          temp_count <= temp_count -"0001";

          end if;

         end process;

          q_out <= temp_count;

end arch_counter ;
```

- Architecture defines the functionality of design.
- Process is sensitive to 'clk', 'and 'reset_n'. Any event on one of the signal invokes the process.
- If-then-else is sequential statement and used inside the process.
- For logic '0' 'reset_n' condition the output 'q_out' is assigned to zero. During normal operation 'reset_n' is active high and counter decrements.
- Counter decrements on positive edge of clock.

Example 5.10 VHDLRTL for four-bit down counter

has active low asynchronous 'reset_n' input, and when it is active low, the status on output line 'q_out' is '000.' During normal operation, 'reset_n' is active high.

The synthesis result is shown in Fig. 5.30 and as shown it has active low reset input 'reset_n.' Output is indicated by the 'q_out' lines and positive edge-triggered clock by 'clk.'

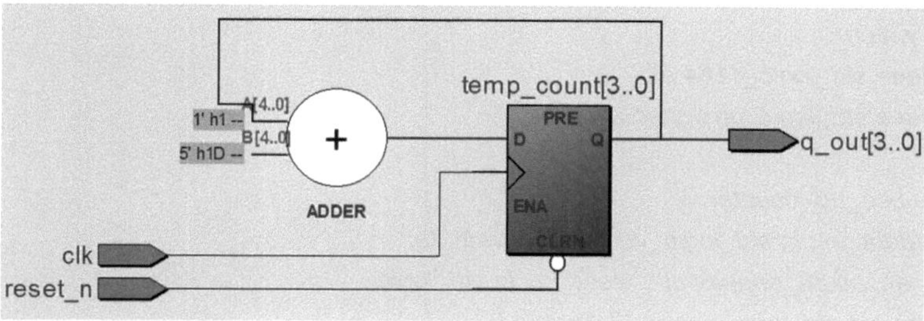

Fig. 5.30 Synthesized four-bit down counter

5.6.3 *BCD Up Counter*

Four-bit BCD up counter can be described by using synthesizable VHDL constructs. Up counter counts from '0000' to '1001' and triggered on the active edge of the clock and initializes to '0000' on the next active edge of the count after reaching to count value '1001.' The timing sequence for the BCD up counter is shown in Fig. 5.31. As shown in the timing sequence the BCD UP counter uses negative edge triggered clock.

Figure 5.31 gives the information about the timing sequence for the up counting. The counter can be designed as synchronous or asynchronous counter.

The counter described in Example 5.11 is presettable counter, and it has the synchronous active high 'load_en' input to sample the four-bit value. The data input is four-bit and indicated as 'data_in.' The 'count_enable' is used to enable the counting on the rising edge of the clock.

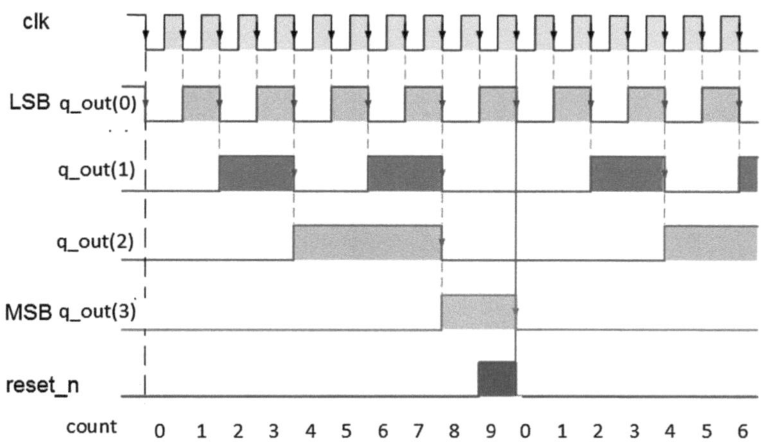

Fig. 5.31 Four-bit BCD up counter

```
library ieee;
use ieee.std_logic_1164.all;
use ieee.std_logic_unsigned.all;

entity bcd_up_counter is
port (data_in : in std_logic_vector (3 downto 0);
load_en, count_enable, clk, reset_in : in std_logic;
q_out : out std_logic_vector (3 downto 0));
end bcd_up_counter;

architecture arch_counter of bcd_up_counter is
signal sig_count : std_logic_vector (3 downto 0);
begin

process (clk, reset_in)
begin
if (reset_in ='1') then
sig_count <= (others => '0');
elsif rising_edge(clk) then
if (load_en = '1' ) then
sig_count <= data_in;
elsif (count_enable = '1') then
 if (sig_count ="1010") then
  sig_count <= (others => '0');
 else
  sig_count <= sig_count + '1';
 end if;
end if;
end if;
end process;

q_out<= sig_count;
end arch_counter;
```

- Architecture defines the functionality of design.
- Process is sensitive to 'clk', 'and 'reset_in'. Any event on one of the signal invokes the process.
- Nested If-then-else is sequential statement and used inside the process.
- For logic '1' 'reset_in' condition the input 'q_out' is assigned to zero. During normal operation 'reset_in' is active low and counter increments.
- For 'load_en' is equal to logic '1' 'data_in' is assigned to output q_out.
- Counter increments to the next value for 'count_enable=1'
- Counter counts from 0 to 9 and triggered on positive edge of clock.

Example 5.11 VHDLRTL for BCD up counter

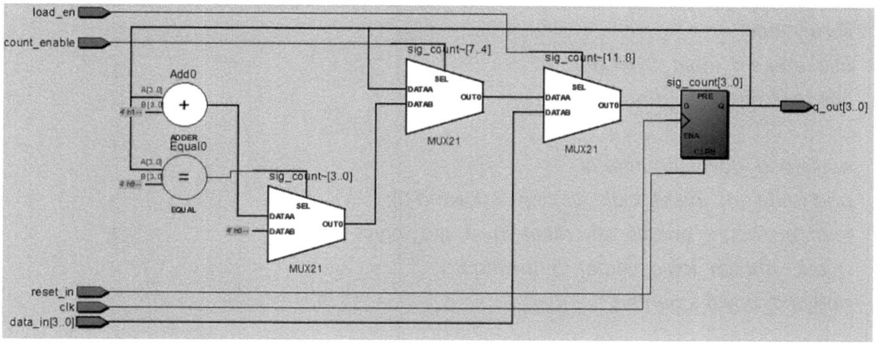

Fig. 5.32 Synthesis result for BCD up counter

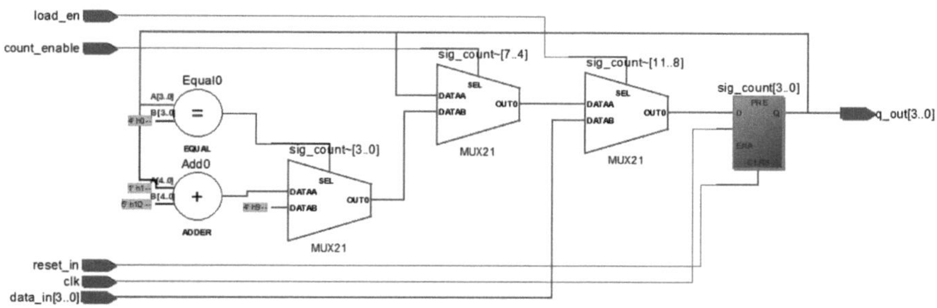

Fig. 5.33 Synthesis result for BCD down counter

Counter has active high asynchronous 'reset_in' input, and when it is active high, the status on output line 'q_out' is '0000.' During normal operation, 'reset_in' is active low.

The synthesis result is shown in Fig. 5.32 and has four-bit data input lines 'data_in,' active high 'load_en,' 'count_enable,' and active high reset input 'reset_in.' Four bit output is indicated by the 'q_out' lines and positive edge-triggered clock by 'clk.'

5.6.4 BCD Down Counter

Four-bit BCD down counter is described by using VHDL and uses the synthesizable constructs. Down counter counts from '1001' to '0000' and triggered on the positive edge of the clock and initializes to '1001' on the next positive edge of the count after reaching to count value '0000.'

The counter described in Example 5.12 is presettable counter, and it has the synchronous active high 'load_en' input to sample the four-bit required value. The

```
library ieee;
use ieee.std_logic_1164.all;
use ieee.std_logic_unsigned.all;

entity bcd_down_counter is
port (data_in : in std_logic_vector (3 downto 0);
load_en, count_enable, clk, reset_in : in std_logic;
q_out : out std_logic_vector (3 downto 0));
end bcd_down_counter;

architecture arch_counter of bcd_down_counter is
signal sig_count : std_logic_vector (3 downto 0);
begin

process (clk, reset_in)
begin
if (reset_in ='1') then
sig_count <= (others => '0');
elsif rising_edge(clk) then
if (load_en = '1' ) then
sig_count <= data_in;
elsif (count_enable = '1') then
 if (sig_count ="0000") then
  sig_count <= "1001";
 else
  sig_count <= sig_count - '1';
 end if;
end if;
end if;
end process;

q_out<= sig_count;
end arch_counter;
```

- Architecture defines the functionality of design.
- Process is sensitive to 'clk', 'and 'reset_in'. Any event on one of the signal invokes the process.
- Nested If-then-else is sequential statement and used inside the process.
- For logic '1' 'reset_in' condition the input 'q_out' is assigned to zero. During normal operation 'reset_in' is active low and counter increments.
- For 'load_en' is equal to logic '1' 'data_in' is assigned to output q_out.
- Counter decrements to the next value for 'count_enable=1'
- Counter counts from 9 to 0 and triggered on positive edge of clock.

Example 5.12 Synthesizable VHDL RTL for the BCD down counter

data input is four-bit and indicated as 'data_in.' The 'count_enable' is used to enable the counting on the rising edge of the clock.

Counter has active high asynchronous 'reset_in' input, and when it is active high, the status on output line 'q_out' is '0000.' During normal operation, 'reset_in' is active low.

The synthesis result is shown in Fig. 5.33 and has four-bit data input lines 'data_in,' active high 'load_en,' 'count_enable,' and active high reset input 'reset_in.' Four bit output is indicated by the 'q_out' lines and positive edge-triggered clock by 'clk.'

5.6.5 BCD Up–Down Counter

BCD up–down counter can be designed by using the synthesizable VHDL constructs for counting depending on the status of the mode input. Mode input is used to indicate up or down counting. Depending on the status of 'up_down' input, the counter increments or decrements. For 'up_down' is equal to logic 1, it performs the up counting; otherwise, it performs the down counting.

The counter described in Example 5.13 is presettable counter, and it has the synchronous active high 'load_en' input to sample the four-bit data. The data input is four-bit and indicated as 'data_in.' The 'count_enable' is used to enable the counting on the rising edge of the clock.

Counter has active high asynchronous 'reset_in' input, and when it is active high, the status on output line 'q_out' is '0000.' During normal operation, 'reset_in' is active low.

The synthesis result is shown in Fig. 5.34 and has four-bit data input lines 'data_in,' active high 'load_en,' 'count_enable,' 'up_down,' and active low reset input 'reset_in.' Four bit output is indicated by the 'q_out' lines and positive edge-triggered clock by 'clk.'

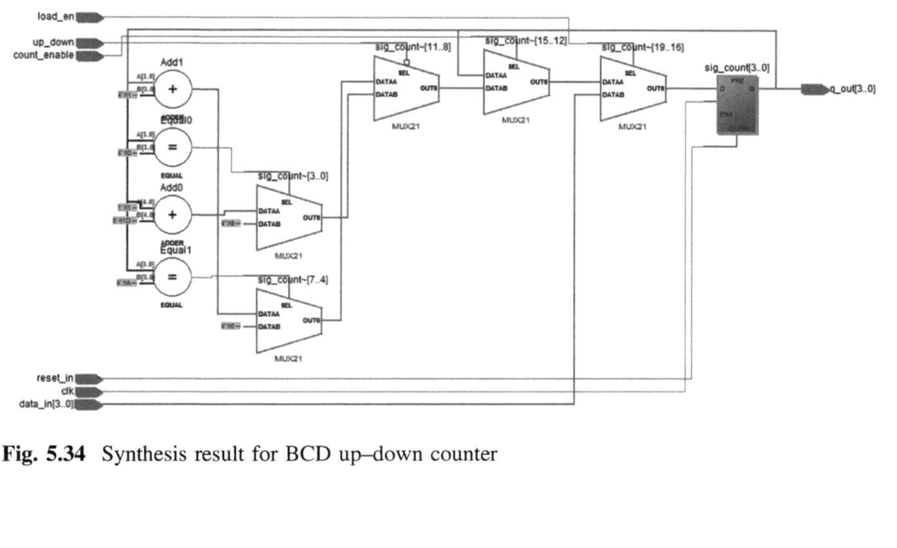

Fig. 5.34 Synthesis result for BCD up–down counter

```
library ieee;
use ieee.std_logic_1164.all;
use ieee.std_logic_unsigned.all;
entity bcd_up_down_counter is
port (data_in : in std_logic_vector (3 downto 0);
load_en, count_enable, up_down, clk, reset_in : in std_logic;
q_out : out std_logic_vector (3 downto 0));
end bcd_up_down_counter;
architecture arch_counter of bcd_up_down_counter is
signal sig_count : std_logic_vector (3 downto 0);
begin
process (clk, reset_in)
begin
if (reset_in ='1') then
sig_count <= (others => '0');
elsif rising_edge(clk) then
if (load_en = '1' ) then
sig_count <= data_in;
elsif (count_enable = '1') then
 if (up_down ='0') then
  if (sig_count ="0000") then
  sig_count <= "1001";
  else
  sig_count <= sig_count - '1';
  end if;
 else
  if (sig_count ="1010") then
  sig_count <= "0000";
  else
  sig_count <= sig_count + '1';
 end if;
end if;
end if;
end if;
end process;
q_out<= sig_count;
end arch_counter;
```

- Architecture defines the functionality of design.
- Process is sensitive to 'clk', 'and 'reset_in'. Any event on one of the signal invokes the process.
- Nested If-then-else is sequential statement and used inside the process.
- For logic '1' 'reset_in' condition the input 'q_out' is assigned to zero. During normal operation 'reset_in' is active low and counter increments.
- For 'load_en' is equal to logic '1' 'data_in' is assigned to output q_out.
- Counter increments or decrements to the next value for 'count_enable=1'
- Depending on the status of up_down input counter increments or decrements.
- For up_down='1' the counter increments and for up_down='0' the counter decrements on the rising edge of clock.

Example 5.13 Synthesizable VHDL RTL for the BCD up–down counter

5.7 Gray Counter

Gray counters are used in the multiple clock domain designs as only one output bit changes on the active clock edge. Gray codes are used in the design of synchronizers. Gray counter is described in the example, and in this, only one bit is changing on the active clock edge with reference to the previous output of the counter. In this, active low reset input is 'reset_n.' When 'reset_n = 0,' the output of counter 'q_out' is assigned to '000.' During normal operation, 'reset_n' is active high.

The RTL using VHDL is described in Example 5.14, and the synthesis result is shown in Fig. 5.35.

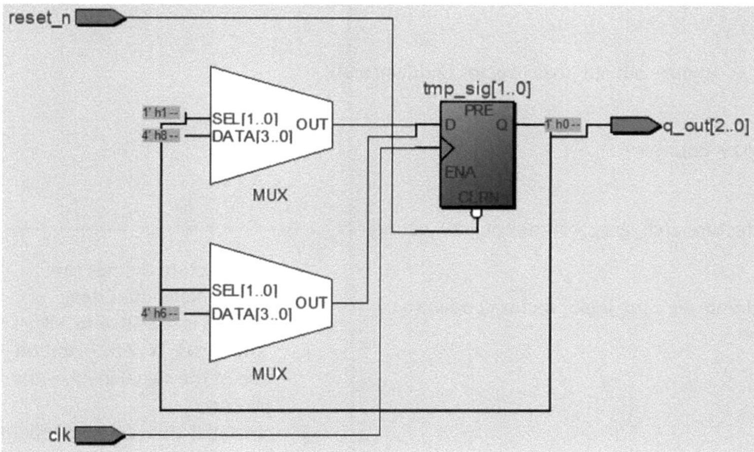

Fig. 5.35 Synthesis result of three-bit gray counter

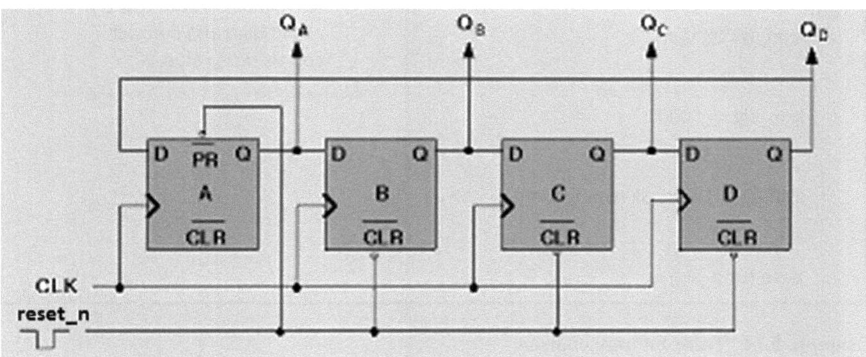

Fig. 5.36 Ring counter internal structure

```
library ieee;

use ieee.std_logic_1164.all;

entity gray_counter is

port ( clk : in std_logic;

     reset_n : in std_logic;

            q_out : out std_logic_vector (2 downto 0));

end gray_counter;

architecture arch_gray_counter of gray_counter is

signal tmp_sig : std_logic_vector (2 downto 0);

begin                                  <-----

 process ( clk, reset_n)

 begin

  if (reset_n='0') then

       tmp_sig <= "000";

       elsif (clk='1' and clk'event) then

        case (tmp_sig) is
```

- Architecture defines the functionality of design.
- Process is sensitive to ‘clk’, ‘and ‘reset_n’. Any event on one of the signal invokes the process.
- Nested If-then-else is sequential statement and used inside the process.
- For logic ‘0’ ’reset_n’ condition the input ‘q_out’ is assigned to “000”. During normal operation ‘reset_n’ is active high and counter counts next value.

Example 5.14 Three-bit gray counter

```
when "000" => tmp_sig <= "000";
        when "001" => tmp_sig <= "001";
        when "010" => tmp_sig <= "011";
        when "011" => tmp_sig <= "010";
        when "100" => tmp_sig <= "110";
        when "101" => tmp_sig <= "111";
        when "110" => tmp_sig <= "101";
        when "111" => tmp_sig <= "100";
        when others => tmp_sig <= null;
        end case;
        end if;
       end process;
       q_out <= tmp_sig;
           end arch_gray_counter;
```

- 'Case' construct is used to describe the design functionality. The design generates the three bit output count as gray code.
- The output is generated on output lines 'q_out'.

Example 5.14 (continued)

5.8 Ring Counter

Ring counters are used in the practical applications to provide the predefined delay. These counters are synchronous in nature and used in the practical applications such as traffic light controller and timers to introduce the certain amount of predefined delay. The internal logic structure using the D flip-flops for four-bit ring counter is shown in Fig. 5.36; as shown, the output of the MSB flip-flop is fed back to the LSB flip-flop input and the counter shifts the data on every active edge of clock signal.

The RTL using VHDL for the four-bit ring counter is described in Example 5.15, and the counter has 'set_in' input to set the input initialization value of '1000' and works on the positive edge of clock signal (Example 5.15).

```
library ieee;
use ieee.std_logic_1164.all;

entity ring_counter is

generic (counter_size : integer := 4);
port (clk : in std_logic;
set_in : in std_logic;
reset_in : in std_logic;
q_out : out std_logic_vector(counter_size 1 downto 0));
end ring_counter;

architecture arch_ring of ring_counter is
signal temp_sig : std_logic_vector(counter_size
begin

process(clk, reset_in, set_in)
begin
if (reset_in = '1') then
temp_sig <= (others => '0');
elsif (set_in ='1') then
temp_sig <= "1000";
elsif (clk='1' and clk'event) then
for k in 1 to (counter_size - 1) loop
temp_sig(k) <= temp_sig(k-1);
end loop;
temp_sig(0) <= temp_sig(counter_size-1);
end if;
end process;

q_out <= temp_sig;

end arch_ring;
```

- Architecture defines the functionality of design.
- Process is sensitive to 'clk', 'set_in' and 'reset_in'. Any event on one of the signal invokes the process.
- Nested If-then-else is sequential statement and used inside the process.
- For logic '1' 'reset_in' condition the input 'q_out' is assigned to zero. During normal operation 'reset_in' is active low.
- For 'set_in' is equal to logic '1' the value "1000" is assigned to output q_out.
- Counter is basically the shift register with the output of LSB fed back to the input of MSB
- These type of counters are used to generate the repeated sequence or used to insert the delay. The counting sequence is 1000,0100,0010,0001---

Example 5.15 VHDLRTL for four-bit ring counter

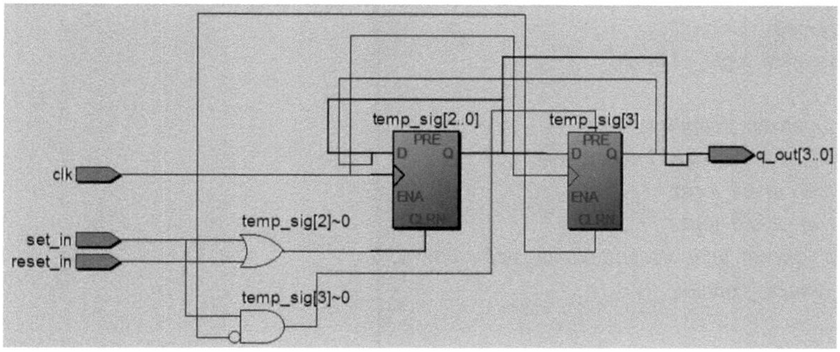

Fig. 5.37 Synthesis result for four-bit ring counter

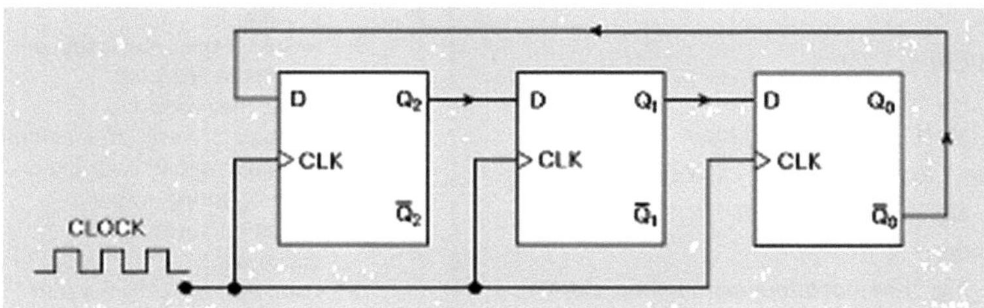

Fig. 5.38 Three-bit Johnson counter

The synthesis result for the ring counter is shown in Fig. 5.37. It uses the additional logic for forcing the asynchronous set_in and reset_in inputs. The logic is not the part of the data path but it is used to control the output of the ring counter.

5.9 Johnson Counter

The Johnson counter is the special type of synchronous counter and designed by using the shift register. This type of counter is also called as twisted ring counter. The internal structure for three-bit Johnson counter is shown in Fig. 5.38. In this type of counter the complement output of LSB flip-flop is fed back to the input of MSB.

The RTL using VHDL for four-bit Johnson counter is shown in Example 5.16. The synthesized logic is shown in Fig. 5.39.

```
library ieee;
use ieee.std_logic_1164.all;

entity johnson_counter is
generic (counter_size : integer := 4);
port (clk : in std_logic;
reset_in : in std_logic;
q_out : out std_logic_vector(counter_size-1 downto 0));
end johnson_counter;

architecture arch_johnson of johnson_counter is
signal temp_sig : std_logic_vector(counter_size
begin

process(clk, reset_in)
begin
if reset_in = '1' then
temp_sig <= (others => '0');
elsif (clk='1' and clk'event) then
for k in 1 to (counter_size - 1) loop
temp_sig(k) <= temp_sig(k-1);
end loop;
temp_sig(0) <= not temp_sig(counter_size-1);
end if;
end process;

q_out <= temp_sig;

end arch_johnson;
```

- Architecture defines the functionality of design.
- Process is sensitive to 'clk' and 'reset_in'. Any event on one of the signal invokes the process.
- Nested If-then-else is sequential statement and used inside the process.
- For logic '1' 'reset_n' condition the input 'q_out' is assigned to zero. During normal operation 'reset_n' is active low.
- Counter is basically the shift register with the complement output of LSB fed back to the input of MSB register.
- These type of counters are used to generate the repeated sequence or used to insert the delay. The counting sequence is 1000,1100,1110,1111,0111, 0011,0001,0000---

Example 5.16 VHDLRTL for four-bit Johnson counter

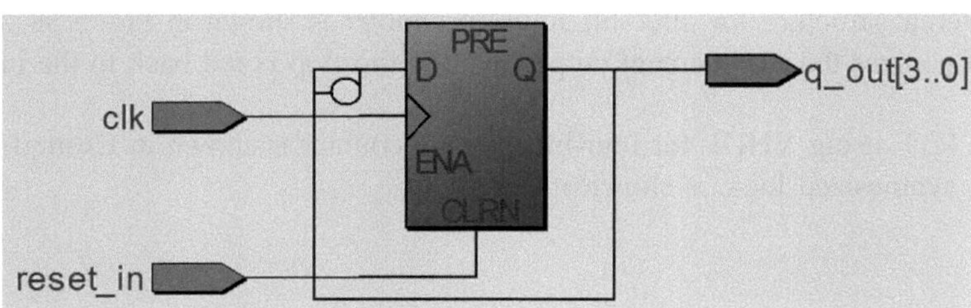

Fig. 5.39 Synthesis result for four-bit Johnson counter

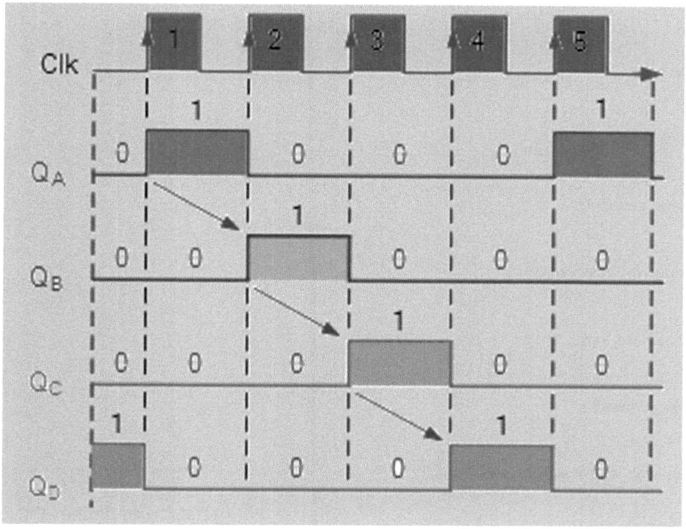

Fig. 5.40 Timing sequence of shift register

5.10 Shift Registers

Shift registers are used in most of the practical applications to perform the shifting or rotation operations on the active edge of clock. The shifter timing sequence with reference to the positive edge of clock signal is shown in Fig. 5.40. As shown in the timing sequence for every positive edge of the clock, the data from LSB shifts by one bit to the next stage, and hence, for the four-bit shift register, it requires four-clock latency to get the valid output data from MSB.

The RTL using VHDL for the Serial Input, Serial Output shift register (SISO) is described in Example 5.17. As described in the example, the data 'serial_in' is shifted on every clock edge to generate the serial output 'serial_out.' To generate the valid serial output for any change on the serial input, the shift register needs four clock pulses.

The synthesis result having four registers for the serial input, serial output shift register is shown in Fig. 5.41.

```
library ieee;
use ieee.std_logic_1164.all;
entity serialin_serialout is
port(clk, serial_in : in std_logic;
serial_out : out std_logic);
end serialin_serialout;
architecture arch_siso of serialin_serialout is
signal tmp_sig : std_logic_vector(3 downto 0);
begin
process (clk)
 begin
 if (clk='1' and clk'event) then
 for i in 0 to 2 loop
 tmp_sig (i+1) <= tmp_sig (i); end loop;
 tmp_sig (0) <= serial_in;
 end if;
 end process;
 serial_out <= tmp_sig (3);
 end arch_siso;
```

- Architecture defines the functionality of design.
- Process is sensitive to 'clk'. Any event on the clock input invokes the process.
- If-then-else is sequential statement and used inside the process.
- The synthesizable 'for' loop is used inside the process and it infers the hardware triggered on the positive edge of clock.
- The serial input is assigned to the input of LSB register.
- The logic infers the design with four registers triggered on the positive edge of the clock input.

Example 5.17 VHDLRTL for serial input, serial output shift register

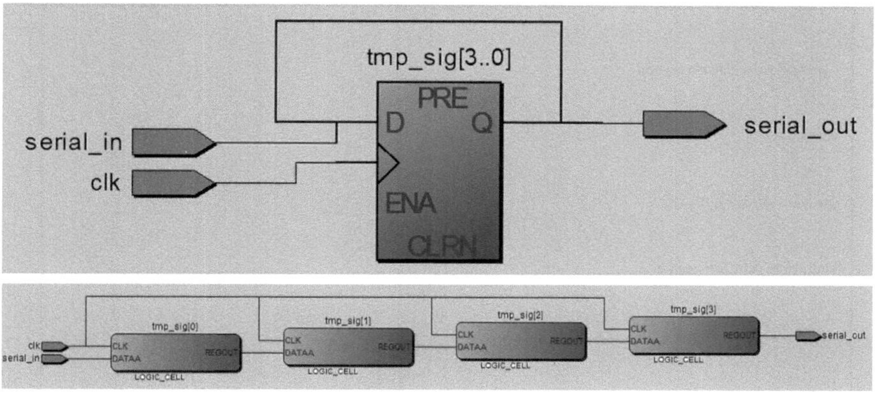

Fig. 5.41 Synthesis result for four-bit shift register

5.10.1 Right and Left Shift Registers

Most of the practical application involves the use of right or left shift of the data. Consider the protocol where requirement is to shift the string on the right side or on the left side by one bit or by multiple bits. In such scenario, the bidirectional (right/left) shift registers are used.

The RTL using VHDL is described in Example 5.18 for bidirectional shift register, and the direction of data is controlled by 'right_left' input. For 'right_left = 1,' the data is shifted toward the left side, and for the 'right_left = 0,' the data is shifted toward the right side.

The synthesis result is shown in Fig. 5.42, and the direction of data transfer is controlled by 'right_left' input. The synthesis result shown consists of four registers with additional combinational logic to control, the data flow direction.

5.10.2 Parallel Input, Parallel Output (PIPO) Shift Register

In most of the processor design applications, the data needs to be transferred in parallel. Consider the four-bit data bus communicating with the external peripheral. If both processor and peripheral operate on the parallel data, then it is essential to transfer the data using parallel input, parallel output logic.

```
--Shift right and shift left register

library ieee;

use ieee.std_logic_1164.all;

entity shift_right_left_register is

port ( serial_in : in std_logic;

     clk , right_left , reset_in :  std_logic;

     q_out : out std_logic_vector ( 3 downto 0) );

end shift_right_left_Register;

architecture arch_register of shift_right_left_Register is

signal sig_tmp : std_logic_vector ( 3 downto 0);

begin

  process ( serial_in, clk, reset_in, right_left)

       begin

        if (reset_in ='0') then       sig_tmp <= "0000";

        elsif ( clk='1' and clk'event) then

         if ( right_left ='1') then

         sig_tmp  <= serial_in & sig_tmp ( 3 downto 1);

           else  sig_tmp  <= sig_tmp ( 2 downto 0) & serial_in;

           end if;        end if;     end process;            q_out<= sig_tmp;  end arch_Register ;
```

- Architecture defines the functionality of design.
- Process is sensitive to 'clk', 'reset_in', 'right_left' and 'serial_in' . Any event on the clock input invokes the process.
- Nested If-then-else is sequential statement and used inside the process.
- For 'right_left' equal to '1' q_out is left shifted and for 'right_shift is equl to '0' the q_out is right shifted.
- The 'reset_in' is an asynchronous input and when logic '0' it is used to initialize the four bit register.
- During normal operation 'reset in' is **active high.**

Example 5.18 VHDLRTL for the right/left shift register

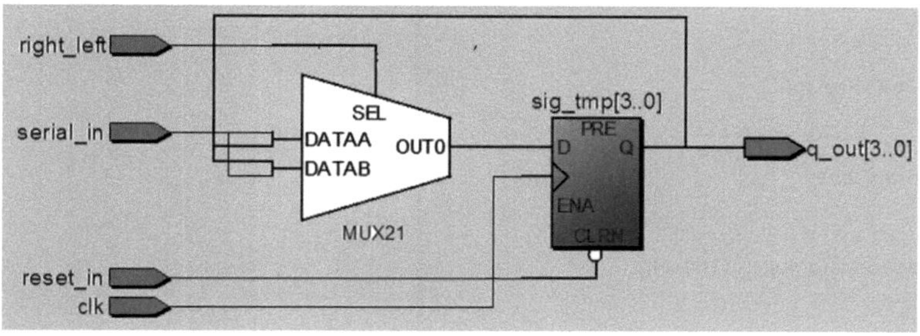

Fig. 5.42 Synthesized logic for bidirectional shift register

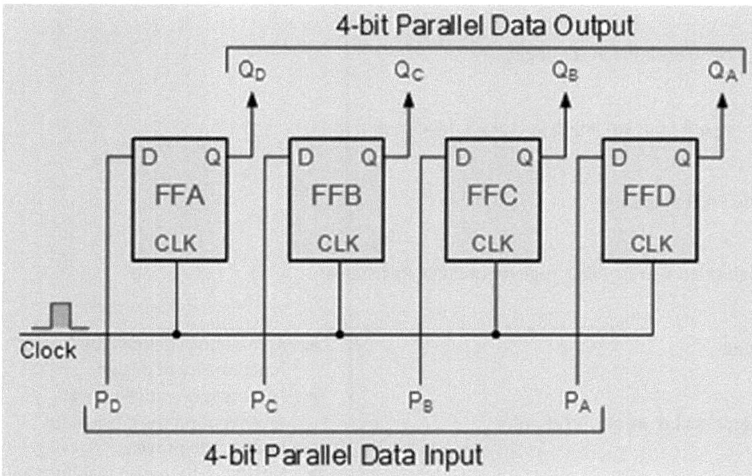

Fig. 5.43 Four-bit PIPO register

In such scenarios, PIPO registers are used. The logic diagram of PIPO four-bit register is shown in Fig. 5.43. Four parallel input lines are named as P_A, P_B, P_C, and P_D, and four-bit parallel output lines are named as Q_A, Q_B, Q_c, and QD. The PIPO register is triggered on the positive edge of clock signal.

The RTL using VHDL is described in Example 5.19.

The synthesis result for the four-bit PIPO register is shown in Fig. 5.44.

```
--PIPO Register
library ieee;
use ieee.std_logic_1164.all;
entity PIPO_Register is
port ( d_in : in std_logic_vector ( 3 downto 0);
    clk , reset_in : in std_logic;
    q_out : out std_logic_vector ( 3 downto 0) );
end PIPO_Register;
architecture arch_PIPO_register of PIPO_Register is
begin
  process ( d_in, clk, reset_in)
     begin
      if (reset_in ='0') then
      q_out <= "0000";
      elsif ( clk='1' and clk'event) then
      q_out <= d_in;   end if;
     end process;  end arch_PIPO_Register ;
```

- Architecture defines the functionality of design.
- Process is sensitive to 'clk'. Any event on the clock input invokes the process.
- Nested If-then-else is sequential statement and used inside the process.
- The parallel input is assigned to output on rising edge of clock.
- The 'reset_in' is an asynchronous input and when logic '0' it is used to initialize the four bit register.
- During normal operation 'reset_in' is active high.

Example 5.19 VHDLRTL for 4-bit PIPO register

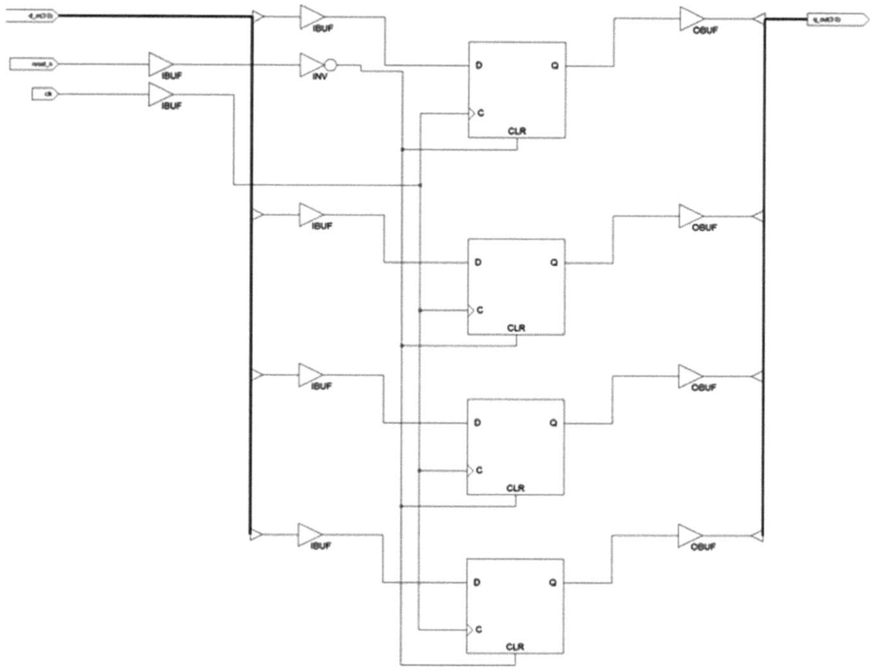

Fig. 5.44 Synthesized logic for four-bit PIPO register

5.11 Asynchronous Designs

In the asynchronous designs, the clock signal is not driven by the common clock source. If the output of LSB flip-flop is given as an input to the subsequent flip-flop, then the design is asynchronous. The issue with the asynchronous design is the cumulative clock to q delay of flip-flop due to the cascading of the stages. Asynchronous counters are not recommended in the ASIC design due to the issue of glitches or spikes, and even the timing analysis for such kind of design is difficult task.

The asynchronous counter design and the memories are discussed in the next subsequent chapter. Even the test benches and verification using VHDL for the RTL design are discussed in Chap. 7.

5.12 Summary

The following are the key points to summarize the sequential logic design:

1. Sequential design elements are latches and flip-flops.

2. Sequential designs are of two types: synchronous and asynchronous.
3. If the two arriving clock inputs are from different sources and has phase difference, then the design is called asynchronous.
4. Latches are level sensitive and not recommended in the ASIC designs. Flip-flops are edge-triggered and are recommended in the ASIC designs.
5. Number of statements inside the if (clk'event and clk = '1') infers those many number of registers.
6. Flip-flop timing parameters are setup time, hold time, and clock to q delay or propagation delay.
7. Gray counters can be designed by using the binary counters with the additional combinational logic.
8. Synchronous counters are recommended in the ASIC or FPGA design as timing analysis will be easy and they are not prone to the glitches.
9. Asynchronous counters are prone to the glitches or spikes and hence not recommended in the ASIC or FPGA designs.
10. Special counters such as ring and Johnson can be designed by using the shift registers.
11. If setup or hold time is violated, then the flip-flop goes into the metastable state.
12. Use the two-stage level synchronizer to pass the data from one of the clock domains to another clock domain.
13. Maximum operating frequency for any design is dependent on the time period of the register-to-register path and setup time of the flip-flop.

Chapter 6
Introduction to PLD

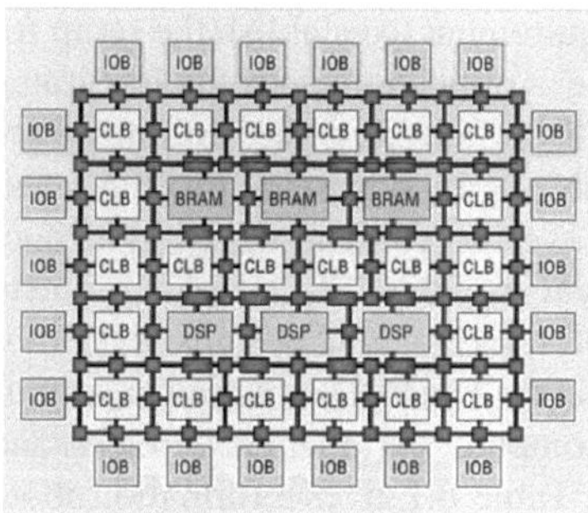

"I have no special talent. I am only passionately curious..." --- Albert Einstein

Understand the concept of PLD and PLD based designs and use the imagination and intelligence to develop the design using PLDs.

Abstract This chapter describes the practical understanding about the PLD architecture and the practical use in the ASIC prototyping and FPGA based design. This chapter is organized in such a way that it explains the PLD evolution and the classification with the detailed architecture. Even this chapter covers the practical scenarios while using the FPGA for prototyping. The architecture for XILINX and Altera is covered with the practical-oriented examples and the synthesis results.

Keywords ASIC · PLD · CPLD · SPLD · PAL · PLA · FPGA · LUT · Register · LFSR · Configuration file · MUX · Registered output · Combinational output · Device utilization · CLB · Slice · Carry chain · PROM · Bit-map · XILINX · Altera · Spartan · Cyclone · Virtex · Stratix · Multiplex clocking

During the past decade, the programmable logic devices (PLDs) are used for the rapid prototyping of ASICs. PLD based designs can be used to detect the bugs during early design cycle and to validate the design in lesser time duration for the given functional specifications. If we consider the era of miniaturization, during the past 50 years, then we can easily conclude that the designs have become very complex. In the present scenario there is need of million gate programmable ASIC for realization of the complex designs. As PLD-based design is more cost-effective

V. Taraate, *PLD Based Design with VHDL*,
DOI 10.1007/978-981-10-3296-7_6

and can be realized in lesser time duration, the PLD market has grown substantially for the quick prototyping of ASICs. This chapter discusses about the evolution of PLDs, the types of PLDs, and the architecture of PLDs. Even this chapter discusses the PLD-specific design guidelines and scenarios.

6.1 History and Evolution of PLDs

In the semiconductor design industry, it is very much required to have the programmable logic devices. The reason is that the device can be repeatedly programmed for the different design specifications. The cost requirement to establish the setup for PLD-based designs is lesser as compared to the Application-Specific Integrated Circuit (ASIC). For the new idea realization, it is not possible to infuse the million-dollar funds at the early stage. So it is always recommended to have less investment while realizing the product idea and even to validate the design functionality. If we consider the past decade, then the real growth of PLD market is due to the requirements of the million-to-billion-gate SOCs. In the SOC designs, the PLDs are extensively used to validate the design functionality. The PLDs are used in various market segments such as automotive, consumer, computer peripherals, wireless, and industrial domains for proof of concept of the ideas. Table 6.1 gives information about the worldwide semiconductor revenue projections till the year 2018.

Even if we consider the shrinking process node below 10 nm, then we can conclude that there is a need of multiple FPGAs in the realization of billion-gate complexity ASICs.

The major advantage of PLDs is they can be Programmed by end user in the field. The first PLD that was invented before 1970 is Programmable Read-Only Memory (PROM). But PROM is one-time-programmable memory. Again, we can differentiate this as mask programmable devices and field-programmable logic devices. The mask programmable logic devices are programmed by vendor using the interconnect and custom mask, whereas the programmable devices are programmed or configured by user depending on the required design specification and complexity.

During the late 1970s, the programmable array logic (PAL) was introduced in the market. **The PAL consists of programmable AND and fixed OR plane**. In the subsequent section, we will discuss the PAL architecture and how the Boolean expressions are realized using the PAL. But they are used to realize low-complexity designs. But during the present decade, the PAL devices are available with a varied size of inputs and outputs. Instead of using the AND–OR array plane, most of the vendors use the NAND–NAND or NOR–NOR structures.

During the early 1970s, the programmable logic array (PLA) was introduced in the market and it has programmable AND and programmable OR structures. During the 1980s, the evolution of PLD happened and the PLDs are classified as simple programmable logic devices (SPLDs) and high-density programmable logic devices (HPLDs). SPLD includes the PROM, PLA, PAL, and GAL. HPLD

Table 6.1 The revenue forecast in semiconductor design

Worldwide semiconductor revenue forecast by market segment

$M	2010	2011	2012	2013	2014	2015	2016	2017	2018	13–18 CAGR %	13/12 Y/Y %
Automotive	23,149	25,344	24,324	25,897	28,325	30,017	32,658	35,242	38,002	8	6
Computer	93,269	86,897	79,501	82,494	88,268	95,631	103,570	111,353	119,243	8	4
Consumer	58,024	53,664	54,177	57,661	63,621	69,515	75,041	80,553	86,382	8	6
Wired	28,444	27,749	28,270	32,453	32,465	36,557	38,533	41,668	45,995	7	15
Wireless	59,891	70,826	72,694	79,454	90,976	101,916	111,400	121,735	134,378	11	9
Industrial	35,544	35,043	33,224	35,082	37,475	39,336	43,384	45,216	48,012	6	6
Total	298,321	299,521	292,190	313,041	341,129	372,972	404,587	435,766	472,012	9	7

Databeans estimates

includes the CPLD and FPGA. CPLD is a complex programmable logic device, and FPGA is a field-programmable gate array.

6.2 Simple Programmable Logic Device (SPLD)

The SPLDs are simple programmable logic devices and used for low-density gate count design. The SPLD can be visualized as the array of AND and OR. Figure 6.1 describes the key functional blocks for the SPLD.

Now, before going through the internal structure of every block to understand the SPLD, let us explore the simple design of full adder using the concept of programmable OR.

The full adder using two half adders and the OR gate is shown in Fig. 6.2. Please refer Chap. 2 for the basic combinational elements.

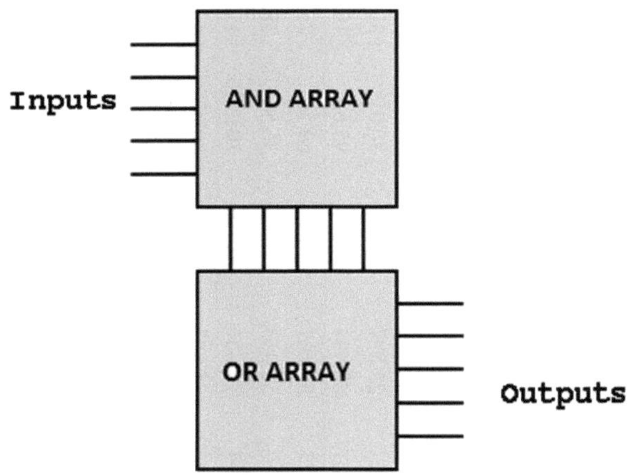

Fig. 6.1 Internal structure of logic array

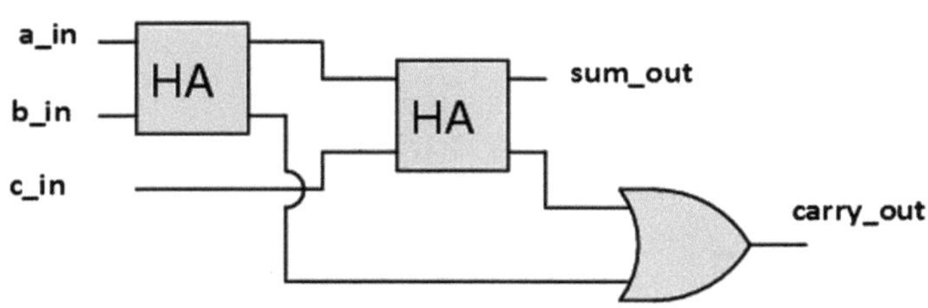

Fig. 6.2 Full adder using the logic gates

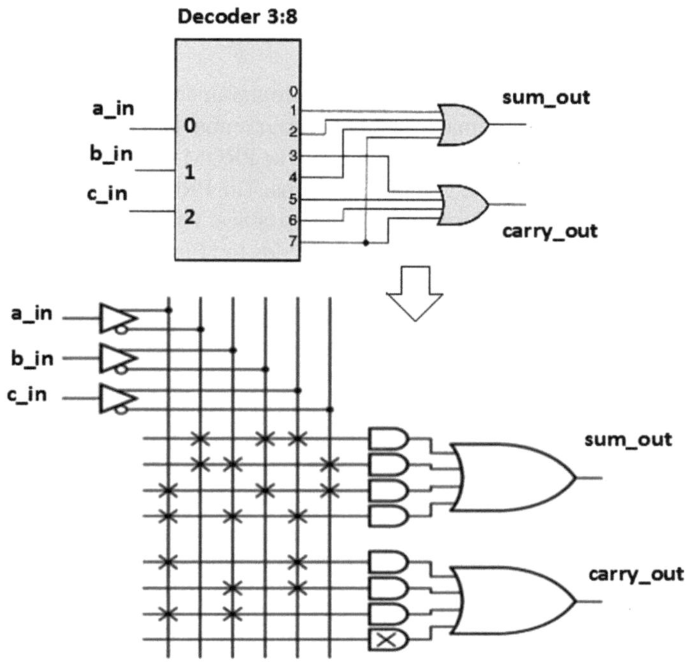

Fig. 6.3 Full adder using the programmable concept

Instead of using the half adders and or gate the full adder can be realized using the fixed OR array and programmable AND array. The same concept is used in the programmable array. Figure 6.3 describes the realization of the full adder using the programmable concept.

As shown in Fig. 6.3 depending on the required min-terms, the AND plane acts as programmable decoder and the OR plane that is fixed is used to generate to programmable outputs 'sum_out' and 'carry_out'. Depending on the programmable or fixed array, SPLDs are classified as follows:

1. Programmable read-only memory (PROM);
2. Programmable array logic (PAL);
3. Programmable logic array (PLA);
4. Generic array logic (GAL).

The subsequent section discusses about the details of the SPLD and the basic architecture for the each SPLD.

6.2.1 *Programmable Read-Only Memory (PROM)*

The PROM was firstly developed as one-time-programmable read-only memory. It is available as one-time-programmable and field-programmable. The field-programmable PROM is EPROM-based and EEPROM-based. The PROM is the array of the read-only cells and extensively used in the computer systems. The PROMs are used to realize the small gate count logic using the concept of lookup tables. The logic can be programmed into the PROM. It is like the lookup table that holds the functionality of the design. In this, the function inputs can be visualized as address lines, the memory array cells are used to hold the information about the functionality and the outputs lines are from the memory cells of the PROM. So in the simple words, we can describe PROM architecture as, the input decoder which is AND array and at outputs are generated from programmable OR array. This allows programming of every output individually for the given set of the inputs. Consider the architecture shown in Fig. 6.4.

As shown in Fig. 6.4, the function inputs are given to the select lines of 3×8 decoder. The decoder acts as fixed AND array, and the output lines of decoder are used to program the OR array. Depending on the fan-out capability of decoder, the number of outputs can be programmed. The function realization using PROM is shown in Fig. 6.5.

As shown in the figure, the three input lines are A2, A1, and A0 and output lines are F2 and F1. The output functions are $F2 = \sum(0,1,2,5,7)$ and $F1 = \sum(1,2,4,6)$ which are realized using the PROM. The fixed AND array uses the decoding logic, and the programmable OR array is used to generate programmable output.

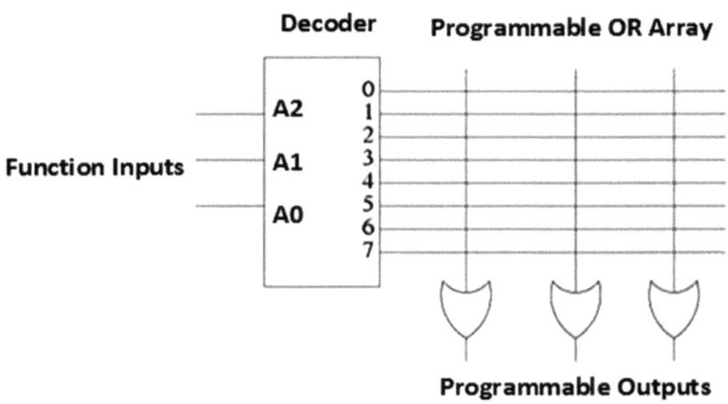

Fig. 6.4 Basic PROM architecture

A2	A1	A0	F2	F1
0	0	0	1	0
0	0	1	1	1
0	1	0	1	1
0	1	1	0	0
1	0	0	0	1
1	0	1	1	0
1	1	0	0	1
1	1	1	1	0

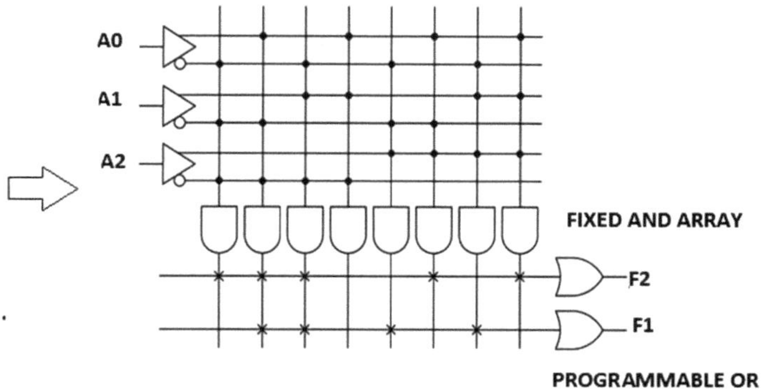

Fig. 6.5 Logic function realization using PROM

6.2.2 *Programmable Array Logic (PAL)*

The PAL uses the programmable AND array and fixed OR array and can be used to design small gate count logic. So these are used to implement the canonical form sum of product Boolean functions using the programmable AND array followed by fixed OR array. So each of the two-level AND–OR terms has the number of inputs which can be programmed. These kinds of PAL are the oldest programmable logic and can be used to generate the combinational output, or the output can be registered or can be fed back internally.

The PAL with the programmable AND array and fixed OR array is shown in Fig. 6.6. This is used to realize the functions F1 and F2. The output functions are $F2 = \sum(4, 5, 6, 7)$, and $F1 = \sum(0.1.2.3)$ are realized using the PAL.

As shown in Fig. 6.6, the outputs are combinational, that is, output is the function of the present input only. The PAL structure can be modified by using the register at the output of the OR array to generate the registered output. Figure 6.7 describes one more example of the macrocell which uses the concept of PAL with the registered output.

As shown in Fig. 6.7, the output can be programmed as registered output or unregistered output. The PAL output passes through the XOR logic which and XOR gate acts as the polarity control. If the output of XOR gate passes through the register the output can be available as the registered output. Output MUX is used to select the registered or combinational output depending on the status of the select line. The loopback of the register output is possible internally and can be used for the internal processing by the PAL.

6.2.3 *Programmable Logic Array (PLA)*

The PLA is more flexible as compared to PAL, and PLA uses the programmable AND and programmable OR arrays. For the logic circuit optimization of the small

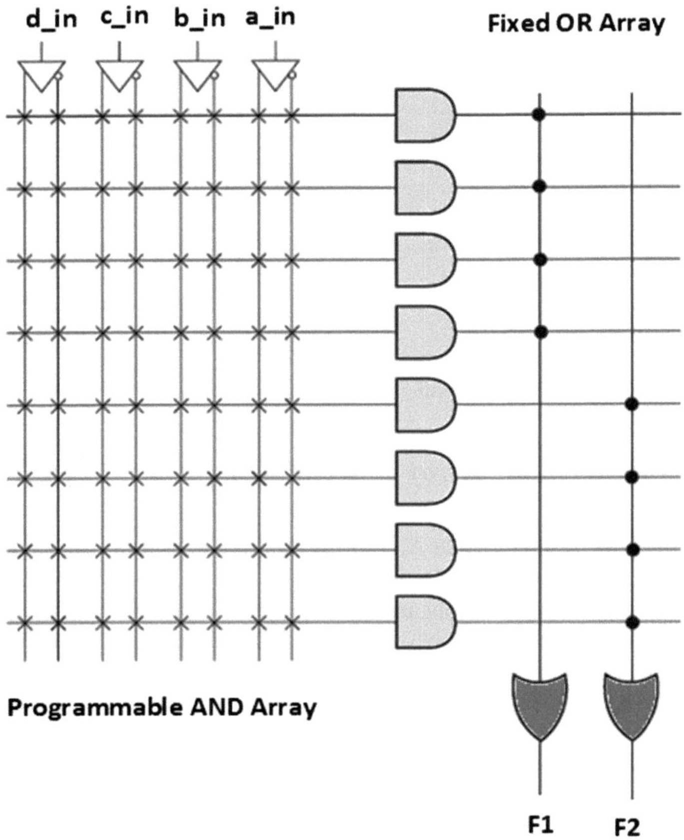

Fig. 6.6 PAL architecture

gate count design, the PLA can be a good choice. Boolean functions can be realized by using the programmable AND followed by programmable OR. The implementation of functions F1 and F2 using PLA is shown in Fig. 6.8. The function implementation for $F2 = \sum(4, 5, 6, 7)$ and $F1 = \sum(0.1.2.3)$ is shown below. As shown the cross indicates the connection.

Figure 6.9 shows the structure of macrocell using the PLA block. The output from PLA can be registered or combinational at the output pad. Even depending on the status of select lines of multiplexer, the output can be configured. The output configuration for the better understanding is shown in Table 6.2.

As shown in Fig. 6.9, the output can be combinational active low or active high. Even the output can be configured as registered active low or registered active high output.

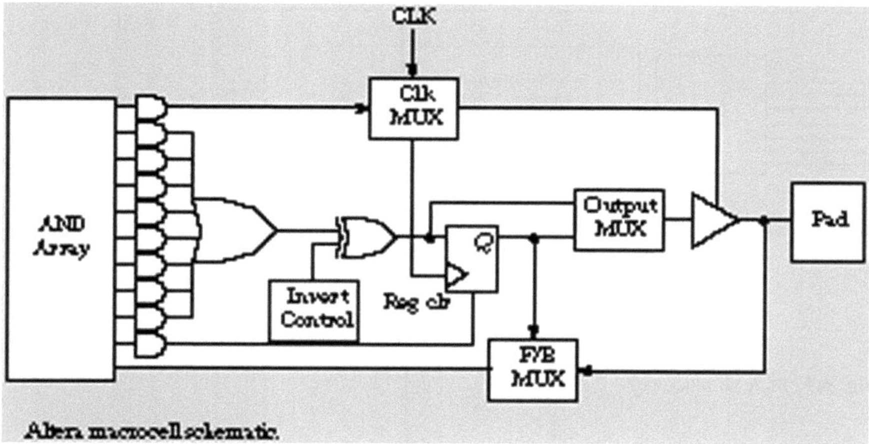

Fig. 6.7 Altera macrocell

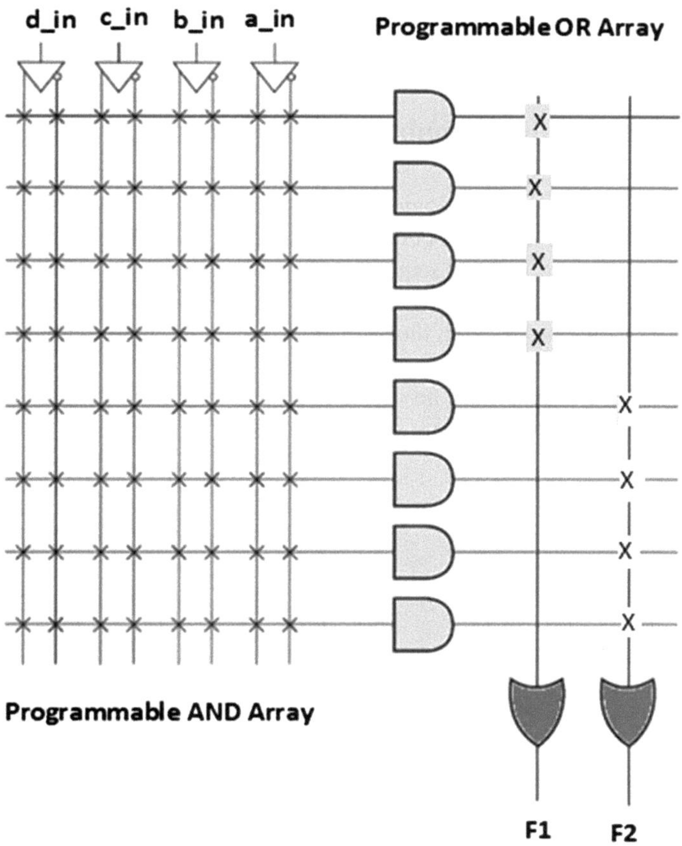

Fig. 6.8 PLA architecture

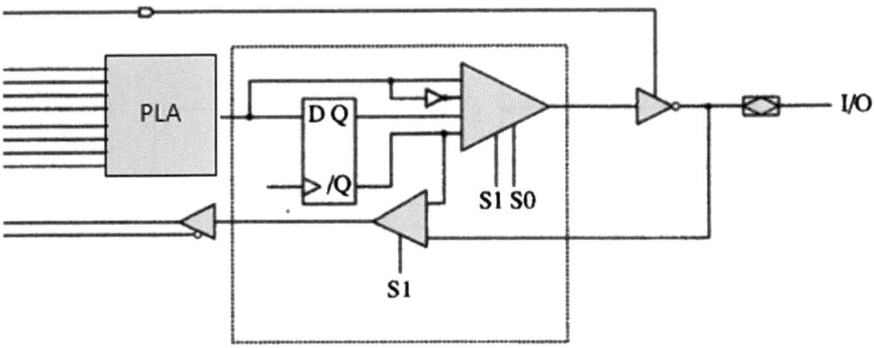

Fig. 6.9 PLA as macrocell

Table 6.2 PLA macrocell output types

S1	S0	Output type
0	0	Combinational active high output
0	1	Combinational active low output
1	0	Registered active high output
1	1	Registered active low output

6.3 Complex Programmable Logic Devices

The complex programmable logic devices (CPLD) are used to realize the small-to-moderate count density controllers using the FSM. Even they can be used to design the combinational and sequential logic of moderate density count designs. The CPLD has evolved by using the concept of PAL-like blocks. The CPLD consists of PAL-like blocks with the routing resources and input/output (IO) blocks to realize the moderate gate count designs.

Each PAL-like block can be treated as simple PLD of few gate count. Figure 6.10 describes the structure of CPLD.

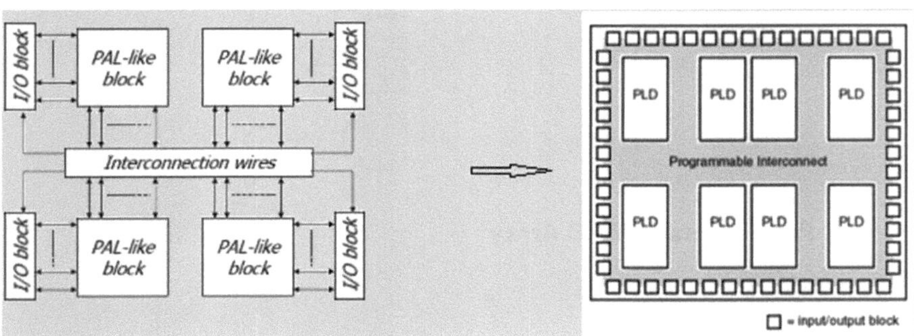

Fig. 6.10 CPLD structure

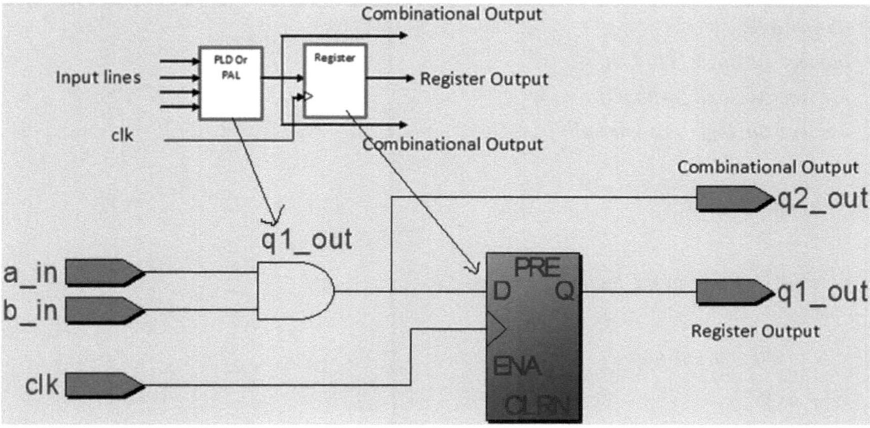

Fig. 6.11 Basic PLD block

Every IO block is used to establish communication between the external world and the PLDs. The multiple PLDs are stacked on the silicon and can be connected using the programmable interconnect. The PLD or PAL block structure can consist of the gate array with the register (number of registers is dependent of the architecture) and be used to generate either combinational or sequential output or both. Figure 6.11 shows the basic PLD block, the logic implementation using the VHDL is shown in Example 6.1.

As shown in Example 6.1, the combinational output 'q2_out' is the AND logic of 'a_in' and 'b_in' and realized using the simple PAL block. The registered output 'q1_out' is sensitive to positive edge of clock 'clk' and realized by use of dedicated register inside the PAL or use of the register from the IO block. The use of registered input and output can improve the design timing and performance even the addition of pipelined logic becomes easy if required.

CPLD is gate-rich logic and has lesser number of sequential cells (registers). The major limitation of the CPLD is small gate count up to few thousand gates, and hence, although having the clean register timing due to small gate count implementation, the multiple CPLDs may be needed to realize the logic which consists of complex design. In such scenario, the best choice is FPGA as it is flip-flop-rich logic. The subsequent session discusses the basic architecture of the field-programmable gate array (FPGA) and the realization of the logic using FPGA.

```
library ieee;
use ieee.std_logic_1164.all;
use ieee.std_logic_arith.all;
use ieee.std_logic_unsigned.all;

entity cpld_logic is

port ( a_in, b_in : in std_logic;

        clk :  in std_logic;

        q1_out, q2_out : out std_logic);

end cpld_logic;

architecture arch_cpld_logic of cpld_logic is

begin

q2_out <= a_in and b_in;

 process ( clk, a_in, b_in)

 begin

      if (clk='1' and clk'event) then

       q1_out <= a_in and b_in;

      end if;

 end process;

end arch_cpld_logic;
```

- Architecture defines the functionality of design.
- Process is sensitive to 'a_in', 'b_in' and 'clk'. Any event on one of the signal invokes the process.
- If-then-else is sequential statement and used inside the process.
- For rising edge of clock 'q1_out' is assigned as 'a_in and b_in'.
- The output 'q2_out' is combinational output and 'q1_out' is registered output.

Example 6.1 Synthesizable VHDL RTL code for the logic realization using PLD

6.4 Field-Programmable Gate Arrays

The field-programmable gate arrays (FPGAs) can be programmed or configured in the field by the user programs and extensively used in the design of complex gate count designs. Even nowadays, FPGAs are used to realize the complex SOC designs and for proof of concept of the ideas. The extensive use of FPGA during this decade is due to the availability of the soft and hard macros required for the processor, DSP, and video processing. Even most of the complex architecture FPGAs support the high-speed interfaces, Ethernet, USB, and AHB protocols.

The basic FPGA architecture can be visualized as a sea of the logic blocks or configurable logic blocks (CLBs), input/output blocks (IOBs), block RAMs (BRAMs), DSP blocks (DSP), and other routing resources. The basic FPGA blocks are shown in Fig. 6.12, and as shown in Fig. 6.12, it consists of

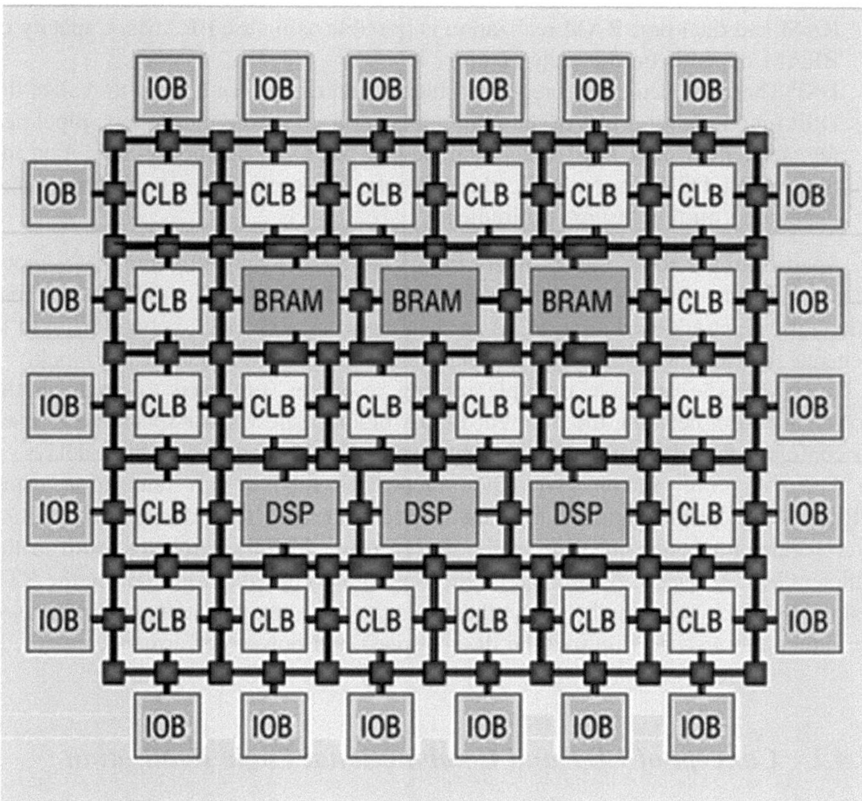

Fig. 6.12 Basic FPGA architecture

1. **CLB** It is used to realize the combinational and sequential logic. Dedicated CLB consists of the number of lookup tables (LUTs) and registers. The combinational function is realized using the LUTs where LUTs have the uniform delay. If logic is not fitted in the single CLB, then the multiple CLBs need to be used and reconfigured depending on the functionality. They are configured by using the vendor-specific program that is configuration or bit-map file.
2. **IOB** It is used to establish the communication between the external world and the CLBs and vice versa. The IOB consists of the bidirectional buffers with the registers. The input can be registered using the IO block, and even the output can be registered using the IO block. The unregistered input and output are possible. Depending on the functional requirements in the design IO can be configured as registered IO or unregistered IO.
3. **BRAM** The FPGA can have the distributed RAM and block RAM (BRAM). Distributed RAM can be realized using the LUTs, and the BRAMs are realized by using the dedicated BRAM blocks which can be programmed by the vendor-specific design tool for the required configuration and size. Single-port RAM and dual-port RAM realization is possible using the BRAMs. Capacity of BRAM depends on the architecture.
4. **DSP** These are dedicated predefined blocks and can be configured to realize the DSP functionality. Most of the DSP application needs the multipliers, pipelined registers, dedicated DSP functional blocks for DSP operations, etc. For the high-speed DSP computation, these blocks are used and can be configured depending on the design requirements.

Apart from the above blocks, every FPGA has the clocking structure that is clock block; Xilinx uses digital clock manager (DCM) with delay-locked loop (DLL), whereas Altera uses phase-locked loop (PLL) as clock network. The clock network is used to generate the clock with uniform clock skew and with glitch and hazard-free clock.

Every FPGA can have multiple routing resources and used to establish the communication between the different FPGA blocks. The vendor-specific tool used to configure the FPGA uses the routing resources with the least routing delays.

The following section gives the practical-oriented design realization using FPGA. As a design engineer, it is always recommended to understand the FPGA architecture for better outcome of the design using VHDL. This can result in the efficient FPGA design. As architecture resources are known before writing the RTL using VHDL, it gives the better synthesis results and even this strategy can be used to reduce the area, to improve the design speed and power performance.

6.4.1 Concept of LUT and Combinational Logic Realization

The LUT concept can be easily understood by using the MUX-based designs. Effectively, the logic function is realized using the LUT. LUT provides the uniform delay, irrespective of the number of inputs for the same design.

Fig. 6.13 Four-input LUT

Consider four-input LUT shown in Fig. 6.13, and it is used to realize the logic function having four inputs and single output.

6.4.2 VHDL Design and Realization Using CLB

As discussed earlier, the basic CLB consists of the LUTs and registers. Depending on the device architecture, the number of LUTs and registers can vary and even the inputs and outputs of LUT structure can vary. For easy understanding, consider the basic CLB architecture shown in Fig. 6.14.

The LUT can be used to realize the logic functions, can be used as distributed RAM, and even used to realize the shift registers. As shown in Fig. 6.14, the output generated is combinational or sequential depending on the configuration set by bitstream file. If the SRAM cell (FF) holds logic '1', then output is combinational logic, and if configuration FF holds the logic '0', then output is registered output.

Consider the RTL using VHDL described in Example 6.2.

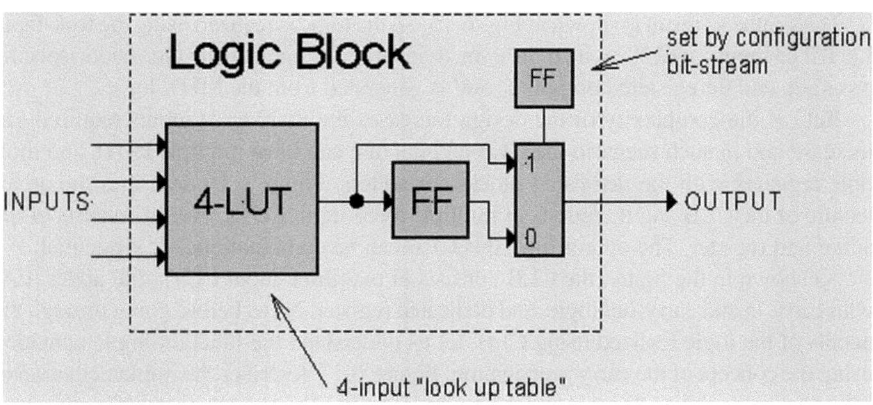

Fig. 6.14 Basic CLB architecture

```
library ieee;
use ieee.std_logic_1164.all;
use ieee.std_logic_arith.all;
use ieee.std_logic_unsigned.all;

entity comb_logic is

port ( a_in, b_in, c_in,d_in: in std_logic;

          q_out: out std_logic);

end comb_logic;

architecture arch_comb_logic of comb_logic is

begin

q_out <= (a_in and b_in) xor (c_in and d_in);

end arch_comb_logic;
```

- Architecture defines the functionality of design.
- The assignment statement is used where the 'q_out' is combinational logic.

Example 6.2 Synthesizable VHDL RTL for the combinational design

The synthesis result is shown in Fig. 6.15. So the logic is realized using the four-input LUT. The inputs of LUT 'a_in, b_in, c_in, d_in' are configured using the vendor-specific program, and unregistered output 'q_out' is generated from the MUX logic.

But as the complexity of the design increases the number of inputs required can increase and in such scenario the CLB architecture can have multiple LUTs and multiple registers with the dedicated blocks for adders. Figure 6.16 describes the architecture of the CLB and it consists of multiple three-input LUTs, even it consists of the adder and register. The output from this CLB can be combinational or sequential.

As shown in the figure, the CLB consists of two three-input LUTs, full adder (FA) with carry-in and carry-out logic, and dedicated register. Now, before going through the details of the logic realized using CLB, let us understand the function implementation using the concept of the carry propagation. Figure 6.17 describes the implementation of full adder using the XOR logic and the multiplexer. In FPGA similar kind of logic can be used to perform the addition using 'A, B, Cin' to generate the 'Cout and Sum' output.

Consider the following structure of CLB which has three-input LUT and register. The output of CLB can be combinational and registered.

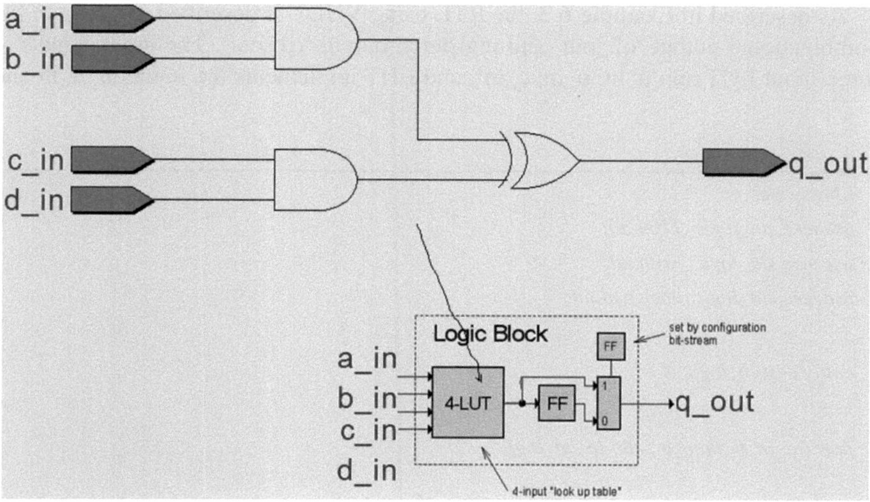

Fig. 6.15 Logic realization using the LUT

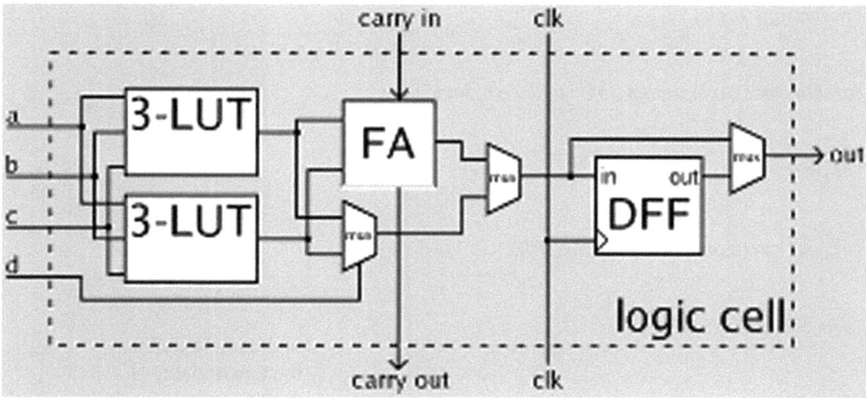

Fig. 6.16 Architecture of CLB with multiple LUTs

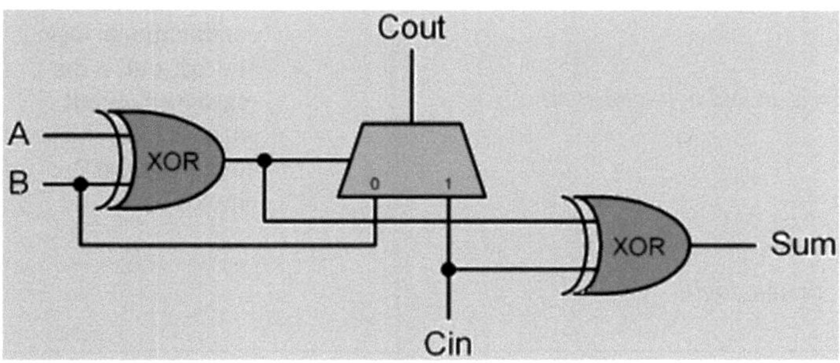

Fig. 6.17 Full adder using the concept of carry propagation

As described in Example 6.3, the RTL using VHDL is described to generate the combinational output 'q2_out' and registered output 'q1_out'. The input signals to three-input LUT are 'a_in, b_in, c_in', and LUT implements the function 'a_in and

```
library ieee;
use ieee.std_logic_1164.all;
use ieee.std_logic_arith.all;
use ieee.std_logic_unsigned.all;

entity design_logic is

port ( a_in, b_in, c_in, clk: in std_logic;

          q1_out, q2_out:  out std_logic);

end design_logic;

architecture arch_design_logic of design_logic is

begin

q2_out <= (a_in and b_in and c_in);

process (clk)

begin

if (clk='1' and clk'event) then

q1_out <= a_in and b_in and c_in;

end process;

end arch_design_logic
```

- Architecture defines the functionality of design.
- The assignment statement is used where the 'q2_out' is combinational logic.
- The 'q1_out' is the registered output which is from the positive edge triggered D flip-flop.

Example 6.3 Synthesizable VHDL RTL for the registered output

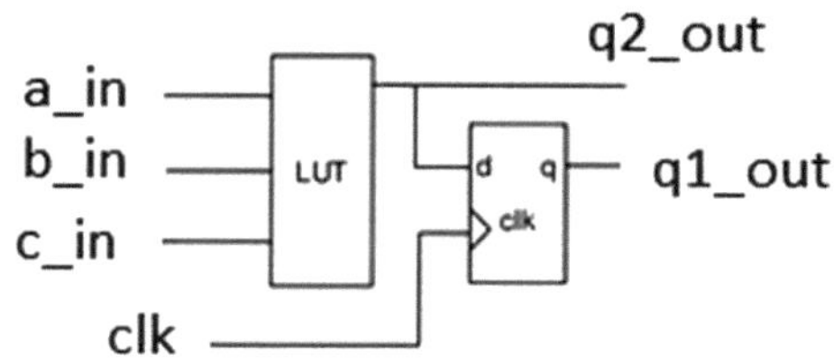

Fig. 6.18 Logic realization using the CLB

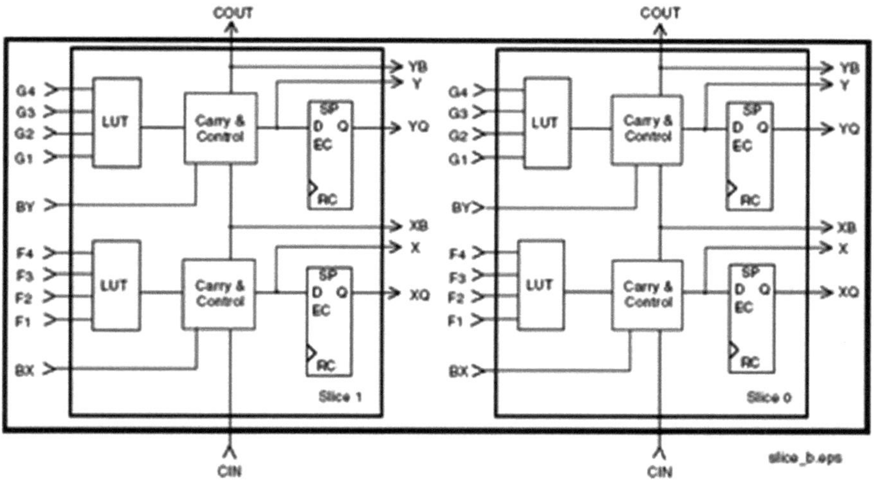

Fig. 6.19 CLB architecture for Virtex series FPGA

b_in and c_in' to get the output 'q2_out'. The registered output 'q1_out' is generated from the D register which is triggered on the positive edge of clock 'clk'.

Figure 6.18 gives information about the synthesis and the implementation using CLB.

As discussed previously, the FPGA architecture is device-specific. The different device series of XILINX/Altera have different CLB blocks. The CLB architecture for Virtex series of XILINX is shown in Fig. 6.19. This consists of two slices, and each CLB has two slices. Slice 1 and slice 0 are identical and used to realize the digital logic with registered output and combinational output.

As shown in Fig. 6.19, every slice consists of two four-input LUT and two registers. Even every slice consists of the carry and control logic to realize the

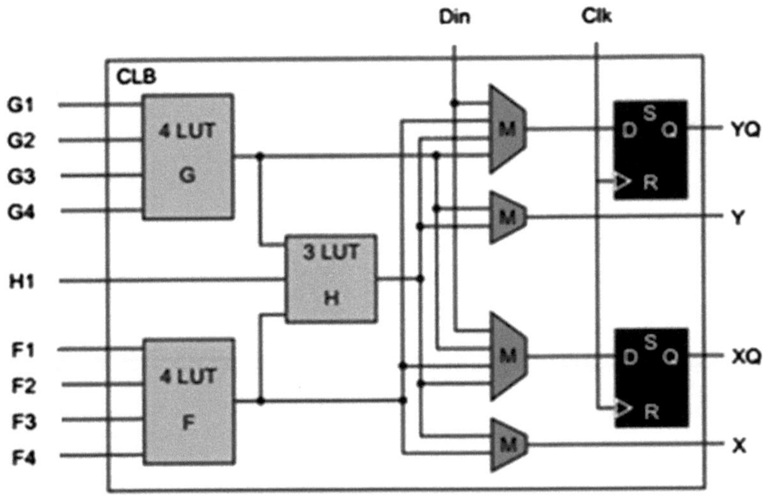

Fig. 6.20 CLB with two four-input LUTs

addition logic. The four-input LUT is used to realize the function with four inputs. By using the routing resources, the slice 1 logic can communicate with the slice 0. Even the output of one CLB can transfer the data to another CLB. Routing is done using the vendor-specific routing algorithms.

The eight-input CLB with the registered output and combinational output is shown in Fig. 6.20. This is used to design the logic function which needs 8 inputs. The CLB can be used to get registered or unregistered output.

The LUT or function generator is named as G, H, and F. The LUT G and F are four-input LUT and LUT H is three-input LUT. Now, consider the design scenario to realize the 8:256 decoder using FPGA. By using the CLB architecture shown in the Fig 6.20 the 8:256 decoder can be realized. But it needs many LUTs, to realize the logic with 256 output lines it needs 3 × 256 LUTs, that is 768 LUTs. The single output can be taken from the 'y' or 'x'. The output is unregistered output.

As discussed in Chap. 4 to implement the decoders, the 'case' construct can be used. If 'case' construct is used then, to realize the combinational logic to generate single output it utilizes F,G and H LUTs. So it is very expensive as per as overall gate count is concern.

In such scenario, the logic duplication can be used and this technique is discussed in Chap. 8. If logic duplication is used, then the overall area in such scenario can be minimized to realize the 8:256 decoder. Instead of using the 8-bit select input, use the 4-bit lower nibble of select lines as input and describe the RTL using VHDL for

4:16 decoder. Use the upper nibble of select lines as the input and describe the RTL using VHDL for 4:16 decoder. Two four as to 16 decoders needs 32 LUTs only . Use the design strategy in such a way that it can minimize the area. Create the AND plane for 256 output lines. use the decoder output lines as inputs lines of AND gates to meet the design functionality. So using the logic duplication, the design needs the 288 LUTs, thus this technique saves area of 768 − 288 = 480 LUTs.

6.4.3 IO Block

The IO blocks are used to communicate with the external world. The basic IO block structure is shown in Fig. 6.21.

As shown in Fig. 6.21, the IO can be configured to transfer the data from the external world to the configurable array. The data is transferred from the input port to PAD and through the input buffer to CLB.

The data can be transferred from the CLB to the outside world through the output buffer, and the buffer enable can be controlled by the programmable

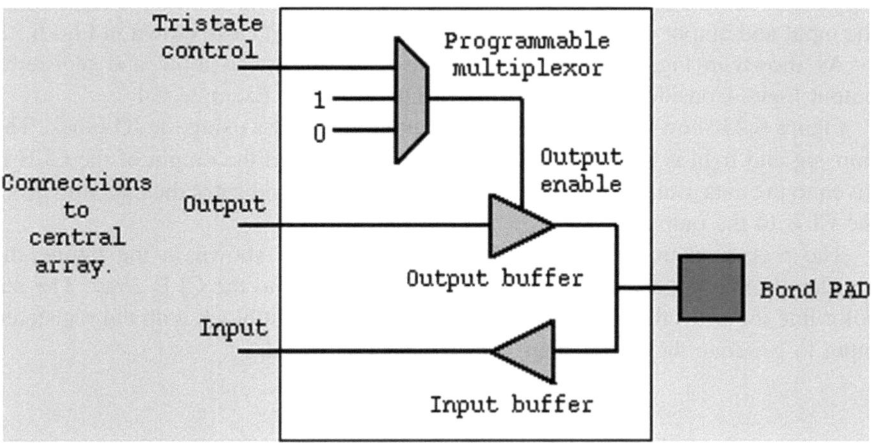

Fig. 6.21 Basic IO block structure

multiplexer.

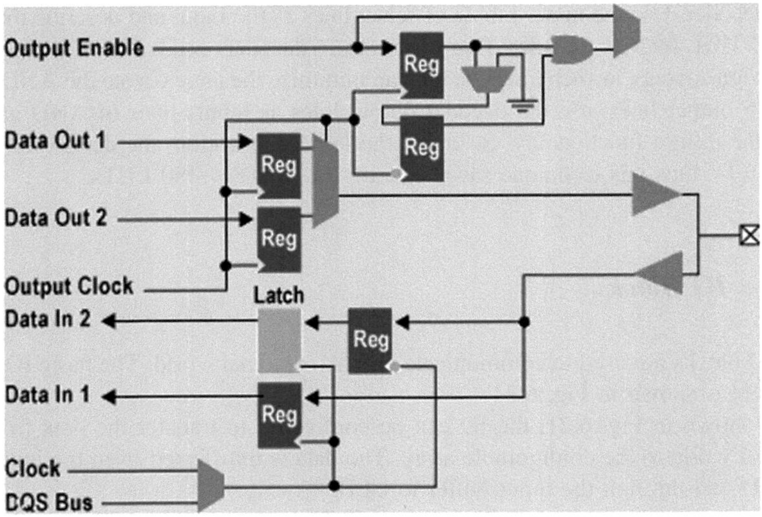

Fig. 6.22 Basic IO block structure used by Altera

In the practical scenario, it is required to have the registered inputs and registered outputs, and under such circumstances, the IO block can have multiple registers in the input and output path. The IO block for the Altera FPGA is shown in Fig. 6.22.

As shown in Fig. 6.22, the IO block uses the registered input and registered output logic. Consider the RTL using VHDL shown in Example 6.4.

Figure 6.23 shows the programmable output 'q_out' by using the IO block. The tmp_sig and b_in is implemented by using the CLB, and the output of the CLB is given to the data_out1 of the register. The red color line indicates the data flow from the CLB to the output buffer.

The registered input is shown in Fig. 6.24, and as shown in the figure, the registered input is generated at Data In 1 and is passed to the CLB array. The red color line indicates the programming of IO block as input block with the registered input to generate the signal 'tmp_sig' as registered input.

6.4.4 Block RAM (BRAM)

BRAM is embedded memory, and the FPGA consists of the single-port and dual-port BRAM. Depending on the architecture of FPGA device, each BRAM consists of the number of static RAM cells. Among them, the few cells are used for the configuration of the memory and the remaining are used for the data storage. The BRAMs are used for the internal storage of the data, to design FIFO, buffers, stacks and can be used to store data required for the FSMs.

Every BRAM has the clock and clock enable, read, and write, and every BRAM is synchronous. If we consider the two-port BRAM, then both ports can be

```
library ieee;
use ieee.std_logic_1164.all;
use ieee.std_logic_arith.all;
use ieee.std_logic_unsigned.all;

entity IO_logic is

port ( a_in, b_in, clk: in std_logic;

        q_out:  out std_logic);

end IO_logic;

architecture arch_IO_logic of IO_logic is

signal tmp_sig : std_logic;

begin

process (clk)

begin

if (clk='1' and clk'event) then

tmp_sig <= a_in;

q_out <= tmp_sig  XOR  b_in;

end process;

end arch_design_logic
```

- Architecture defines the functionality of design.
- The assignment statement is used where the 'q_out' is registered output.
- The 'tmp_sig' is the intermediate signal which is registered input

Example 6.4 Synthesizable VHDL RTL for registered input and registered output

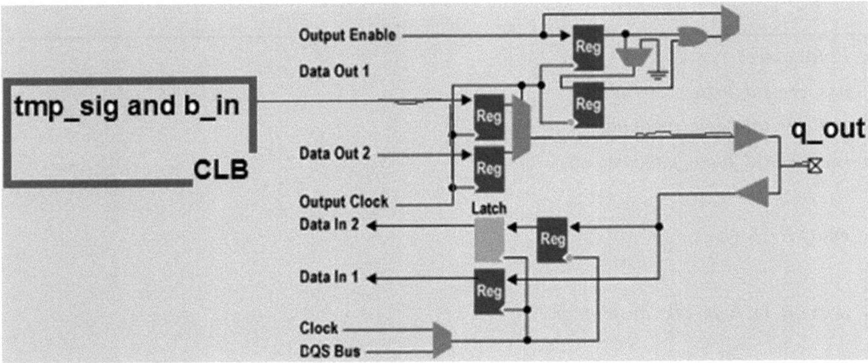

Fig. 6.23 IO configured as registered output

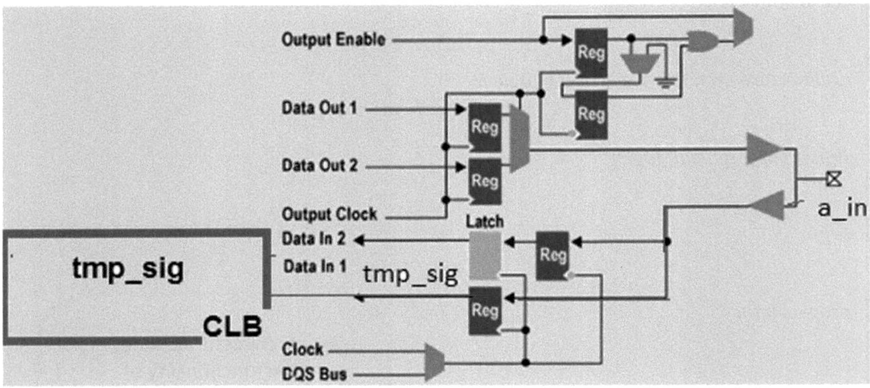

Fig. 6.24 IO configured as registered input

interchangeably used and can be controlled for the synchronous read–write operation. If we consider Spartan-3 devices, then it has BRAM which works at 200 MHz operating frequency. The BRAM single-port and dual-port structure is shown in Fig. 6.25.

As shown in Fig. 6.25, the BRAM consists of the reconfigurable memory, address lines, write enable, clk, data input and data output lines. The RTL using VHDL for the inference of the BRAM is described in Example 6.5, and the synthesis result for the 16 $\times$ 2 BRAM is shown in Fig. 6.26.

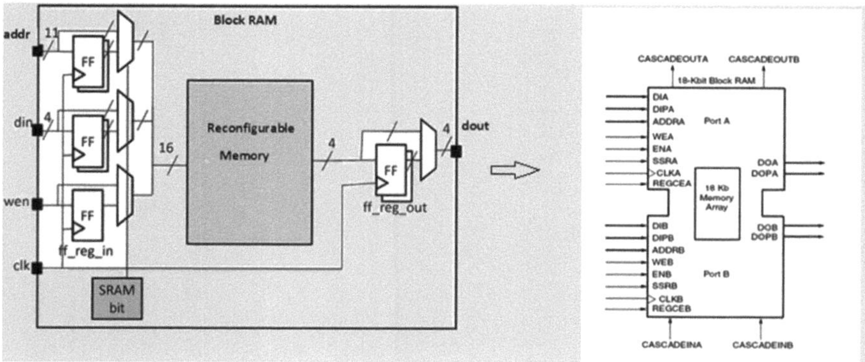

Fig. 6.25 BRAM structure

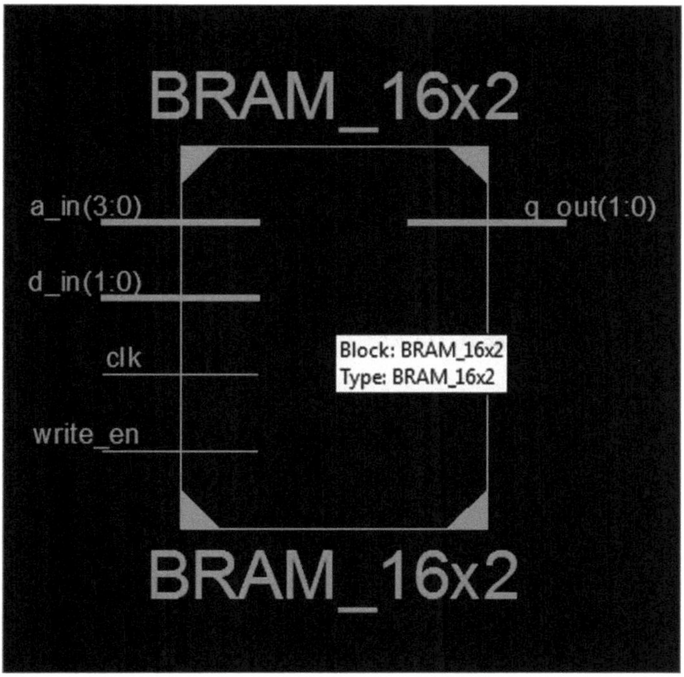

Fig. 6.26 Synthesis result BRAM

```
library  ieee;

use ieee.std_logic_1164.all;

use ieee.std_logic_unsigned.all;

entity BRAM_16x2 is

port (q_out: out std_logic_vector(1 downto 0);

write_en : in std_logic;

clk: in std_logic;

d_in: in std_logic_vector(1 downto 0);

a_in: in std_logic_vector(3 downto 0));

end BRAM_16x2;

architecture BRAM_arch of BRAM_16x2 is

component RAM16x1S is

port (O : out std_logic;

D : in std_logic;

A3, A2, A1, A0 : in std_logic;

WE, WCLK : in std_logic); end component;
```

- Single port BRAM of size 16X2 is described using the component of BRAM 16x1

Example 6.5 Synthesizable VHDL RTL using BRAM component

```
begin

U0 : RAM16x1S

port map (O => q_out(0), WE => write_en, WCLK => clk, D

=> d_in(0), A0 => a_in(0), A1 => a_in(1), A2 => a_in(2), A3 => a_in(3));

U1 : RAM16x1S

port map (O => q_out(1), WE => write_en, WCLK => clk, D

=> d_in(1), A0 => a_in(0), A1 => a_in(1), A2 => a_in(2),A3 => a_in(3));



end BRAM_arch;;
```

- Component instantiation is used and the RAM16x1s is instantiated twice
- This will generate the BRAM 16x2.

Example 6.5 (continued)

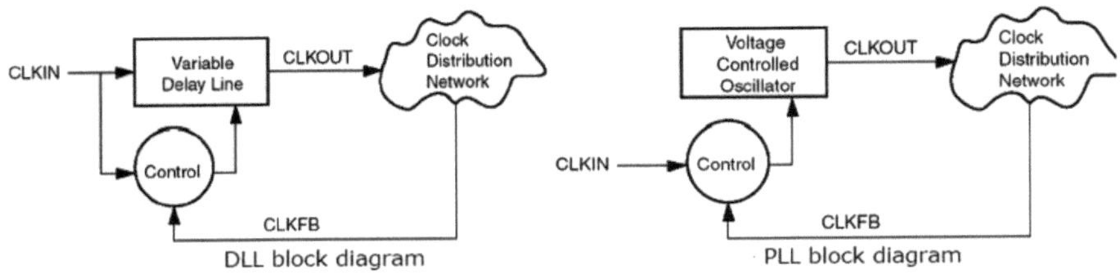

Fig. 6.27 Clock management for FPGA

6.4.5 Clocking Resources

The clock management in the XILINX FPGA uses the DCM, and in the Altera FPGA, it uses the PLL. The clock management is used to generate the clock with the uniform clock skew. Even the clock should be free from glitches and hazards. Figure 6.27 shows the clock management structure using the DLL for Xilinx FPGAs and by using PLL for the Altera FPGA.

The clock management plays the important role in the architecture of the FPGA. It is essential that all the blocks in the design should work synchronously, and hence, providing the uniform clock skew or zero delay across clock network is essential.

The DLL uses the variable delay line with the clock distribution network to provide the clock and to route the clock signals to the internal registers. To adjust the delay, the control logic is used to sample the input clock 'CLKIN' and feedback clock 'CLKFB'. DLL is used to adjust the phase shift of the input clock and feedback clock. The PLL uses the voltage-controlled oscillator (VCO) instead of the delay line to adjust the 360° or 0° phase shift between the input clock and the feedback clock. The control logic in the PLL consists of the filter and the phase detector and is used to generate the desired clock frequency in the lock range.

In most of the practical scenarios, the multiple clocks need to be generated depending on the design requirements. Under such scenarios, the clock management with the clock tree with the uniform clock skew plays an important role. Even the clock tree should be able to propagate the clocks with the zero propagation delays. Consider the simple Example 6.6 described by using the VHDL.

The synthesis result shows the clock selection logic using MUX. As shown in the figure, depending on the 'sel_in' status, one of the clocks is assigned to 'clk_out' (Fig. 6.28).

```
library ieee;
use ieee.std_logic_1164.all;
use ieee.std_logic_arith.all;
entity clk_muxing is
port (sel_in: in std_logic;
clk_slow, clk_fast: in std_logic;
clk_out: out std_logic);
end clk_muxing;
architecture arch_clk_muxing of clk_muxing is
begin
P1: process (clk_slow, clk_fast, sel_in)
begin
if (sel_in = '1') then
clk_out <= clk_fast;
else
clk_out <= clk_slow;
end if; end process; end arch_clk_muxing;
```

- 'The selection of the fast clock 'clk_fast' or the slower clock 'clk_slow' is described by using the if-then-else statement.
- For 'sel_in' equal to one the fast clock is output from MUX
- For the 'sel_in' equal to zero the slow clock is output from the MUX.

Example 6.6 Synthesizable VHDL RTL with multiplex clocking

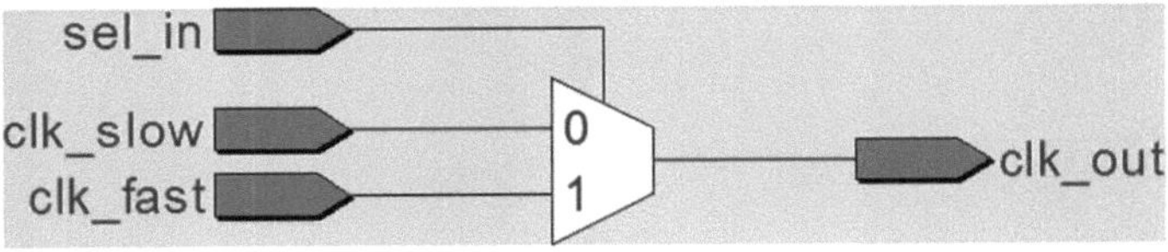

Fig. 6.28 Synthesis result for multiplex clocking

6.4.5.1 Data and Clock Paths and Use of Clock Buffers

The global clock buffers can be inserted to improve the overall fan-out of the clock tree. The following VHDL code describes the instantiation of the global clock buffer (BUFG) in the clock path. It is recommended that the data path and the clock path logic should be separate. Figure 6.29 shows the synthesis result for Example 6.7.

6.4.6 DSP Blocks and Multipliers

For the high computational design and for the improved performance, the modern FPGAs have the dedicated resources as multipliers and DSP blocks. Chapter 9 discusses the use of the multipliers, barrel shifters while prototyping the design. The major DSP applications are filtering, compression, FFT, DFT, encoding, and decoding of the input streams. These operations needs the dedicated resources which can support the pipelining and execution in the shorter time duration. To support this, the DSP blocks are used as dedicated resource in the modern FPGAs.

Figure 6.30 gives information about the dedicated DSP block and can be used efficiently to design some DSP processing algorithms like multiply and accumulate (MAC). As shown in the figure, the DSP block has the input register, multiplier,

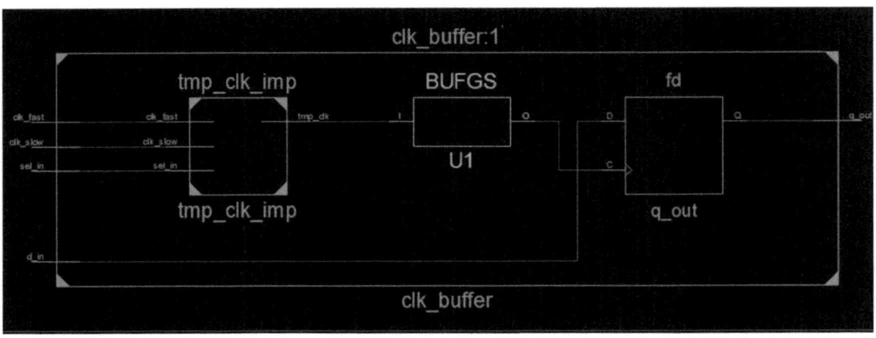

Fig. 6.29 Synthesis result for the use of BUFG

MAC, summation block, and output register. To improve the design performance, the DSP block has the pipelined registers.

The multiplier block is used to perform the multiplication on signed, unsigned, and floating-point numbers, and the architecture is shown in the following figure.

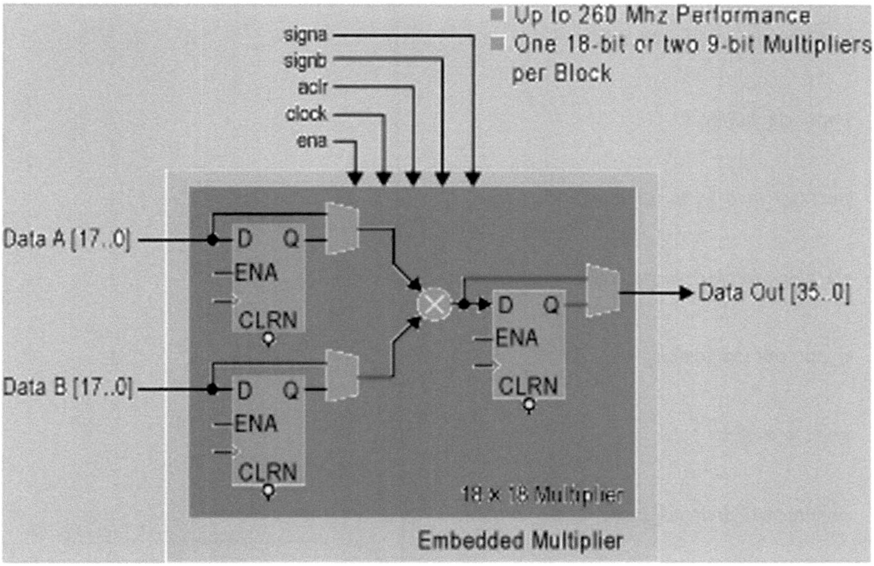

6.4.7 Routing Resources and IO Standards

For the local routing inside the CLB and for the global routing between the CLBs, different types of routing resources are used. The interconnects for the devices are arranged in the form of horizontal and vertical lines. The interconnection lines are single length, double length, and long lines. Single-length lines are used within the CLB and are used for the shorter distance routing. They are flexible enough and used for the faster routing, but when it passes through the switch matrix, it has some delay depending on the length of the line.

In case of double-length line, as each line is two times the single-length line, they are used for routing two CLBs. Long lines are used as routing resource for the full FPGA chip. They are used for high fan-out nets.

Many FPGA vendors support the different IO standards, and they are described in Table 6.3.

The design flow for the FPGA designs is discussed in Chap. 9 with the complex designs and prototyping using modern FPGAs.

```
library ieee;
use ieee.std_logic_1164.all;
use ieee.std_logic_arith.all;
entity clk_buffer is
port (sel_in, d_in: in std_logic;
clk_slow, clk_fast: in std_logic;
q_out: out std_logic);
end clk_buffer;
architecture arch_clk_buffer of clk_buffer is
signal tmp_clk: std_logic;
signal clk_bufg: std_logic;
component BUFGS
port (I: in std_logic;
O: out std_logic);
end component;
```

- For the device Xilinx "XC3s100e the global clock buffer component 'BUFGS' is declared and used to instantiate the clock.
- Depending on the requirement of the slower of faster clock in the design the clock is passed to trigger the register.

Example 6.7 Synthesizable VHDL using BUFG for multiplex clocking

```
begin

P1: process (clk_slow, clk_fast, sel_in)

begin

if (sel_in = '1') then

tmp_clk <= clk_fast;

else

tmp_clk <= clk_slow;

end if;

end process;

U1: BUFGS port map (I => tmp_clk, O => clk_bufg);

P2: process (clk_bufg)

begin

if (clk_bufg ='1' and clk_bufg'event)then

q_out <= d_in;

end if;

end process;  end arch_clk_buffer;
```

- Process 'p1' is for the clock path and describes the selection of the fast or slow clock depending on the status of select input.
- BUFGS is instantiated and input the BUFGS is 'tmp_clk' and output from BUFGS is 'clk_bufg'.
- Depending on the rising edge on the 'clk_bufg' the data input 'd_in' is passed to 'd_out'
- Process 'P2' is data path for the design.

Example 6.7 (continued)

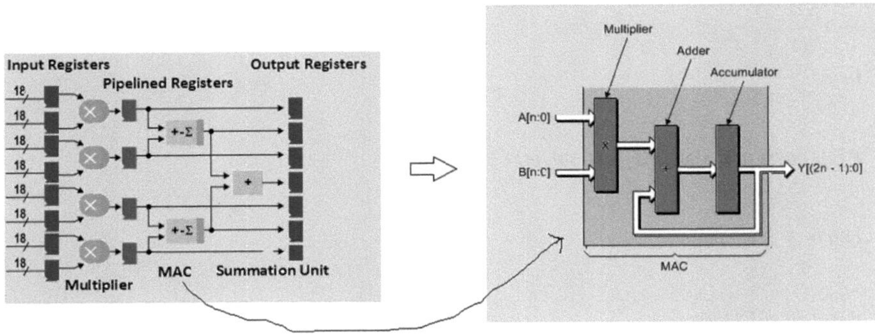

Fig. 6.30 DSP block

Table 6.3 IO standards

IO standard	Long form	Description
LVTTL	Low-voltage TTL	A general-purpose IO standard
LVCMOS	Low-voltage CMOS	A general-purpose IO standard
HSTL	High-speed transceiver logic	A general-purpose high-speed IO standard supported for 1.5 V bus by IBM
SSTL	Subseries terminated logic	The general-purpose memory bus standard
PCI	Peripheral component interface	It uses LVTTL input buffers and supports the PCI bus applications at 33 and 66 MHz
AGP	Advanced graphics port	This standard supports the graphics application and works at 3.3 V

6.5 Practical Scenarios and Guidelines

While using the FPGAs, the designers need to take care of the design guidelines. Most of the design guidelines for the PLD based design are explained in Chap. 8. The major focus of this section is to have the practical-oriented information and guidelines for the synchronous and asynchronous designs, clocks, resets, and the use of the synchronizers during the design.

6.5.1 Reset Strategy

Most of the times, the designers are confused whether to use the synchronous reset or synchronous resets in the design.

6.5.1.1 Synchronous Reset

The synchronous resets are recommended in most of the applications as the reset logic is part of the data path, and the reset is sampled on the active edge of the clock. The logic in the data path of register using synchronous reset is shown in the Fig. 6.31.

The RTL using VHDL for the design using synchronous reset is described in Example 6.8.

The synthesis result for the synchronous reset is shown in Fig. 6.32.

In the synchronous reset, the reset logic is part of the data path and reset signal is sampled on the active clock edge, and hence there is no need of the synchronizer.

By using the synchronous reset strategy the glitches are filtered out and it shown in the Fig. 6.33.

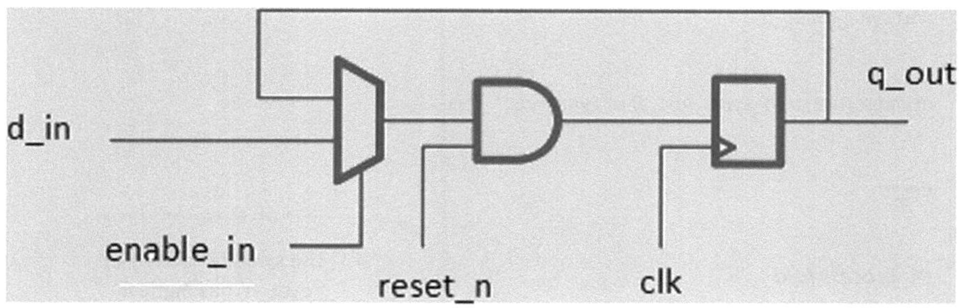

Fig. 6.31 Synchronous reset logic

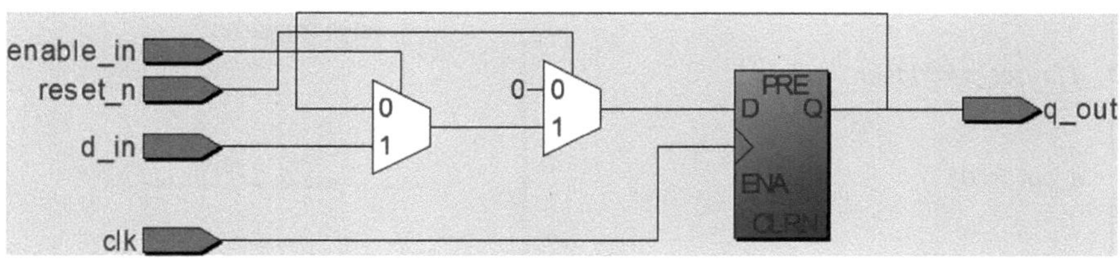

Fig. 6.32 Synthesis result for D flip-flop using synchronous reset and enable

```
library ieee;
use ieee.std_logic_1164.all;
use ieee.std_logic_arith.all;
entity sync_reset_ff is
port (d_in, clk, enable_in, reset_n: in std_logic;
q_out: out std_logic);
end sync_reset_ff;
architecture arch_sync_rest_ff of sync_reset_ff is
begin
P1: process (clk)
begin
if (clk='1' and clk'event) then
 if ( reset_n='0') then
 q_out <='0';
 elsif ( enable_in ='1') then
 q_out <= d_in;  end if; end if;end process; end arch_sync_rest_ff;
```

- The synchronous reset 'reset_n' has highest priority as compare to the 'enable_in' input.
- The synchronous reset logic is part of the data path.
- The flip-flop is rising edge triggered.

Example 6.8 Synthesizable VHDL using the synchronous reset and enable

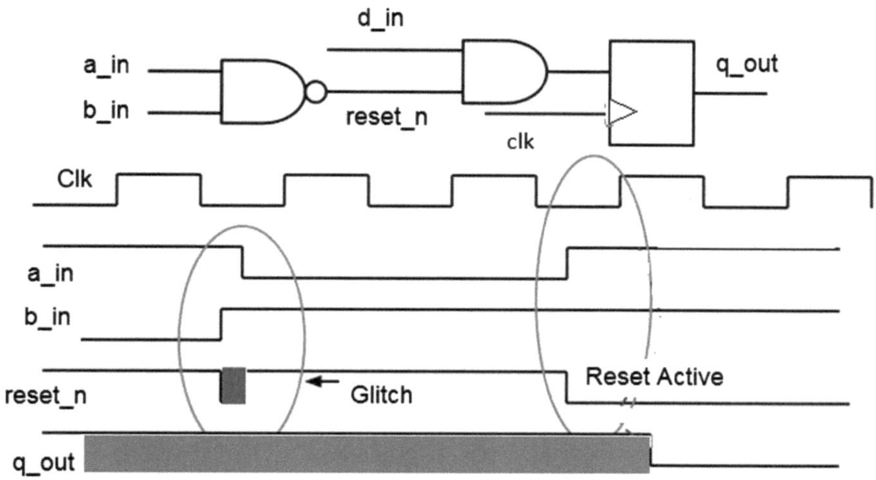

Fig. 6.33 Reset glitch filtering

6.5.1.2 Asynchronous Reset

The asynchronous reset signal is sampled at any time irrespective of the active clock edge. In the asynchronous reset strategy, the reset logic is not a part of the data path.

While designing using asynchronous resets, designer needs to take care of the reset assertion and reset deassertion. The reset recovery and removal time play the significant role in such type of the strategy.

The reset recovery time is the minimum amount of time required where reset signal should be active before the arrival of the active clock edge. If at a time clock and reset will change, then the register goes into the metastable state. To avoid this, the asynchronous reset can be synchronized internally using the two-stage level synchronizer.

Figure 6.34 gives information about the reset recovery time. If reset signal arrives before the active edge of clock and remains stable, then the timing is met.

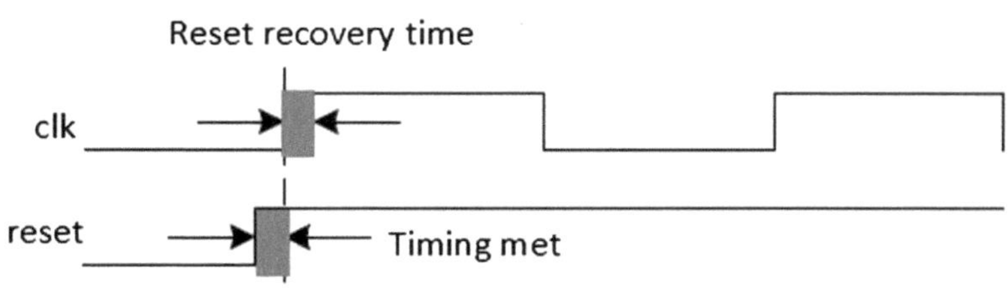

Fig. 6.34 Timing sequence for the reset recovery time

Figure 6.35 gives the information about the timing violation as the reset signal makes changes at the clock edge. The design goes into the metastable state.

The reset removal time is the amount of time required for deassertion of the reset signal. Figure 6.36 shows the timing sequence for the reset removal.

The asynchronous reset can be synchronized internally using the two-stage level synchronizer and shown in Fig. 6.37.

As shown in Fig. 6.37, the two-stage synchronizer is used to generate the asynchronous reset signal to the register. It uses the clock as 'clk' and reset signal as 'master_reset'. During the normal operation, the two-stage synchronizer generates the logic '1' at the reset input of the register during the valid reset time duration the 'master_reset' input is active low and the level synchronizer generates the active low output to reset the register. Refer Chap. 5 for the RTL using VHDL which uses the asynchronous reset.

6.5.2 Asynchronous Versus Synchronous Designs

As discussed already in Chap. 5, the designers need to have a good understanding of the asynchronous and synchronous designs. In the asynchronous designs, as

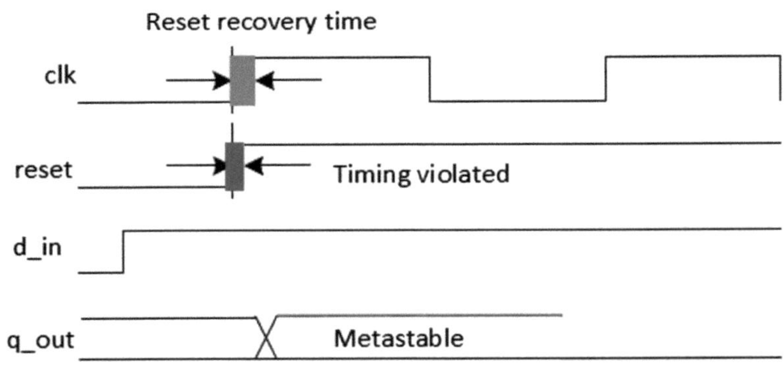

Fig. 6.35 Reset recovery time violation and metastable output

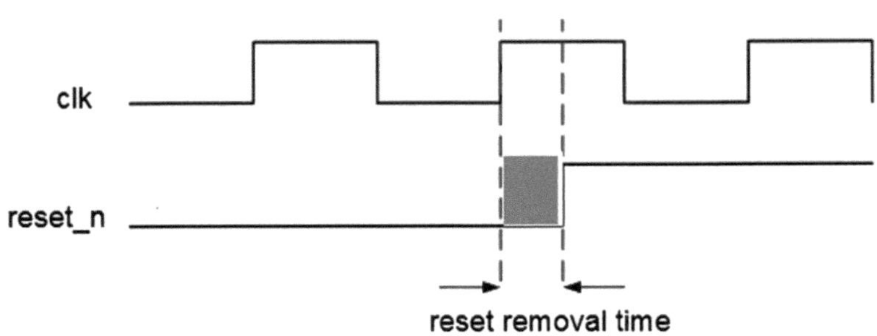

Fig. 6.36 Timing sequence for the reset removal

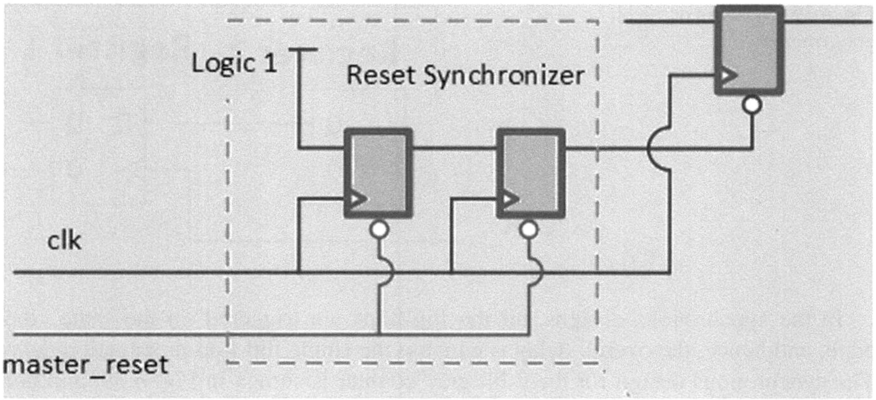

Fig. 6.37 Two-stage level synchronizer for the reset

clock signal is not common for all the registers, the cumulative delay slows down the design performance. In most of the practical scenarios, it is not recommended to use the asynchronous designs.

The asynchronous design using JK flip-flops and timing sequence is shown in Fig. 6.38.

As shown in Fig. 6.38, to get the output 'q-out', it needs the delay of 4 times the propagation delay of the single flip-flop.

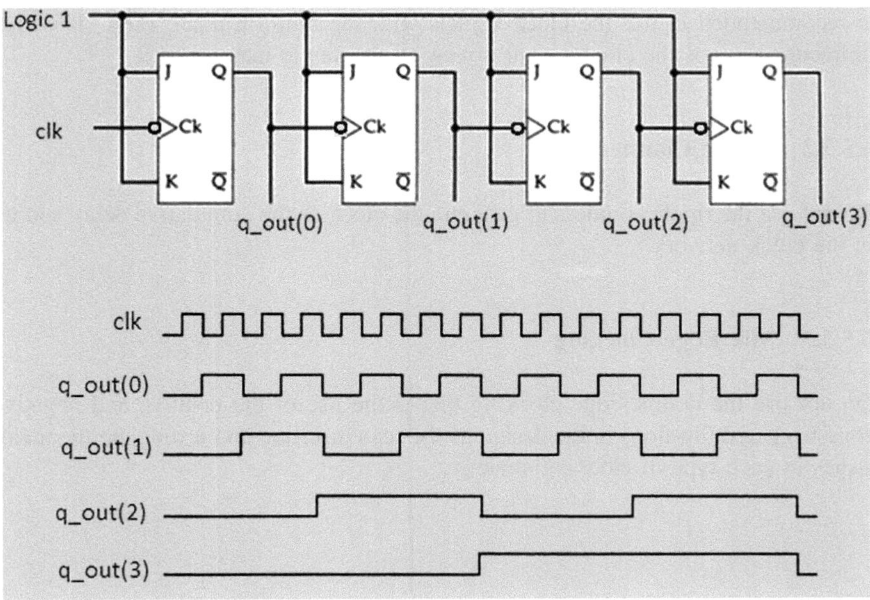

Fig. 6.38 Asynchronous counter design and timing sequence

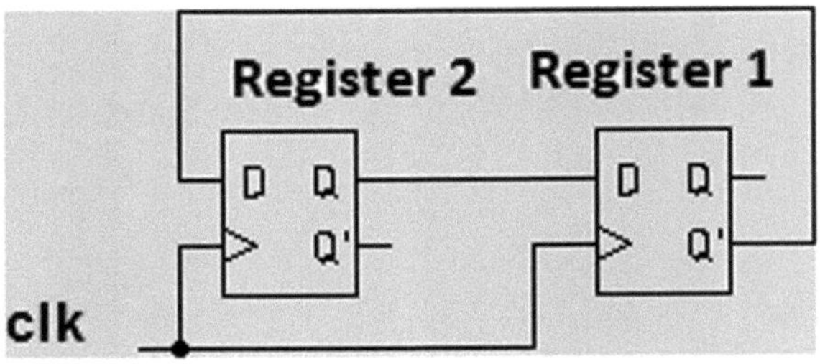

Fig. 6.39 Synchronous 2-bit gray counter

In the synchronous designs, all the flip-flops are triggered on the same clock edge, and hence, the overall delay is equal to the single flip-flop propagation delay. The synchronous design for the 2-bit gray counter is shown in Fig. 6.39, and both the registers uses the same clock source driven by the master clock.

6.5.3 Clocking Strategies

The few important clocking strategies are listed below.

6.5.3.1 Single Master Clock

The clocking strategy plays an important role in the design. In the FPGA designs, it is recommended to use the clock signals with the uniform clock skew. It is recommended to use the clock signal driven by the single master clock.

6.5.3.2 Ripple Counters

Do not use the ripple counters to generate the clock as the cumulative delay add up in the clock network.

6.5.3.3 Mix Edge Clocking

Do not use the double-edge clocking that is the use of the positive and negative edge-triggered flip-flops in the design as the scan insertion and testing are the major issues in such type of clocking strategy.

6.5.3.4 Gated Clocks

Use the gated clocks to reduce the dynamic power dissipation in the design. Gated clock can introduce the glitches in the generated clock, and hence, it is recommended to use the clock gating cells with the latch enable mechanism. Please refer Chap. 8 for the clock gating mechanism and RTL description using VHDL.

6.6 Summary

The following are the key points to summarize this chapter.

1. SPLDs are used for design of small gate count designs.
2. SPLDs are classified as PROM, PAL, and PLA.
3. CPLDs are used to realize the moderate gate count designs with the better timing performance.
4. FPGAs are programmed by the user program at field.
5. FPGA architecture consists of CLBs, IOBs, routing resources, clocking resources, and programmable interconnects.
6. Modern FPGAs consist of the CLBs, IOBs, routing resources, clocking resources, programmable interconnects, DSP blocks, multipliers, processor, etc.
7. It is not recommended to use the ripple counters for the clock generation.
8. Use the clock gating cell to reduce the power dissipation in the design.
9. Synchronous counters are recommended in the ASIC design as timing analysis will be easy and they are not prone to the glitches.
10. Asynchronous counter logic is prone to glitches or spikes and hence not recommended in the ASIC designs.
11. Use the BRAM instead of the distributed RAM. But BRAM can have one clock latency as compared to distributed RAM.
12. Clock management is accomplished in the FPGAs using the DLL or PLL. Xilinx uses the DLL, and Altera uses the PLL.
13. Clock management is used for the uniform clock skew and for the clock propagation with the zero delay.

Chapter 7
Design and Simulation Using VHDL Constructs

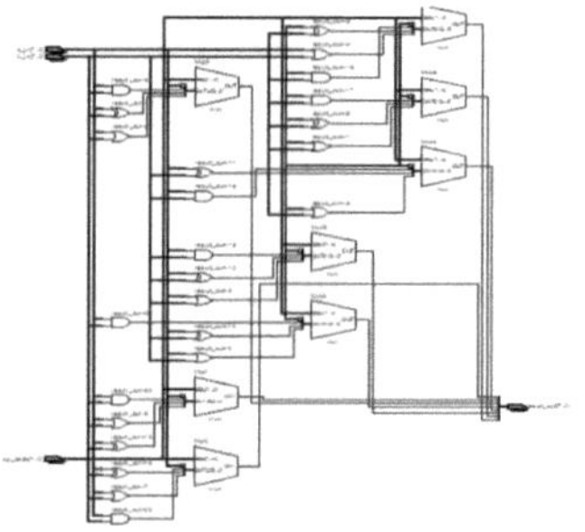

> *"The significant problems we have cannot be solved at the same level of thinking with which we created them...."* --- **Albert Einstein**
>
> Design simulation needs different thinking . Use the VHDL constructs to simulate the design.

Abstract This chapter discusses the VHDL constructs and their use during the design verification. The constructs such as subprogram, procedures, functions, TEXTIO, and file handling are discussed in this chapter with the practical examples. Even this chapter gives basic understanding of design simulation using the VHDL constructs. How to write an efficient testbench and how to carry out the presynthesis simulation are explained in this chapter with the simulation results. This chapter even discusses the use of the packages and file handling.

Keywords Block · Subprogram · Procedure · Functions · Files · TEXTIO · Simulation · Verification · Testbench · File handling · Package · Device under test · Design under verification · Presynthesis simulation · Generate · Binary counter · Attributes

As discussed in the previous chapters, VHDL is efficiently used to code the functionality of the design. VHDL has powerful concurrent and sequential constructs and can be used during the design, simulation or verification of the design. For smaller designs with few input and output ports, it is easy to manually force the inputs to check

---

The original version of this chapter was revised: The incorrect information have been corrected. The erratum to this chapter is available at 10.1007/978-981-10-3296-7_12

V. Taraate, *PLD Based Design with VHDL*,
DOI 10.1007/978-981-10-3296-7_7

the functional correctness of the design. For the complex designs with more number of inputs and outputs, it becomes time-consuming to force the inputs to check the behavior of the design. Even the chances of error are higher with the manual forcing.

The functional correctness of the design can be checked by writing another VHDL code that is testbench. Testbench can be written efficiently using the VHDL constructs to force the inputs for different time instances. The subsequent section discusses the use of the VHDL constructs during simulation.

7.1 Simulation Using VHDL

The required inputs can be forced by using the stimulus generator, and the output can be observed. For the complex designs the functions, packages, TEXTIO with file handling can be used efficiently to write the testbenches and even to simulate the results. For the complex designs, file handling can play the important role and can be used to store the results. Figure 7.1 shows the simulation setup for the design using VHDL.

As shown in Fig. 7.1, the stimulus generator (testbench) can drive the required input signal to the design under test (DUT). DUT is also called as design under verification. Stimulus generator is used to check the functional correctness of the design. In the practical scenario, the self-checking testbenches using the stimulus generator, monitor, and checkers can be used to check the design functionality. Following section discusses the use of the VHDL constructs during simulation.

7.1.1 Testbench for 4:1 MUX

As discussed in the previous chapters, the multiplexers are combinational elements. In the multiplexer (MUX), at a time one of the inputs is selected and passed to the output. Example 7.1 describes the 4:1 MUX using VHDL RTL.

Multiplexer has single-bit inputs 'a_in, b_in, c_in, d_in' and 2-bit select input 'sel_in.' Output of multiplexer is 'y_out.' Example 7.2 describes the forcing of the inputs and select lines of 4:1 MUX using VHDL construct. The simulation result for 4:1 MUX is shown in Fig. 7.2.

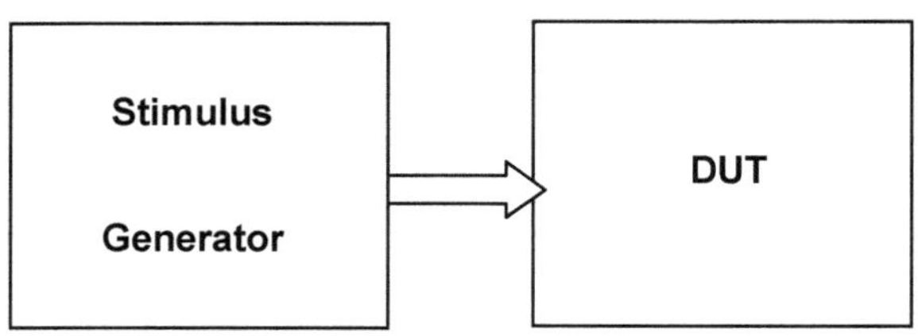

Fig. 7.1 Simulation using VHDL

```
library ieee;
use ieee.std_logic_1164.all;
entity mux_4to1 is
  port (
        a_in, b_in,c_in,d_in: in std_logic;
        sel_in : in std_logic_vector(1 downto 0);
    y_out: out std_logic);
 end mux_4to1;
architecture arch_mux_4to1 of mux_4to1 is
begin
comb_p1:  process ( a_in, b_in,c_in,d_in, sel_in)
 begin
  case ( sel_in) is
      when "00" => y_out<= a_in;
      when "01" => y_out<= b_in;
      when "10" => y_out<= c_in;
      when "11" => y_out<= d_in;
      end case;
end process comb_p1;
end arch_mux_4to1;
```

- Architecture defines the functionality of design.
- Combinational Process 'comb_p1' is sensitive to the input changes at 'a_in', 'b_in' in', 'd_in' and 'sel_in'.
- Case construct is used to infer the parallel logic.
- Depending on the status of 2-bit 'sel_in' the 'y_out' is assigned.
- An output is either 'a_in', 'b_in','c_in' or 'd_in' at a time.

Example 7.1 Synthesizable VHDL RTL for 4:1 MUX

```
library ieee;use ieee.std_logic_1164.all;

entity testbench_mux_4to1 is

end;

architecture arch_testbench of testbench_mux_4to1 is

 component mux_4to1

  port (

      a_in, b_in,c_in,d_in:in std_logic;

      sel_in :in std_logic_vector(1 downto 0);

   y_out: out std_logic);

 end component;

 signal sel_in: std_logic_vector(1 downto 0);

 signal a_in,b_in,c_in,d_in,y_out: std_logic;

begin

 sel_in <= "00", "01" after 20 ns, "10" after 40 ns,

  "11" after 60 ns, "XX" after 90 ns,

  "00" after 110 ns;

 a_in <= 'X', '0' after 5 ns, '1' after 10 ns;

 b_in <= 'X', '0' after 20 ns, '1' after 30 ns;

 C_in<= 'X', '0' after 40 ns, '1' after 60 ns;

 d_in <= 'X', '0' after 100 ns, '1' after 110 ns;  U: mux_4to1 port map ( a_in, b_in, c_in, d_in, sel_in, y_out);

end arch_testbench;
```

- The component 'mux_4to1' is declared inside the architecture.
- The name of component should be same as that of name of VHDL entity.

- To bind the individual ports to carry out the simulation the temporary signals are declared using 'signal'
- The 'sel_in' is forced to different values "00',"01", "10", "11" and "XX" for the different time durations.
- Inputs 'a_in', 'b_in', 'c_in', 'd_in' are forced to different logic levels '0' or '1' at various time instances.
- The instance of the design under test is 'mux_4to1' and using 'port map' the input and output ports are mapped.

Example 7.2 Testbench for 4:1 MUX

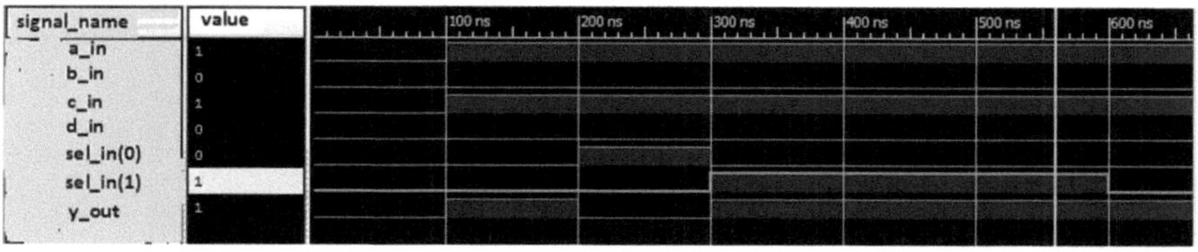

Fig. 7.2 Simulation result for 4:1 MUX

7.1.2 Testbench for 4-Bit Binary up Counter

As discussed in Chap. 5, the counters are used to count the specific steps depending on the clock input. Counters are sequential logic circuits. Example 7.3 describes the 4-bit binary counter using VHDL.

The testbench using VHDL constructs to force the clk and reset value is described in Example 7.4, and simulation results are shown in Fig. 7.3.

7.2 Functions

A function call is in the form of an expression that returns a value. A function call is subprogram which consists of function declaration and sequential statements. Functions are used to describe the algorithm or the required behavior. The functions are used to return the complex-type or scalar-type values.

The function calls can be pure or impure. In the pure function calls, it returns the default or the same type of values as that of parameter type. In case of impure functions, it may return different types of values. Impure functions can update the objects that are out of scope, but pure functions will not be able to update objects that are out of scope.

Function declaration has two main parts and they are function declarations and function body.

- **Function Declaration**: It consists of name of function, parameter list, and type of the values returned by the function. The function declaration can start with an optional reserved word pure or impure; it denotes the character of the function. Without any reserved word, the function is assumed as pure.
- **Function Body**: It contains variables, types, constants, local declarations of nested subprograms, files, aliases, attributes, groups, and sequence of statements to perform the required algorithm.

It is important to note that function body may not contain a wait statement or a signal assignment. But subprograms (functions and procedures) can be nested. Function can be recursive.

```
library ieee;
use ieee.std_logic_1164.all;
use ieee.std_logic_arith.all;
use ieee.std_logic_unsigned.all;

entity counter is
port (clk : in std_logic;
     reset_n :in std_logic;
     count_out : out std_logic_vector(3 downto 0)
    );
end counter;

architecture arch_counter of counter is

signal tmp_count : std_logic_vector(3 downto 0) :=(others => '0');

begin

count_out <= tmp_count;

seq_p1: process(clk,reset_n)
begin

if(reset_n='0') then

 tmp_count <=(others => '0');

elsif(clk'event and clk='1') then

   if(tmp_count = "1111") then
     tmp_count <="0000";
   end if;
    tmp_count<= tmp_count+'1';

end if;

end process seq_p1;

end arch_counter;
```

- Architecture defines the functionality of design.
- Sequential Process ‘seq_p1’ is sensitive to the input changes at ‘clk’ and ‘reset_n’.
- ‘if-then-else’ construct is used to infer the priority logic.
- For the ‘reset_n=’0’’ the output ‘count_out’ is equal to “0000”. During counting ‘reset_n=’1’’.
- For active edge of the clock input ‘clk’ the ‘count_out’ is incremented by one.
- When ‘count_out’ reaches to “1111” counter output is assigned to “0000”.

Example 7.3 Synthesizable VHDL RTL for 4-bit binary up counter

```
library ieee;
use ieee.std_logic_1164.all;
use ieee.std_logic_arith.all;
use ieee.std_logic_unsigned.all;
entity testbench_counter is
end testbench_counter;
architecture arch_tb of testbench_counter is
  component counter
  port(
      clk : in std_logic;
      reset_n : in std_logic;
      count_out : out std_logic_vector(3 downto 0)
      );
  end component;
  signal clk : std_logic := '0';
  signal reset_n : std_logic := '1';
  signal count_out : std_logic_vector(3 downto 0);
```

- The component 'counter' is declared inside the architecture.
- The name of component should be same as that of name of VHDL entity.
- For the port binding the temporary signals are declared using 'signal' and of type 'std_logic'.

Example 7.4 Testbench for 4-bit binary up counter

```
U: counter PORT MAP (
    clk => clk,
    reset_n => reset_n,
    count_out => count_out      );
clk_p1 :process
begin
    clk <= '0';
    wait until clk_period/2;
    clk <= '1';
    wait until clk_period/2;
end process;
reset_p2: process
begin
    wait for 12 ns;     reset_n <='0';
    wait for 5 ns;      reset_n <='1';
    wait for 25 ns;     reset_n <= '0';
    wait for 3 ns;      reset_n <= '0';
    wait;
end process; end arch_tb;
```

- Using one to one mapping the design under test input and output ports are mapped.

- Process 'Clk_p1' is without sensitivity list.
- The process is used to generate the stimulus at 'clk' input.

- Process 'reset_p2' is without sensitivity list.
- The process is used to generate the stimulus at 'reset_n' input.
- The 'reset_n' input is forced to different logic levels '0' or '1'

Example 7.4 (continued)

Syntax

function *name_function (parameters)* ***return*** *type;*

function *name_function (parameters)* ***return*** *type* ***is***

function declarations

begin

sequential statements;

end function *name_function;*

Function Examples

function *Function_real (a_in,b_in,c_in: real)* ***return*** *real;*

The function declared is called as Function_real, the function has three parameters a_in, b_in and c_in of real type and this returns the real value.

function *"*" (a_in,b_in: integer_value)* ***return*** *integer_value;*

The function uses the operator as function name and used to defines a algorithm for multiplication

function *addition* ***(signal*** *sig_in1,sig_in2: real)* ***return*** *real;*

In the above function signals are used as input parameters. Signals are denoted by the reserved word signal.

type data_int **is file of** natural;
function file_end (**file** file_name: data_int) **return** boolean;

The function is used to check the end of the file and it consists of natural numbers. The parameters are Boolean type declarations.

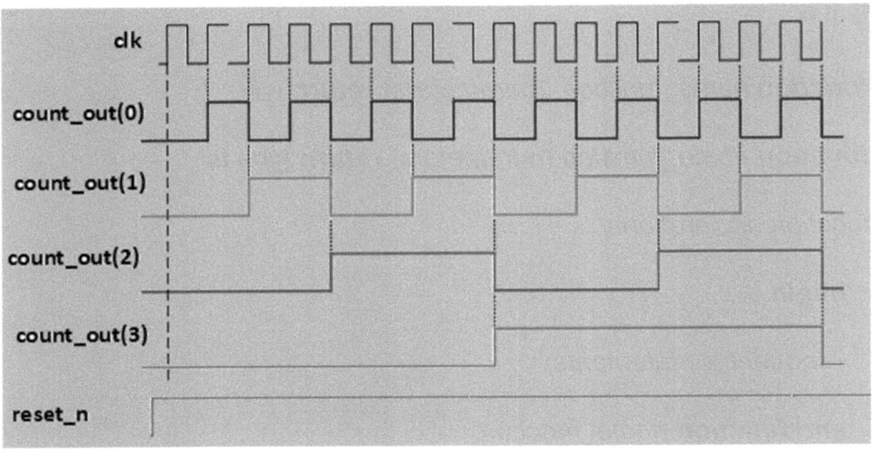

Fig. 7.3 Simulation result for the 4-bit binary up counter

Function to return complement
Function Body to return the complement of bit vector

```
function complement (para_value: in bit_vector(0 to 7)) return bit_vector is
    begin
      case para_value  is
        when "00000000" => return "11111111";
        when "11111111" => return "00000000";
        when others => return "00000000";
      end case;
    end complement;
```

The sequential construct ‘case’ is used in the function body. The parameter is ‘para_value’ and/or ‘bit_vector’ type, the function returns the value of same type.

Function for multiplication

```
function multiplication (constant a_in, b_in,c_in: real) return real is
    begin
      return a_in*b_in**4+b_in;
    end multiplication;
```

The parameters declared are of type real, and after the execution of the expression in the function body, it returns the real value.

Function to count minimum value

```
function min_value (con-
    stant a_in,b_in,step_size,left_b,right_b: in real) return real is
    variable count, min, temp_v: real;
    begin
      count:= left_b;
      max:=min_value(a_in,b_in, count);
      Loop1:
       while count >= right_b loop
         temp_v:=min_value(a_in,b_in, count);
         if temp_v < min then
           min:=temp_v;
         end if;
         count := count-step_size;
       end loop Loop1;
      return min;
    end min_value;
```

The parameters declared are constants and of the real type. The value of a_in and b_in is passed when function is called. The left_b and right_b are declared and used to define the range to search the minimum value.

The function body uses the sequence of statements to search for the minimum value, the function returns the minimum value from the range.

Impure function

```
variable number: integer := 0;
    impure function imp_fucntion (a_in: Integer) return integer is
    variable count: integer;
    begin
      count := a_in * number;
      number := number + 1;
      return count;
    end imp_function;
```

The declared function is of impure type, and the parameter a_in is of integer type, and when the function is executed, it returns the count value. The returned value by function may be of different type, and hence, it is called as an impure function.

7.3 Packages

Packages are used to share the design objects with the different kinds of VHDL designs. Packages consist of the following declarations:

- Subprogram
- Attributes
- Aliases
- Types
- Files
- Components

Package is declared by using the following syntax:

package *package_name is*

subprogram_declaration | subprogram_body

| type_declaration | subtype_declaration

| constant_declaration | shared_variable_declaration

| file_declaration | alias_declaration

| use_clause | group_template_declaration

| group_declaration

end package package_name ;

Package body consists of the functional information of the procedures and functions. The functional information may be visible to many other designs.

Package body *package_name is*

subprogram_declaration | subprogram_body

| type_declaration | subtype_declaration

| constant_declaration | shared_variable_declaration

| file_declaration | alias_declaration

| use_clause | group_template_declaration

| group_declaration

end package body package_name ;

Consider the design scenario to perform the addition (XOR) of two operands 'a_in' and 'b_in.' The RTL using VHDL is described in Example 7.5. Synthesis result is shown in Fig. 7.4.

7.3.1 Package Use in Design

The package 'add_package' is declared in the VHDL program, and by using; use work.add_package.all; it is accessed in the main VHDL program. The package declaration and package body are shown in the Example 7.6, to perform the XOR operation.

```
library ieee;
use ieee.std_logic_1164.all;
library work;
use work.add_package.all;
entity seq_logic is
port (clk : in std_logic;
    a_in : in s1_pck;
    b_in : in s1_pck;
    y_out: out s1_pck    );
end seq_logic;
architecture arch_seq_logic of seq_logic is
begin
process(clk)
begin
if(clk'event and clk='1') then
y_out<=add_op(a_in,b_in);
end if;
end process;
end arch_seq_logic;
```

- The sequential logic is positive edge triggered.
- It performs the addition (xor) of two operands.
- The package used is 'add_package'.
- Package is included using use work.add_package.all;
- The ports clk is of 'std_logic'
- The ports 'a_in', 'b_in' and 'y_out' are of type s1_pck.

Example 7.5 Use of package for addition operation

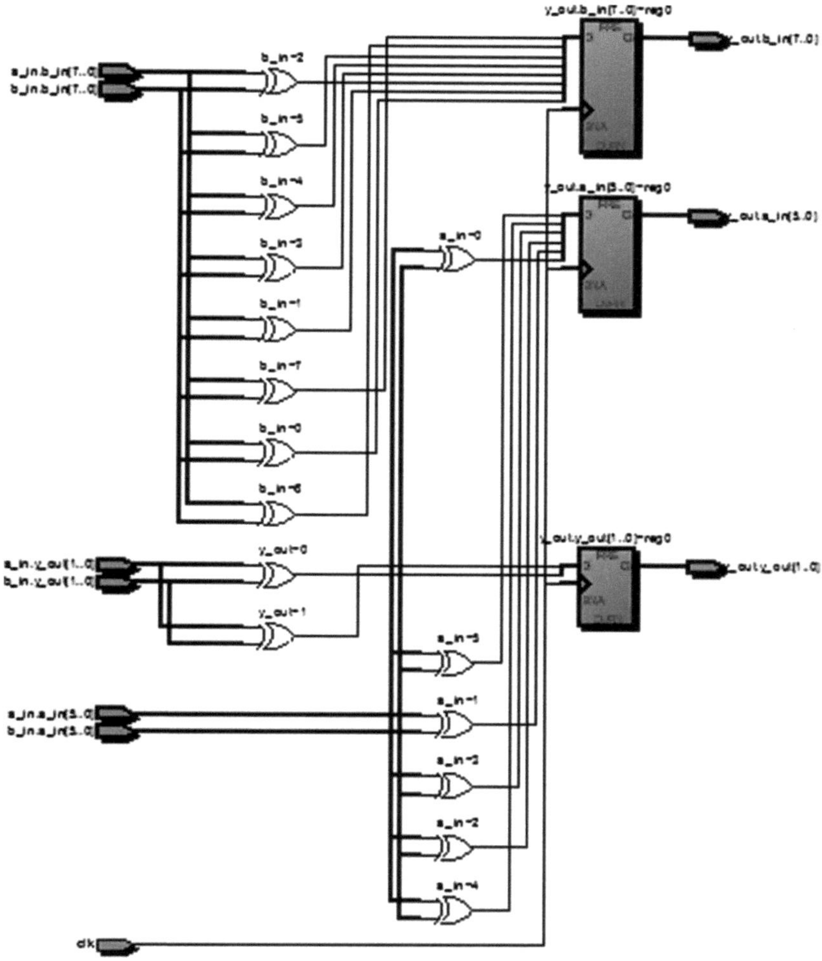

Top-level Entity Name	seq_logic
Family	MAX II
Total logic elements	16 / 240 (7 %)
Total pins	49 / 80 (61 %)
Total virtual pins	0
UFM blocks	0 / 1 (0 %)
Device	EPM240T100C3
Timing Models	Final

Fig. 7.4 Synthesis result for the addition operation using package

```
library IEEE; use ieee.std_logic_1164.all; use ieee.std_logic_arith.all;
package add_package is
type s1_pck is
  record
    a_in :std_logic_vector(5 downto 0);
    b_in :std_logic_vector(7 downto 0);
    y_out :std_logic_vector(1 downto 0);
  end record;
function add_op (a_in : s1_pck; b_in: s1_pck) return s1_pck;
end add_package;
package body add_package is
function  add_op (a_in : s1_pck; b_in: s1_pck) return s1_pck is
variable sum : s1_pck;
begin
sum.a_in:=a_in.a_in xoR b_in.a_in;
sum.b_in:=a_in.b_in xoR b_in.b_in;
sum.y_out:=a_in.y_out xoR b_in.y_out;
return sum;
end add_op;  end add_package;
```

- And The function 'add_op' is used in the package.
- Package body consists of the statements and used to perform the addition on two operands.
- Function 'add_op' returns the sum.

Example 7.6 Package declaration for addition

7.4 Attributes

Attributes in VHDL are used to return information about the signal. The attributes are of type signal attributes, array attributes, and type attributes. Attributes consist of the (') quote followed by the name of the attribute. In the array manipulations, the attributes are used.

7.4.1 Signal Attribute

These are used to return the true value on the event. These attributes return a Boolean value. Following is the example of the signal attribute

```
if(clk='1' and clk'event) then

q_out<= data_in; else

q_out <= '0';

end if;
```

As shown in the above code, the clk'event is used to return the true value, that is, to find the positive edge of the clock 'clk.'

Consider another example of signal attribute to find the negative edge of the clock.

```
if(clk='0' and clk'event) then

q_out<= data_in; else

q_out <= '0';

end if;
```

7.4.2 Array Attribute

These types of attributes are used in the array manipulations and array access. Consider a scenario where the particular range of array needs to be accessed. Under such scenario, the name_array'range attribute can be used. This is used to find whether the signal is zero or not.

Consider the following example for the array attribute

```
signal tmp_sig :std_logic_vector(31 downto 0):=(others =>'0');

signal value_sig :std_logic_vector(31 downto 0):=(others =>'0');

if(value_sig /= value_sig'range => '0')) then

        tmp_sig <= value_sig;

end if;
```

As shown in the above example, the array attribute value_sig'range is used. These types of attributes are used for the long signals.

7.5 File Handling

Most of the complex designs using VHDL may need the larger number of inputs and outputs. In such scenarios, it becomes difficult to write the testbench. Even it becomes difficult to read the testbench code. The better way is to use the files. The input can be stored in the text file and can be read from the text file. The results can be even stored in the output file. The following section discusses the file handling using VHDL.

7.5.1 Use of Files in Design Simulation

During the simulation of the VHDL design if it is required, that the data written in one of the files need to be copied in another file, then under such circumstances, the file handling can be efficiently used. In such scenario, data can be stored in one of the input files and the data can be copied in the other output files.

Consider Example 7.7; in this, the input data is stored in 'file1.txt' and the 'file2.txt' is used to hold the output data.

As shown in Example 7.7, the process 'read_p1' is used to read the data till the end of the line in input file 'file1.txt'. An another process 'write_p2' is used to write the data in another file 'file2.txt.'

```
library ieee;

use ieee.std_logic_1164.all;

use ieee.std_logic_arith.all;

use ieee.std_logic_unsigned.all;

use std.textio.all;

entity file_handling is

end file_handling;

architecture arch_file_handling of file_handling is

signal clk,end_file : bit := '0';

signal   data_read : real;

signal   data_save : real;

signal   line_number : integer:=1;
```

Example 7.7 VHDL code for the file read and write

```
begin
clk <= not (clk) after 2 ns;
read_p1:process
  file input_file   : text is in "file1.txt";
 variable line_no_in   : line;
  variable data_read_1   : real;
begin
wait until (clk = '1' and clk'event);
if (not endfile(input_file)) then
readline(input_file, line_no_in;
  read(line_no_in, data_read_1);
data_read <=data_read_1;  .
else
end_file <='1';
```

Example 7.7 (continued)

7.5.2 *TEXTIO*

Library consists of the predefined packages. If the requirement is to access the predefined packages from standard library, then 'TEXTIO' can be used. To use the 'TEXTIO,' declare *use ieee.std_logic_TEXTIO.all;*

To read and write the ASCII files, the packages should consist of the functions and procedures. The 'TEXTIO' uses the files, where a line is a carriage return

```
end if;
end process read_p1;
write_p2 : process
   file     output_file  : text is out "file2.txt";
   variable  line_no_out : line;
begin
wait until (clk ='0' and clk'event);
if(end_file='0') then
write(line_no_out, data_read, right, 20, 16);
writeline(output_file, line_no_out);
line_number <= line_number + 1;
else
null;
end if;
end process write_p2;
end arch_file_handling;
```

Example 7.7 (continued)

terminated by text string. The package defines a number of types that can be used with text files. A variable of type 'line' is defined to hold a line of text. The 'line' is the basic unit upon which 'TEXTIO' operates.

```
library ieee,std;
use ieee.std_logic_1164.all;
use ieee.std_logic_textio.all;
use std.textio.all;
entity text_io is
end text_io;
architecture arch_text_io of text_io is
begin
file_p1: process is
 file input_file : text open read_mode is "input_data_values";
 file output_file : text open write_mode is "output_data_values";
 variable output_line : line;
 variable input_line : line;
```

Example 7.8 VHDL code for the TEXTIO

Consider Example 7.8, where the XOR operation is performed by using the 'TEXTIO.' Files 'input_file' and 'output_file' are used and operated in the read mode and write mode respectively.

By using the 'TEXTIO,' the result for the different input values is shown in Fig. 7.5.

```
variable a_in,b_in,c_out : std_logic;

begin

while not endfile(input_file) loop

readline(input_file, input_line);

read(input_line, a_in);

read(input_line, b_in);

c_out := a_in xor b_in;

write(output_line, c_out);

writeline(output_file, output_line);

end loop;

assert false report "simulation is over" severity warning;

wait;

end process;

end arch_text_io;
```

Example 7.8 (continued)

```
input_values output_values

000

101

011

110
```

Fig. 7.5 Results using TEXTIO

7.6 Summary

The following are key points to summarize this chapter:

1. VHDL has powerful constructs to carry out simulation. For complex designs, use file handling.
2. Functions can be of pure or of impure type.
3. Pure function can return default value of same type as that of the parameters.
4. Impure function can return value and may be of different type as that of the parameters.
5. Packages are used extensively to pass the information about the design objects to other VHDL designs.
6. Testbench is used to force the values to the design under verification, and it acts like stimulus generator.

Chapter 8
PLD-Based Design Guidelines

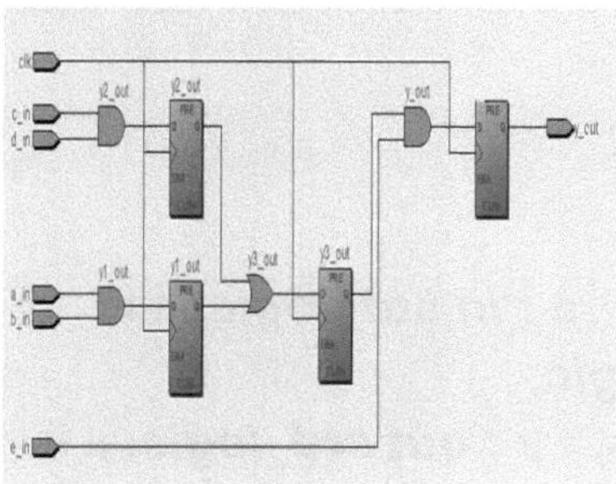

"**Once we accept our limits, we go beyond them.**
" --- Albert Einstein

Try to use the guidelines to overcome the issues during design cycles. Strong design guidelines can lead to the efficient design.

Abstract This chapter describes the design guidelines for ASIC and FPGA designs. The coding and design guidelines are useful in the RTL design cycle and recommended to be used for the efficient performance of the design. The design guidelines such as resource sharing, pipelining, logic duplications, grouping, use of signals and variables, gated clock, and clock enable logic are discussed in this chapter. Designers are requested to use these guidelines for area, speed, and power improvement in the design.

Keywords ASIC · PLD · Signals · Variables · FPGA · Grouping · Pipelining · Logic duplication · Area minimization · Speed improvement · Power · Constraining design · Parallel logic · Priority logic · Bidirectional IO · Clock gating · Clock enable · LUT · Latch · Sensitivity list · Registered output · Tri-state unintentional latches

During the design using programmable ASIC, the use of guidelines is important. For efficient design, coding and design guidelines are used in the industries. Every organization has their own coding guidelines and used during the RTL design cycle for ASIC and FPGA designs. The design and optimization guidelines are used for the performance improvement of the design, and these guidelines are covered with the practical scenarios in the subsequent sessions.

The original version of this chapter was revised: The incorrect information have been corrected. The erratum to this chapter is available at 10.1007/978-981-10-3296-7_12

V. Taraate, *PLD Based Design with VHDL*,
DOI 10.1007/978-981-10-3296-7_8

Among these coding guidelines, few of them are naming conventions, use of the registered outputs, complete sensitivity list, and use of the signals and variables.

8.1 Naming Conventions

Every organization has their own style in describing the naming conventions while writing the RTL code for the given design functionality. The naming conventions improve the readability of code. Even the good naming conventions can give information about the functional intent of the declarations used in the VHDL design.

For Example :

1. Instead of declaring input as, ***a : in std_logic;*** it is better to use ***a_in : in std_logic;***
2. Instead of declaring output as, ***y : out std_logic;*** it is better to use ***y_out : out std_logic;***
3. Signal can be declared as : ***tmp_sig : std_logic;***
4. varibale can be declared as : ***tmp_var : std_logic;***

8.2 Use of Signals and Variables

Most of the times, it is essential to use either the signals or variables in the VHDL code RTL design using VHDL. These are used to assign the values depending on the simulation time stamp or based on the simulator time tick. Simulator uses the time stamp to update the variable and signal values. The signals and variables are used for the interconnection, and the purpose of using signal or variable is depending upon the design functional requirements.

The major difference between the signals and variables is that signals are updated on the next-simulation time stamp or at the end of the process, whereas the variables are updated instant immediately during the same simulation time stamp. For more information, please go through the Chap. 3.

Example 8.1 describes the use of signal. Signals are global to the architecture and hence used in all the sequential processes across the architecture. Signals are updated at the end of the process.

The synthesis result is shown in Fig. 8.1 and as shown in the result, the y1_out and y2_out both are assigned as b_in XNOR c_in. In this a_in, input is not used and connected to ground. The reason is that synthesis tool updates the last assignment to d_in for evaluating the expression of y1_out and y2_out.

```
library ieee;
use ieee.std_logic_1164.all;
use ieee.std_logic_unsigned.all;
entity design_signal  is
port ( a_in, b_in, c_in : in std_logic;
y1_out, y2_out: out std_logic);
end design_signal;
architecture arch_signal of design_signal  is
signal  d_sig : std_logic;
begin
process (a_in, b_in , c_in)
begin
d_sig <= a_in; -- this assignment will be ignored
y1_out <= c_in xnor d_sig;
d_sig <= b_in; -- this assignment overrides the previou
y2_out <= c_in xnor d_sig;
end process;
end arch_signal;
```

- Architecture defines the functionality of design.
- Process is sensitive to 'a_in', 'b_in' and 'c_in'. Any event on one of the signal invokes the process.
- The 'd_sig' is global signal and assigned twice in the architecture.
- As the same signal is assigned twice the last assignment will be updated and first assignment is ignored.

Example 8.1 Synthesizable VHDL code for the signal assignments

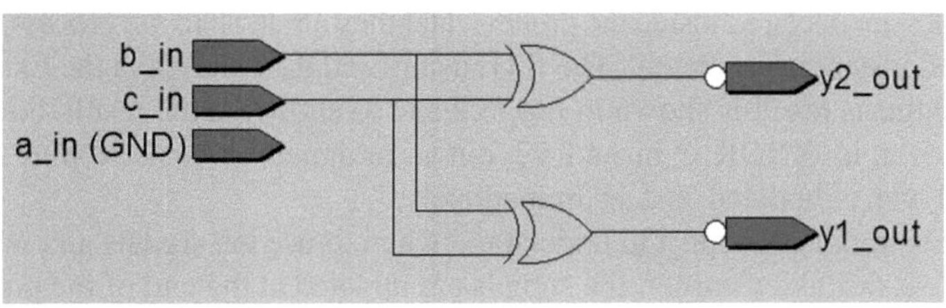

Fig. 8.1 Synthesis result for signal assignment

```
library ieee;
use ieee.std_logic_1164.all;
use ieee.std_logic_unsigned.all;
entity design_variable is
port ( a_in, b_in, c_in : in std_logic;
y1_out, y2_out: out std_logic);
end design_variable;
architecture arch_variable of design_variable is
begin
process (a_in, b_in , c_in)
variable d_var : std_logic;
begin
d_var := a_in; -- this assignment will not be ignored
y1_out <= c_in xnor d_sig;
d_var:= b_in; -- this assignment will not override the p
y2_out <= c_in xnor d_sig;
end process;
end arch_variable;
```

- Architecture defines the functionality of design.
- Process is sensitive to 'a_in', 'b_in' and 'c_in'. Any event on one of the signal invokes the process.
- The variable 'd_var' is declared inside the process.
- The 'd_var' is assigned to a_in and updated immediately
- The 'd_var' is assigned to b_in and updated immediately.

Example 8.2 Synthesizable VHDL using variable

Variables are declared inside the process, and they are local to the process. Variables are updated instant immediately. The RTL using VHDL is shown in the Example 8.2.

The synthesis result is shown in Fig. 8.2 and as shown in the result, the y1_out is assigned as a_in XNOR c_in and y2_out is assigned as b_in XNOR c_in. The variable d_var is updated instant immediately.

Note It is important aspect to understand when to use the signals and when to use variables. As discussed earlier, the signals are updated at the end of the process after delta delay, and variables are updated instant immediately. So if the assigned value needs to be used during the same simulation time stamp, then use variables otherwise use the signals. Another important point is, if the global declaration is required then use the signal so that it can be accessed throughout the architecture. Remember that VHDL-87 does not support the shared variables.

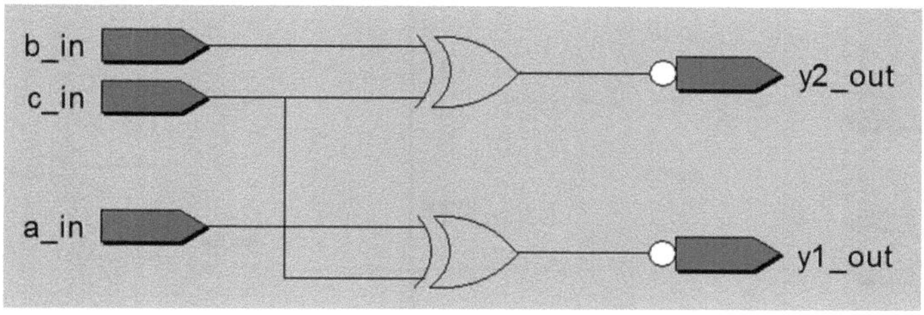

Fig. 8.2 Synthesis result for the VHDL code using variable

8.3 Grouping in Design

To improve the design performance, the grouping of terms or expressions can be used. This can be visualized as the expression with the use of parenthesis. Consider Example 8.3 shown. In this figure, the output y_out is assigned as a_in + b_in − c_in − d_in. Without grouping, the synthesis will generate a logic using cascaded network.

```
library ieee;
use ieee.std_logic_1164.all;
use ieee.std_logic_arith.all;
use ieee.std_logic_unsigned.all;

entity without_grouping is

port ( a_in, b_in, c_in, d_in : in std_logic_vector(1 downto 0);

     y_out : out std_logic_vector(1 downto 0) );

end without_grouping;

architecture arch_without_gropuing of without_grouping is

begin

  y_out <= a_in + b_in -c_in -d_in;

end arch_without_gropuing;
```

- Architecture defines the functionality of design.
- Architecture uses the concurrent assignment and 'y_out' is assigned as 'a_in + b_in –c_in-d_in'
- This generates the cascaded logic.

Example 8.3 Synthesizable VHDL without use of parenthesis

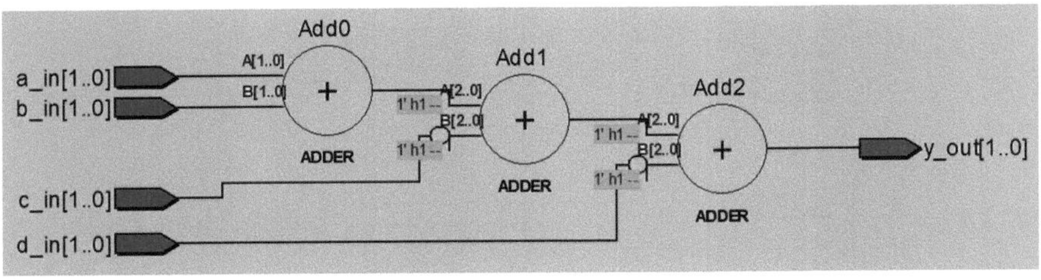

Fig. 8.3 Synthesis result for the VHDL code without use of parenthesis

The synthesis result is shown in Fig. 8.3; as shown in the result, it generates three cascaded adders. If every adder has the propagation delay of 1 ns then the overall delay is 3 ns.

The VHDL code described in Example 8.3 can be modified by the use of parenthesis. The modified code is shown in Example 8.4 and it uses the expression as y_out <= (a_in + b_in) – (c_in + d_in);

The synthesis result is shown in Fig. 8.4 and it uses the parallel logic due to use of the parenthesis. As a_in and b_in are combined using parenthesis, c_in and d_in are combined using parenthesis so it generates the two adders and one subtractor. The subtraction operation is implemented using 2's complement addition. If the delay of every adder is 1 ns, then the overall propagation delay is 2 ns. This technique is used to improve the design performance.

8.4 Guidelines for Use of Tri-State Logic

In most of the practical scenarios, the tri-state logic needs to be used to design the buses. Tri-state has three values logic '0', logic '1', and high impedance 'z'. The tri-state buses are used to communicate with the other design modules. Example 8.5 describes the tri-state logic. It is recommended to use the tri-state logic at the top level in the design to avoid the bus contentions. Instead of using the tri-state logic, it is recommended to use the Mux-based logic with the enable.

Figure 8.5 shows the synthesis result for the tri-state logic and the logic can be used to pass the data when 'enable_in' is equal to logic '1'. For logic '0' enable input, the output of tri-state logic is high impedance that is potential-free contact.

```
library ieee;
use ieee.std_logic_1164.all;
use ieee.std_logic_arith.all;
use ieee.std_logic_unsigned.all;

entity with_grouping is

port ( a_in, b_in, c_in, d_in : in std_logic_vector(1 downto 0);

        y_out : out std_logic_vector(1 downto 0) );

end with_grouping;

architecture arch_with_gropuing of with_grouping is

begin

  y_out <= (a_in + b_in) -(c_in + d_in) ;

end arch_with_gropuing;
```

- Architecture defines the functionality of design.
- The concurrent assignment statement is used inside the architecture.
- The 'y_out' is assigned to (a_in + b_in) – (c_in+d_in);

Example 8.4 Synthesizable VHDL code using parenthesis

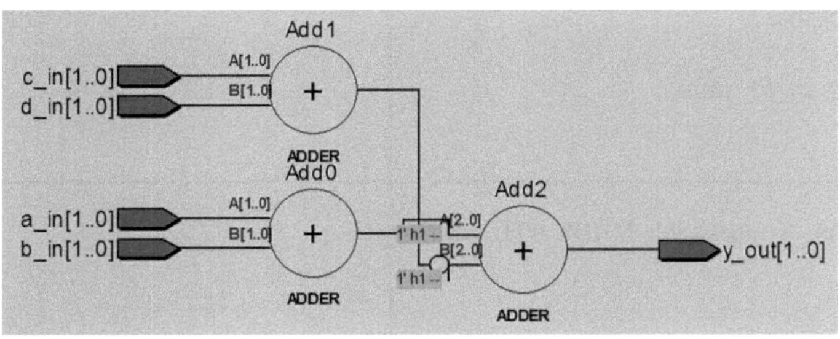

Fig. 8.4 Synthesis result for VHDL code using parenthesis

```
library ieee;
use ieee.std_logic_1164.all;
use ieee.std_logic_arith.all;
use ieee.std_logic_unsigned.all;

entity tri_state is

port ( a_in, enable_in : in std_logic;

       y_out : out std_logic);

end tri_state;

architecture arch_tri_state of tri_state is

begin

  process ( a_in, enable_in)

     begin

 if ( enable_in='1') then

 y_out <= a_in;

 else

 y_out <='Z';

end if;

end process;

end arch_tri_state;
```

- Architecture defines the functionality of design.
- The process is sensitive to 'enable_in', 'a_in'.
- The 'y_out' is assigned to a_in for enable_in ='1'
- For enable_in ='0' y_out is assigned to high impedance state.

Example 8.5 Synthesizable VHDL RTL for tri-state logic

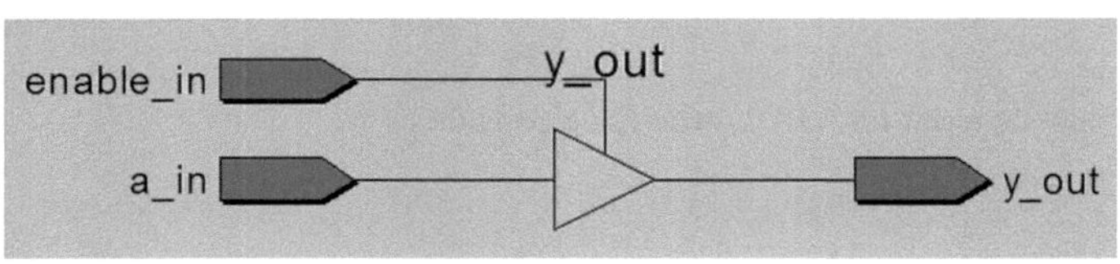

Fig. 8.5 Synthesis result for the tri-state logic

8.5 Arithmetic Resource Sharing

In most of the practical design scenarios, the common resources can be shared by using the fundamental concepts of logic design. For example, if adders are used and consuming the more area, then the area can be reduced by sharing the common adder as resource. This technique plays important role in the reduction of area by minimizing the required gate count during synthesis. Instead of using more number of adders, it is better practice to use more number of multiplexers in the design. Consider the VHDL code described in Example 8.6 for the following truth Table 8.1. As described in the VHDL code, the output needs to be assigned depending on the condition of the select input. For 'sel_in = 1' the output 'y_out' is assigned to 'a_in + b_in' and for the 'sel_in = 0' an output 'y_out' is assigned to 'c_in + d_in'.

The synthesis result for the arithmetic logic without using the concept of resource sharing is shown in Fig. 8.6. As shown in Fig. 8.6, the logic uses two adders and single multiplexer. The adders are used in the data path to perform the addition. The output of multiplexer is controlled by 'sel_in' input and for the 'sel_in' input as logic '1' it generates an output which is addition of 'a_in' and 'b_in'. For the logic '0' condition of 'sel_in' it generates an output as addition of 'c_in' and 'd_in'.

The generated logic has issue, as both adders are performing operations at the same time so unnecessarily it is wastage of power. The result data after performing the additions waits at the input lines of multiplexers and depending on the status of select line, the output is assigned. So this kind of technique is less efficient and has more gate count and leads to more power dissipation. To overcome this limitation, the resource sharing is used where the common resources can be shared by pushing the adder forward to the multiplexers. So using resource sharing more multiplexers are used and less number of adders and this leads to the significant area reduction.

As discussed earlier, the common resource required that is adder can be shared by using the multiplexer chain at the input and that can be achieved by pushing the adder at the output. Table 8.2 gives information about the strategy used for sharing the common resources.

By modification in the VHDL code, the resource sharing can be achieved. The modified RTL using VHDL is described in Example 8.7 and uses the temporary signals as 'sig_1' and 'sig_2'. For logic '0' status on the select line 'sel_in' the 'sig_1' holds the 'c_in' input and 'sig_2' holds the 'd_in' input value. For logic '1' status on the select line 'sel_in' the 'sig_1' holds the 'a_in' input and 'sig_2' holds the 'b_in' input value.

The synthesis result for the Example 8.7 is shown in Fig. 8.7. As shown in the figure, the logic is realized by using the single adder and two multiplexers. As one adder consumes lesser area, the design is efficient and has less gate count and lesser power. In this logic, only one operation is performed at a time.

```
library ieee;
use ieee.std_logic_1164.all;
use ieee.std_logic_unsigned.all;
use ieee.std_logic_arith.all;
entity resource_sharing is
port (a_in,b_in,c_in,d_in: in std_logic_vector (1 downto 0);
sel_in: in std_logic;
y_out: out std_logic_vector ( 1 downto 0));
end resource_sharing;
architecture arch_without_sharing of resource_sh
begin
process (a_in, b_in, c_in, d_in, sel_in)
begin
if (sel_in='1') then
y_out <= a_in +  b_in;
else
y_out <= c_in +  d_in;
end if;
end process;
end arch_without_sharing;
```

- Architecture defines the functionality of design.
- Process is sensitive to 'a_in', 'b_in' and 'sel_in'. Any event on one of the signal invokes the process.
- If-then-else is sequential statement and used inside the process.
- For true 'sel_in' condition the input 'a_in+b_in' is assigned to 'y_out'.
- For false 'sel_in' condition the input 'c_in+d_in' is assigned to 'y_out'

Example 8.6 Synthesizable VHDL code for arithmetic logic without resource sharing

Table 8.1 Truth table for the arithmetic logic

sel_in	y_out
0	c_in + d_in
1	a_in + b_in

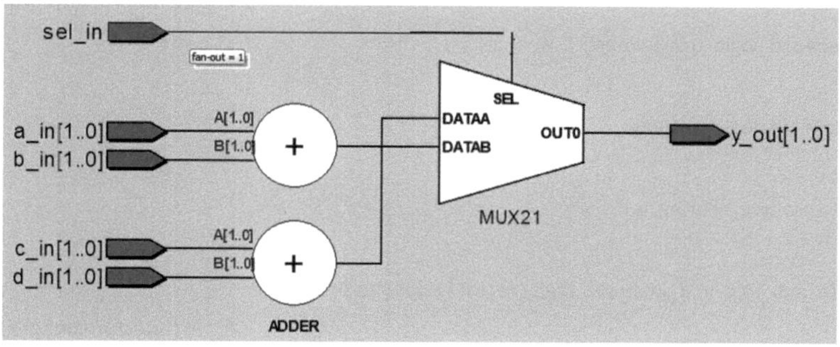

Fig. 8.6 Synthesis result for the VHDL code without resource sharing

Table 8.2 Truth table for the arithmetic logic

sel_in	sig_1	sig_2	y_out
0	c_in	d_in	c_in + d_in
1	a_in	b_in	a_in + b_in

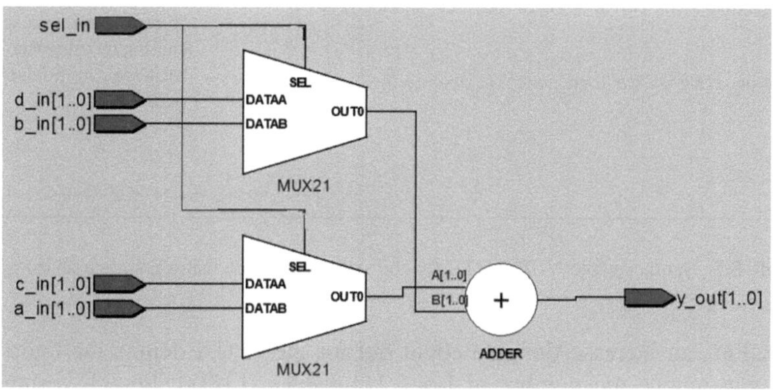

Fig. 8.7 Synthesis result for the VHDL code using resource sharing

8.6 Logic Duplications

The duplication of the logic plays an important role in the design of the digital circuits. The logic duplication depending on the scenario can increase the gate count or can reduce the gate count. In the ASIC design realization, the logic

```
library ieee;

use ieee.std_logic_1164.all;

use ieee.std_logic_unsigned.all;

use ieee.std_logic_arith.all;

entity resource_sharing is

port (a_in,b_in,c_in,d_in: in std_logic_vector (1 downto 0);

sel_in: in std_logic;

y_out: out std_logic_vector ( 1 downto 0));

end resource_sharing;

architecture arch_with_sharing of resource_sharing_1 is

signal sig_1, sig_2 : std_logic_vector (1 downto 0);

begin
```

- Architecture defines the functionality of design.
- The temporary signals 'sig_1' and 'sig_2' are used to hold the value depending on the assignment.
- The declared signals are global to the architecture and used to enable the resource sharing by using the MUX logic in the data path.

Example 8.7 Synthesizable VHDL RTL for the arithmetic logic using resource sharing

duplication can increase the gate count but for the FPGA deigns the logic duplication can reduce the number of Look Up Tables (LUTs). So this technique is scenario specific.

Example 8.8 is the description of the 4:16 decoder using VHDL and the code uses the case construct. As 'case' construct is used, it infers the parallel logic and if the design is realized using the FPGA which has 4 input and single output LUT then it uses two LUTs for every output. The reason being the 'enable_in' is one more input and hence two cascaded LUTs are required to generate single-bit decoder output.

```
P1: process (a_in, b_in, c_in, d_in, sel_in)

begin

if (sel_in='1') then

sig_1 <= a_in ;

sig_2 <= b_in;

else

sig_1 <= c_in ;

sig_2<= d_in;

end if;

end process;

P2: process ( sig_1, sig_2)

begin

 y_out <= sig_1 + sig_2;

end process;

end arch_with_sharing;
```

- Process is sensitive to 'a_in', 'b_in', 'c_in', 'd_in' and 'sel_in'. Any event on one of the signal invokes the process.
- If-then-else is sequential statement and used inside the process.
- For true 'sel_in' condition the input 'b_in' is assigned to 'sig_2' and input 'a_in' is assigned to 'sig_1'.
- For false 'sel_in' condition the input 'd_in' is assigned to 'sig_2' and input 'c_in' is assigned to 'sig_1'.

- Process is sensitive to 'sig_1' and 'sig_2'. Any event on one of the signal invokes the
- The assignment is used inside the process and output 'y_out' is assigned to addition of 'sig_1' , 'sig_2'

Example 8.7 (continued)

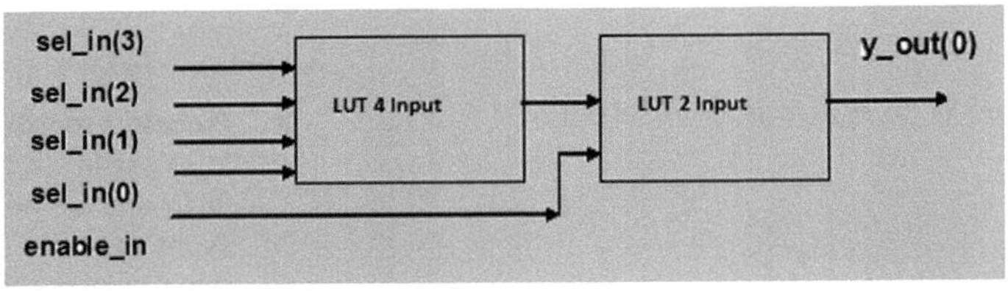

Fig. 8.8 Decoder single output realization using FPGA

As shown in Fig. 8.8, two cascaded LUTs having uniform delay are used to realize the logic at one of the output of decoder. For the 16 output lines, the realization using FPGA uses 32 LUTs without the use of logic duplication.

The RTL using VHDL code described in Example 8.8 can be modified by using the logic duplication where two 2:4 decoders can drive the 16 AND gates. So in such scenario the FPGA realization uses the 8 LUTs for the decoders and 16 LUTs for generating the outputs. Thus this technique reduces almost 8 LUTs as compare to logic without the use of logic duplication.

The modified VHDL design using the logic duplication technique is described in Example 8.9. The synthesis result is shown in Fig. 8.9.

By this technique, the VHDL code length increases but this technique can reduce the usage of number of LUTs while implementing by using FPGA.

8.7 Multiple Driver Assignments

Most of the time during the design using programmable ASICs, if the same signal is assigned in the different processes then it gives the multiple driver assignment error. Most of the EDA tools generate the error as 'Error: Can't resolve multiple constant drivers for net "name_net" in the file_name.vhd'. The scenario is described in Example 8.10.

As described in Example 8.10 the signal 'y_reg' is assigned in the process 'P1' and process 'P2'. So during compilation the EDA tool gives the error as multiple driver assignments.

The RTL using VHDL is described in Example 8.10, is modified to resolve the multiple driver error. The error is resolved by using the intermediate signal 'y1_reg' in the second process. As same signal is not assigned in the process P1 and process P2, it does not have the compilation issue, and hence, there is no any error for the multiple driver assignment. The synthesizable RTL using VHDL is described in Example 8.11.

The synthesis result for Example 8.11 is shown in Fig. 8.10.

```
library ieee;

use ieee.std_logic_1164.all;

entity without_logic_duplication is

port (sel_in  : in std_logic_vector(3 downto 0);

enable_in : in std_logic;

y_out : out std_logic_vector ( 15 downto 0));

end without_logic_duplication;

architecture arch_without_logic_duplication of without_logic_duplication is

begin

process(sel_in, enable_in)

begin

if (enable_in ='1') then

   case ( sel_in) is

       when "0000" => y_out <= "0000000000000001";

       when "0001" => y_out <= "0000000000000010";

       when "0010" => y_out <= "0000000000000100";

       when "0011" => y_out <= "0000000000001000";
```

- Architecture defines the functionality of design.
- Process is sensitive to 'sel_in', and 'enable_in'. Any event on one of the signal invokes the process.
- If-then-else is sequential statement and used inside the process.
- For true 'enable_in' condition the decoder is enabled to generate active high output.

Example 8.8 Synthesizable VHDL RTL for 4:16 decoder without logic duplication

```
when "0100" => y_out <= "0000000000010000";
      when "0101" => y_out <= "0000000000100000";
      when "0110" => y_out <= "0000000001000000";
      when "0111" => y_out <= "0000000010000000";
          when "1000" => y_out <= "0000000100000000";
      when "1001" => y_out <= "0000001000000000";
      when "1010" => y_out <= "0000010000000000";
      when "1011" => y_out <= "0000100000000000";
      when "1100" => y_out<="0001000000000000"
      when "1101" => y_out <= "0010000000000000";
      when "1110" => y_out <= "0100000000000000";
      when "1111" => y_out <= "1000000000000000";
      end case;
else
   y_out <= (others =>0);
end if;
end process;
```

- The case construct is used to describe the parallel logic.
- Depending on the status tus on 'sel_in' input lines it assigns one of the output line as active high.
- For 'enable_in' input as active high the decoder is enabled to force one of the output line as active high. For 'enable_in' as active low all the output lines are forced to active low as decoder is disabled.

Example 8.8 (continued)

```
library ieee;

use ieee.std_logic_1164.all;

entity with_logic_duplication is

port (sel_in  : in std_logic_vector( 3 downto 0);

enable_in : in std_logic;

y_out : out std_logic_vector ( 15 downto 0));

end with_logic_duplication;

architecture arch_with_logic_duplication of with_logic_duplication is

signal y0_reg, y1_reg : std_logic_vector ( 3 downto 0);

begin

process(sel_in(1 downto 0), enable_in)

begin

if (enable_in ='1') then

    case ( sel_in(1 downto 0)) is

        when "00" => y0_reg <= "0001";

        when "01" => y0_reg <= "0010";
```

- Architecture defines the functionality of design.
- Intermediate signals y0_reg and y1_reg are defined to hold the value.
- Process is sensitive to 'sel_in(1 downto 0)', and 'enable_in'. Any event on one of the signal invokes the process.
- If-then-else is sequential statement and used inside the process.
- For true 'enable_in' condition the decoder is enabled to 4 bit generate output y0_reg.

Example 8.9 Synthesizable VHDL RTL using logic duplication

```
 when "10" => y0_reg <= "0100";
      when "11" => y0_reg <= "1000";
      end case;
else
   y0_reg <= "0000";
end if;
end process;
process(sel_in(3 downto 2), enable_in)
begin
if (enable_in ='1') then
   case ( sel_in(3 downto 2)) is
      when "00" => y1_reg <= "0001";
      when "01" => y1_reg <= "0010";
      when "10" => y1_reg <= "0100";
      when "11" => y1_reg <= "1000";
      end case;
```

- Process is sensitive to 'sel_in(3 downto 2)', and 'enable_in'. Any event on one of the signal invokes the process.
- If-then-else is sequential statement and used inside the process.
- For true 'enable_in' condition the decoder is enabled to 4 bit generate output y1_reg.
- For false 'enable_in' condition the decoder is disabled to generate 4 bit output y1_reg as logic '0000'

Example 8.9 (continued)

```
else
   y1_reg <= "0000";
end if;
end process;
y_out(0) <= y0_reg(0) and y1_reg(0);
y_out(1) <= y0_reg(1) and y1_reg(0);
y_out(2) <= y0_reg(2) and y1_reg(0);
y_out(3) <= y0_reg(3) and y1_reg(0);
y_out(4) <= y0_reg(0) and y1_reg(1);
y_out(5) <= y0_reg(1) and y1_reg(1);
y_out(6) <= y0_reg(2) and y1_reg(1);
y_out(7) <= y0_reg(3) and y1_reg(1);
y_out(8) <= y0_reg(0) and y1_reg(2);
y_out(9) <= y0_reg(1) and y1_reg(2);
y_out(10) <= y0_reg(2) and y1_reg(2);
y_out(11) <= y0_reg(3) and y1_reg(2);
```

- Effectively the decoder with output line y1_reg(0) is used as the enable input of AND gate to generate the four output lines y_out(0) to y_out(3) depending on the status of y0_reg(0) to yo_reg(3)

- Effectively the decoder with output line y1_reg(1) is used as the enable input of AND gate to generate the four output lines y_out(4) to y_out(7) depending on the status of y0_reg(0) to y0_reg(3)
- Effectively the decoder with output line y1_reg(2) is used as the enable input of AND gate to generate the four output lines y_out(8) to y_out(11) depending on the status of y0_reg(0) to y0_reg(3).

Example 8.9 (continued)

```
y_out(12) <= y0_reg(0) and y1_reg(3);

y_out(13) <= y0_reg(1) and y1_reg(3);

y_out(14) <= y0_reg(2) and y1_reg(3);

y_out(15) <= y0_reg(3) and y1_reg(3);

end arch_with_logic_duplication;
```

➤ Effectively the decoder with output line y1_reg(3) is used as the enable input to generate the four output lines y_out(12) to y_out(15) depending on the status of y0_reg(0) to y0_reg(3)

Example 8.9 (continued)

8.8 Inferring Latches

Most of the time during the RTL design phase, it has been observed that, the undesired functional behavior due to inference of the latches. The reason for the unintentional latches is the missing 'else' condition from the 'if then else' statement or missing 'when others' from the 'case' construct. Example 8.12 describes the scenario for the unintended latch inference. The intended design functionality is to generate output 'y_out' as 'a_in and b_in' for 'enable_in = 1' and to assign 'y_out = 0' for the 'enable_in = 0'. As 'else' clause is missing it infers unintended latch.

The synthesis result for the Example 8.12 is shown in Fig. 8.11.

The RTL using VHDL described in Example 8.12 can be modified by using the 'else' clause to get the intended design functionality. The modified VHDL RTL is described in Example 8.13. Synthesis result is shown in Fig. 8.12.

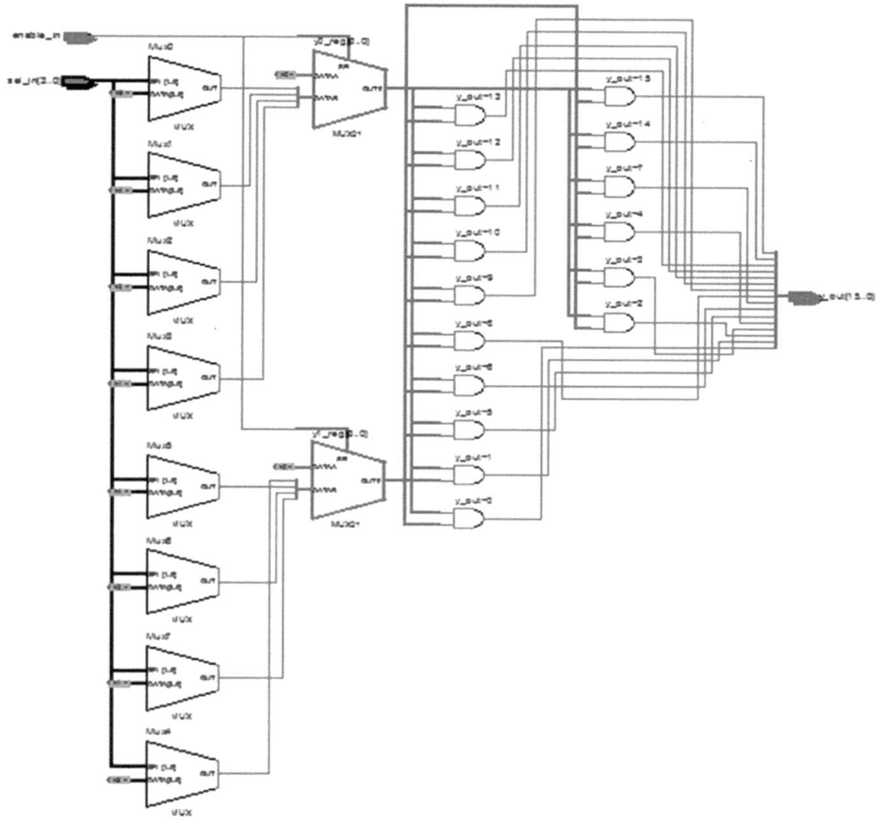

Fig. 8.9 Synthesis result for the VHDL RTL using logic duplication

8.9 Use of If Then Else Versus Case Statements

As discussed in Chap. 4, the sequential statements ‘if then else’ and ‘case’ are used inside the process and they are used to design the combinational or sequential logic. Example 8.14 describes the functionality of the design using ‘case.’ The ‘case’ construct generates the parallel logic. The synthesis result is shown in Fig. 8.13.

The synthesis result for Example 8.14 is shown in Fig. 8.13. As shown, it infers the parallel logic and hence the lesser propagation delay as compare to implementation using ‘if then else’.

The synthesizable VHDL using the ‘if then else’ construct is shown in Example 8.15. As described in Example 8.15, due to use of the nested ‘if then else’

```
--Error: Can't resolve multiple constant drivers for net "y_reg" at multiple_driver.vhd
--Code with the multiple drivers
library ieee;
use ieee.std_logic_1164.all;
entity multiple_driver is
port (a_in, b_in, c_in, d_in : in std_logic;
clk : in std_logic;
y_out : out std_logic);
end multiple_driver;
architecture arch_multiple_driver of multiple_driver is
signal y_reg : std_logic;
begin
P1: process(clk, a_in, b_in)
begin
if (clk='1' and clk'event) then
y_reg <= a_in and b_in;
end if;
end process;
P2: process (clk, c_in, d_in)
 begin
if (clk = '1' and clk'event) then
y_reg <= y_reg or (c_in and d_in );
end if;
end process;
y_out <= y_reg;
end arch_multiple_driver;
```

Error : Can't resolve multiple constant

drivers for net "y_reg" at multiple_driver.vhd

- Process 'P1'is sensitive to 'clk', 'a_in', and 'b_in'. Any event on one of the signal invokes the process.
- If-then-else is sequential statement and used inside the process.
- For rising edge of the clock the y_reg is assigned to "a_In and b_in". in the first process 'P1'.
- Process 'P2'is sensitive to 'clk', 'c_in', and 'd_in'. Any event on one of the signal invokes the process.
- For rising edge of the clock the y_reg is assigned to "y_reg or (c_in and d_in)" in the second process 'P2'.
- This code gives compilation error due to assignment of the same signal twice in the two different processes.

Example 8.10 Synthesizable VHDL RTL with multiple drivers

statement it infers the priority logic. Priority logic is having more propagation delay due to cumulative effect of the individual stage propagation delay.

The synthesis result for the VHDL code using nested if then else is shown in Fig. 8.14 and as shown it infers the priority logic. The input 'a_in' has the highest priority as compared to any other input. The input 'd_in' has the least priority.

```
library ieee;
use ieee.std_logic_1164.all;
entity multiple_driver is
port (a_in, b_in, c_in, d_in : in std_logic;
clk : in std_logic;
y_out : out std_logic);
end multiple_driver;
architecture arch_multiple_driver of multiple_driver is
signal y_reg : std_logic;
signal y1_reg : std_logic;
begin
P1: process(clk, a_in, b_in)
begin
if (clk='1' and clk'event) then
y_reg <= a_in and b_in;
end if;
end process;
P2:process (clk, c_in, d_in)
 begin
if (clk = '1' and clk'event) then
y1_reg <= y_reg or (c_in and d_in );
end if;
end process;
y_out <= y1_reg;
end arch_multiple_driver;
```

- Process is sensitive to 'a_in', 'clk' and 'b_in'. Any event on one of the signal invokes the process.
- If-then-else is sequential statement and used inside the process.
- For rising edge of clk the 'y_reg' is assigned as 'a_in and b_in' in the process P1.
- For rising edge of clk the y1_reg is assigned to 'y_reg or (c_in and d_in).
- The example doesn't have multiple drivers.

Example 8.11 Synthesizable VHDL RTL without multiple drivers

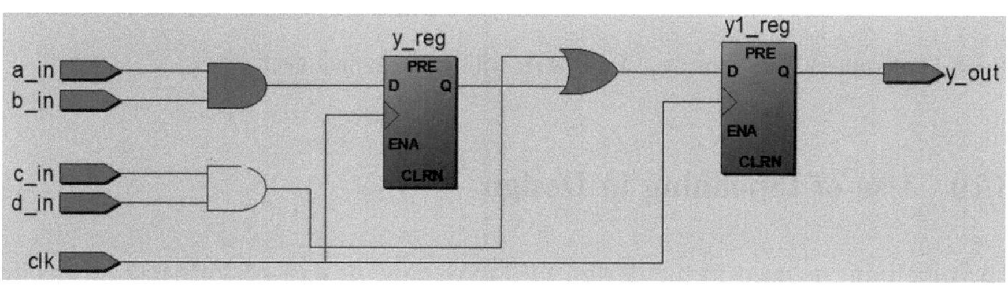

Fig. 8.10 Synthesis result for the logic without multiple drivers

```
library ieee;
use ieee.std_logic_1164.all;

entity latch_inference is
port (a_in, b_in: in std_logic;
enable_in : in std_logic;
y_out : out std_logic);
end latch_inference;

architecture arch_latch_inference of latch_inference is
begin

process(enable_in, a_in, b_in)
begin
if (enable_in='1') then
y_out <= a_in and b_in;
end if;
end process;

end arch_latch_inference;
```

- Process is sensitive to 'a_in', 'b_in' and 'enable_in'. Any event on one of the signal invokes the process.
- For true value of 'enable_in' the 'y_out' is assigned to 'a_in and b_in'.
- For false value of 'enable_in' it has not stated what to do?
- So as 'else' clause is missing it infers logic with latch

Example 8.12 Synthesizable VHDL RTL with missing 'else'

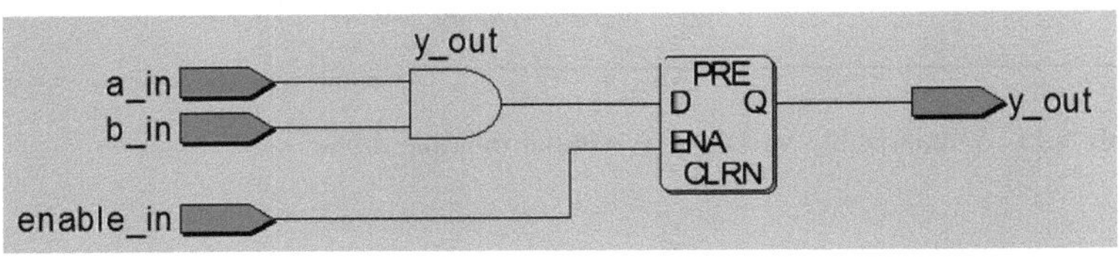

Fig. 8.11 Synthesis result for the VHDL RTL with unintentional latch

8.10 Use of Pipelining in Design

The pipelining is used in the design to improve the design performance. Consider the scenario that the design has the combinational logic between two registers and the delay of combinational logic is more. In such scenario, the combinational logic

```
library ieee;
use ieee.std_logic_1164.all;

entity wo_latch_inference is
port (a_in, b_in: in std_logic;
enable_in : in std_logic;
y_out : out std_logic);
end wo_latch_inference;
architecture arch_wo_latch_inference of wo_latch_inference is

begin
process(enable_in, a_in, b_in)
begin
if (enable_in='1') then
y_out <= a_in and b_in;
else
y_out<='0';
end if;
end process;

end arch_wo_latch_inference;
```

- Process is sensitive to 'a_in', 'b_in' and 'enable_in'. Any event on one of the signal invokes the process.
- For true value of 'enable_in' the 'y_out' is assigned to 'a_in and b_in'.
- For false value of 'enable_in' 'y_out' is assigned to logc'0'
- So it infers the combinational logic.

Example 8.13 Synthesizable VHDL RTL without missing else

can be spitted by adding one more register in the design. The technique is used to improve the overall design timing and performance of the design at the cost of one cycle latency. Considering the design described in Example 8.16, the synthesis result is shown in Fig. 8.15.

As shown in Fig. 8.15, the register-to-register path has AND logic gate followed by the OR logic gate. So it has the maximum combinational delay. If delay of every gate is 1 ns, the combinational delay in the register-to-register path is 2 ns.

This delay has significant impact on the design speed. To improve the design performance, the combinational delay can be reduced by adding the pipelined

```
library ieee;
use ieee.std_logic_1164.all;
use ieee.std_logic_unsigned.all;
entity mux_case is
port (sel_in: in std_logic_vector(1 downto 0);
a_in, b_in,c_in,d_in: in std_logic;
y_out: out std_logic);
end mux_case;
architecture arch_mux_case of mux_case is
begin
process (sel_in, a_in, b_in, c_in , d_in)
begin
case (sel_in) is
 when "00" => y_out <= a_in;
 when "01" => y_out <= b_in;
 when "10" => y_out <= c_in;
 when "11" => y_out <= d_in;
 when others => y_out <= null;

end case;

end process;
end arch_mux_case;
```

- Architecture defines the functionality of design.
- Process is sensitive to 'a_in', 'b_in' , 'c_in' and 'd_in'. Any event on one of the signal invokes the process.
- The 'case' construct is used and it infers the parallel logic.

Example 8.14 Synthesizable VHDL using case

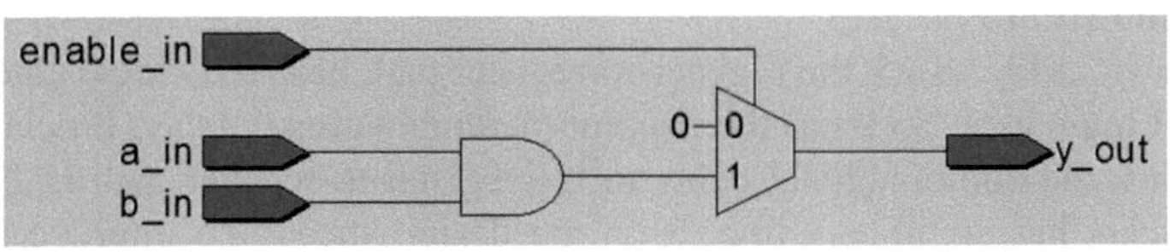

Fig. 8.12 Synthesis result for the VHDL RTL with 'if then else'

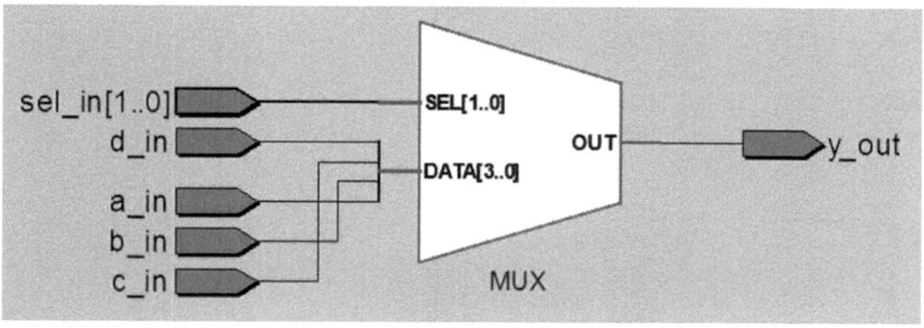

Fig. 8.13 Synthesis result for the VHDL RTL using case

```
library ieee;
use ieee.std_logic_1164.all;
use ieee.std_logic_unsigned.all;
entity mux_if is
port (sel_in: in std_logic_vector(1 downto 0);
a_in, b_in,c_in,d_in: in std_logic;
y_out: out std_logic);
end mux_if;
architecture arch_mux_if of mux_if is
begin
process (sel_in, a_in, b_in, c_in , d_in)
begin
if (sel_in="00") then
   y_out <= a_in;
elsif (sel_in="01") then
   y_out <= b_in;
elsif (sel_in="10") then
   y_out <= c_in;
elsif (sel_in="11") then
   y_out <= d_in;
else
    y_out <= '0';
end if;
end process;
end arch_mux_if;
```

- Architecture defines the functionality of design.
- Process is sensitive to 'a_in', 'b_in','c_in' and 'd_in'. Any event on one of the signal invokes the process.
- The nested 'if-then-else' is used and it infers the priority logic.
- All the conditions are covered in this and it infers the combinational logic.

Example 8.15 Synthesizable VHDL RTL using nested if then else

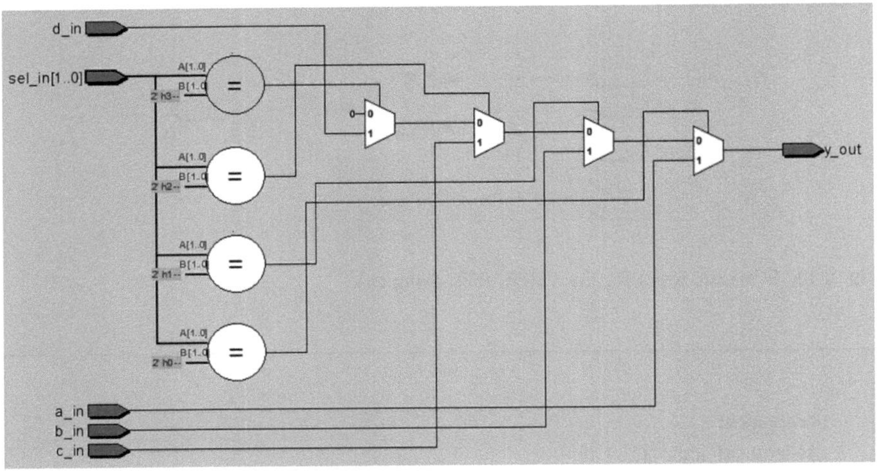

Fig. 8.14 Synthesis result for VHDL RTL using nested if then else

register by retaining the same design functionality. The modified VHDL code is described in Example 8.17 and the synthesis result is shown in Fig. 8.16.

The synthesis result for Example 8.17 is shown in Fig. 8.16. As shown it uses the four registers using the pipelining whereas the Example 8.16, uses only three registers.

Due to pipelining, the design has the less combinational delay in the register-to-register path and has better performance as compared to the design without pipelining. Commonly used techniques to improve the design performance using the pipelining concept are register balancing and register optimization. Depending on the requirement of the hierarchical designs or flattened design, these techniques can be used during the RTL design and synthesis phase.

8.11 Multiple Clock Domain and Data Passing

The complex ASIC designs or design using FPGA can have single clock domain or multiple clock domains. A single clock domain design does not have the issue of data integrity or data convergence. But if the design has multiple clocks then the real issue is the data passing from one of the clock domains to another clock

```
library ieee;
use ieee.std_logic_1164.all;
use ieee.std_logic_unsigned.all;
entity without_pipelining is
port (a_in, b_in,c_in,d_in,e_in, clk: in std_logic;
y_out: out std_logic);
end without_pipelining;
architecture arch_without_pipelining of without_pipelining is
signal y1_out, y2_out : std_logic;
begin
process (clk, a_in, b_in,c_in,d_in, e_in)
begin
if (clk='1' and clk'event) then
y1_out <= a_in and b_in;
y2_out <= c_in and d_in;
y_out <= (y1_out or y2_out) and e_in;
end if;
end process;
end arch_without_pipelining;
```

- The process is sensitive to 'clk', 'a_in','b_jn','c_in, 'd_in'and 'e_in'.
- On rising edge of the clock input the signal 'y1_out' is assigned to 'a_in and b_in'
- On rising edge of the clock input the signal 'y2_out' is assigned to 'c_in and d_in'
- On rising edge of clock an output 'y_out' is assigned to '(y1_out or y2_out) and e_in'.

Example 8.16 Synthesizable VHDL RTL without use of pipelining

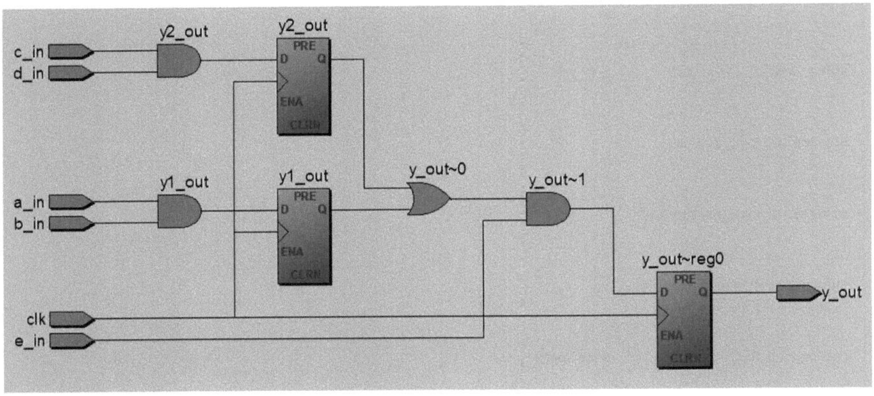

Fig. 8.15 Synthesis result for the VHDL RTL without using pipelining

```
library ieee;
use ieee.std_logic_1164.all;
use ieee.std_logic_unsigned.all;
entity with_pipelining is
port (a_in, b_in,c_in,d_in,e_in, clk: in std_logic;
y_out: out std_logic);
end with_pipelining;
architecture arch_with_pipelining of with_pipelining is
signal y1_out, y2_out,y3_out : std_logic;
begin
process (clk, a_in, b_in,c_in,d_in, e_in)
begin
if (clk='1' and clk'event) then
y1_out <= a_in and b_in;
y2_out <= c_in and d_in;
y3_out <= (y1_out or y2_out);
y_out <= y3_out and e_in;
end if;
end process;
end arch_with_pipelining;
```

- Process is sensitive to 'clk', 'a_in', 'b_in','c_in','d_in'and 'e_in'. Any event on one of the signal invokes the process.
- On the rising edge of clock the y1_out, y2_out, y3_out and y_out areassigned.
- Due to use of 'y3_out' signal it infers one more register.
- Total number of registers inferred by this code are four.

Example 8.17 Synthesizable VHDL RTL using the pipelining

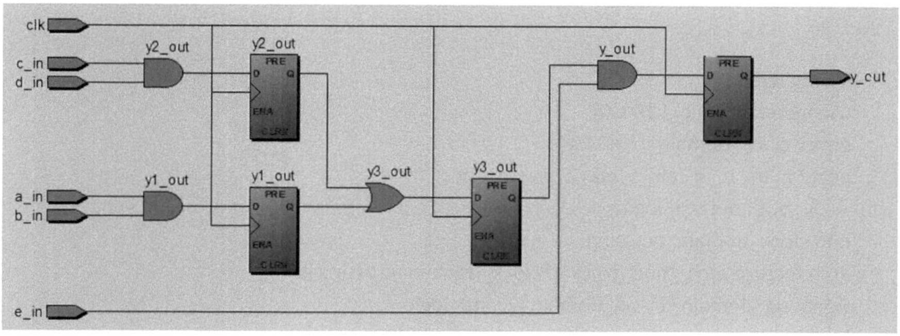

Fig. 8.16 Synthesis result for VHDL RTL using pipelining

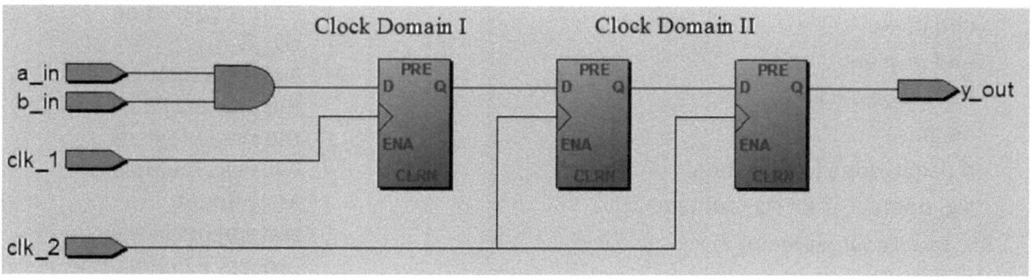

Fig. 8.17 Synthesis result for the VHDL RTL using multiple clocks

domain. To avoid the metastability and the data integrity issues, the data can be passed from one of the clock domain to another by using the two-stage- or multi-stage-level synchronizers.

Example 8.18 describes the multiple clock domain design scenario. But in the practice there can be separate design for clock domain one and clock domain two.

The synthesis result is shown in Fig. 8.17 and as shown while passing the data from clock domain one to the clock domain two, two-level synchronizer is used. The two-level synchronizer output is valid legal state although the first flip-flop in the second clock domain goes into the metastable state.

```
library ieee;
use ieee.std_logic_1164.all;
entity clock_domain_crossing is
port ( a_in , b_in , clk_1, clk_2 : in std_logic;
    y_out : out std_logic);
end clock_domain_crossing;
architecture arch_mult_clock of clock_domain_crossing is
signal sig_domain_1 , sig_domain_2 : std_logic;
begin
P1: process (clk_1)
begin
if rising_edge (clk_1) then
sig_domain_1 <= a_in and b_in;
end if;
end process;
P2: process (clk_2)
begin
if rising_edge (clk_2) then
sig_domain_2 <= sig_domain_1;
y_out <= sig_domain_2;
end if;
end process;
end arch_mult_clock;
```

> - Two different processes P1, P2. Process P1 is triggered on the clk_1, process P2 is triggered on clk_2.
> - Single assignment in the process P1 infers the single register whereas multiple assignment statements in the process P2 infers the two registers.

Example 8.18 Synthesizable VHDL RTL for multiple clock domains

8.12 Bidirectional IO

During the design, the bidirectional IO is used to pass the data from the design to the external world or vice versa. As discussed earlier in Chap. 3, the port can be declared as 'in', 'out', and 'inout'. The RTL using VHDL for the bidirectional IO is shown in Example 8.19. The synthesis result is shown in Fig. 8.18.

The synthesis outcome of Example 8.19 is shown in Fig. 8.18. As shown the four-bit output line is 'y_out' and four-bit bidirectional line is 'y_inout'

```
library ieee;
use ieee.std_logic_1164.all;
entity bidirectional_register is port (
data_in : in std_logic_vector (3 downto 0);
clk,enable_in : in std_logic;
y_out : out std_logic_vector (3 downto 0);
y_inout : inout std_logic_vector (3 downto 0));
end bidirectional_register;
architecture arch_bidirectional_register of bidirectional_register is
signal y_reg : std_logic_vector (3 downto 0);
signal yio_reg : std_logic_vector (3 downto 0);
begin
P1: process(clk,data_in)
begin
if (clk='1' and clk'event) then
y_reg <= data_in;
end if;
end process;
P2: process (y_reg,enable_in)
 begin
if (enable_in = '1') then
yio_reg <= y_reg ;
else
yio_reg <= (others=>'Z');
end if;
end process;
y_inout <= yio_reg;
y_out <= y_inout;
end arch_bidirectional_register;
```

- Process P1 is sensitive to the 'clk' and 'data_in'.
- Process P2 is sensitive to 'y_reg' and 'enable_in'
- The bidirectional IO is 'y_inout' and the value to this are assigned by using the intermediate signal 'yio_reg'

Example 8.19 Synthesizable VHDL RTL for bidirectional IO

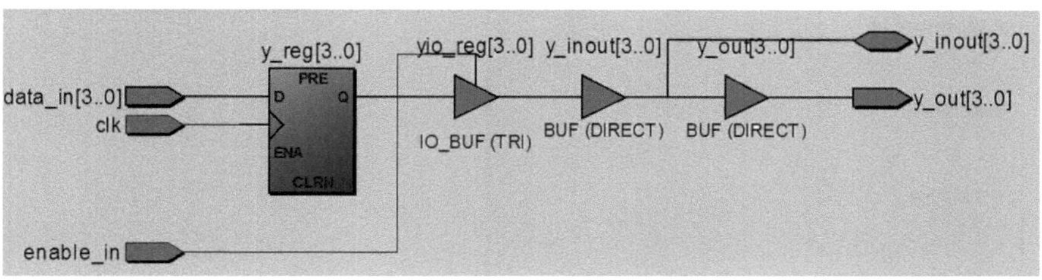

Fig. 8.18 Synthesis result for the VHDL RTL using bidirectional IO

```
library ieee;
use ieee.std_logic_1164.all;
use ieee.std_logic_unsigned.all;
entity gated_clock is
port (data_in, clk, load_en, clock_en: in std_logic;
y_out: out std_logic);
end gated_clock;
architecture arch_gated_clock of gated_clock is
signal clock_gate: std_logic;
begin
clock_gate <= (clk and clock_en);
process (load_en, clock_gate)
begin
if (clock_gate='1' and clock_gate'event) then
if (load_en='1') then
y_out <= data_in;
end if; end if;
end process;
end arch_gated_clock;
```

- Clock enable 'clock_en'signal is used to enable the clock.
- The gated clock 'clock_gate' is created by using 'clock_en and clk'.
- The process is sensitive to the 'clock_gate'and 'load_en'
- For the rising edge of the 'clock_gate' the 'y_out' is assigned as 'data_in' for 'load_en='1'.

Example 8.20 Synthesizable VHDL RTL using clock gating

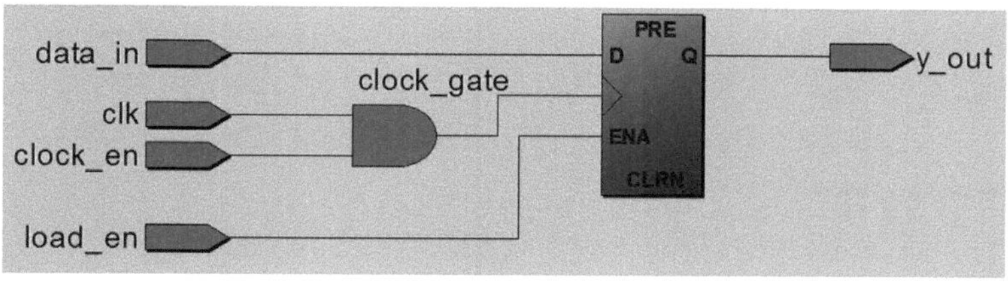

Fig. 8.19 Synthesis result for VHDL RTL using clock gating

8.13 Gated Clock

The clock is hungry net in the design. Due to clock toggling, the design has more dynamic power dissipation. The power dissipation can be reduced by using the clock-gating cells. The design using the clock-gating concept is described in Example 8.20. The synthesis result is shown in Fig. 8.19.

As shown in the synthesis outcome, the clock input of the register is controlled by using the 'clock_gate', where 'clock_gate' signal is generated by using AND logic. But such type of gating strategy is prone to the glitches. To avoid the glitches, it is recommended to use the clock-gating cells.

8.14 Design with Clock Enable

The sequential design can have the additional enable signal. Depending on the enable signal status, the input data can be transferred to the output. Example 8.21 describes the synthesizable VHDL using the enable input and the synthesis result is shown in Fig. 8.20.

As shown in the synthesis outcome the clock enable is generated and used in the enable path of the flip-flop.

More guidelines related to the practical scenarios and their interpretation in the practical ASIC prototyping are discussed in the next subsequent chapters. For the FPGA device-specific guidelines, please refer the Chap. 6.

```
library ieee;
use ieee.std_logic_1164.all;
use ieee.std_logic_unsigned.all;
entity clock_enable is
port (data_in, clk, load_en, clock_en: in std_logic;
y_out: out std_logic);
end clock_enable;
architecture arch_clock_enable of clock_enable is
signal clock_enable : std_logic;
begin
clock_enable <= load_en and clock_en;
process (clk, data_in, clock_enable)
begin
if (clk='1' and clk'event) then
if (clock_enable='1') then
y_out <= data_in;
end if;end if;
end process;
end arch_clock_enab
```

- The clock enable signal 'clock_enable' is generated by using AND of 'load_en', 'clock_en'
- The process is sensitive to 'clk', 'data_in' and 'clock_enable'.
- If 'clock-enable' is logic '1' and clk is rising edge then the 'data_in' is passed to the output 'y_out'

Example 8.21 Synthesizable VHDL RTL using clock enable

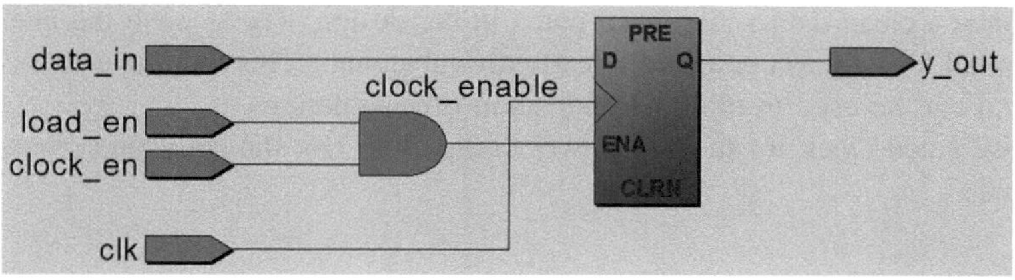

Fig. 8.20 Synthesis result for the VHDL RTL using clock enable

8.15 Summary

The following are the key points to summarize the design guidelines

1. During the RTL design phase, use the naming conventions suggested in the coding guidelines.
2. To avoid the simulation and synthesis mismatch, use all the required signals in the sensitivity list.
3. Do not use 'Buffer' as during synthesis; it creates the problem. Use 'inout' with the suitable intermediate signal for looping back of the signals.
4. Use the 'case' construct to infer the parallel logic and use 'if then else' construct to infer the priority logic.
5. For better timing and constraining design, use the registered input and output.
6. Do not use the glue logic between different modules, instead of that combine the glue logic in the module.
7. Use pipelining for the improved design performance.
8. Use more number of multiplexers as compare to adder. Adder consumes more area as compared to the multiplexers.
9. Use tri-state logic at the top level or model the tri-state behavior using the suitable multiplexing logic with enable input.
10. Use the grouping of the terms using parenthesis to reduce the overall propagation delay.
11. Describe all the conditions in the 'case' construct and 'if then else' construct to avoid inference of the unintentional latches.
12. Latches are inferred in the design if 'else' condition is not covered. Even if all the conditions in the 'case' construct are not covered, then it infers unintentional latches.
13. Use the two- or multi-level synchronizer to pass the data from one of the clock domains to the another clock domain.
14. Use the logic duplication technique to improve the overall design performance. Depending on the scenario, logic duplication can increase the gate count or can reduce the gate count.

15. Have a clean data and control paths in the design. Try to push the late arrival signal forward as compared to early arrival signals. This will have better timing and can be used to eliminate the setup time violations.
16. Use gated clock for the low power dissipation. Use the dedicated clock-gating cell.

Chapter 9
Finite-State Machines

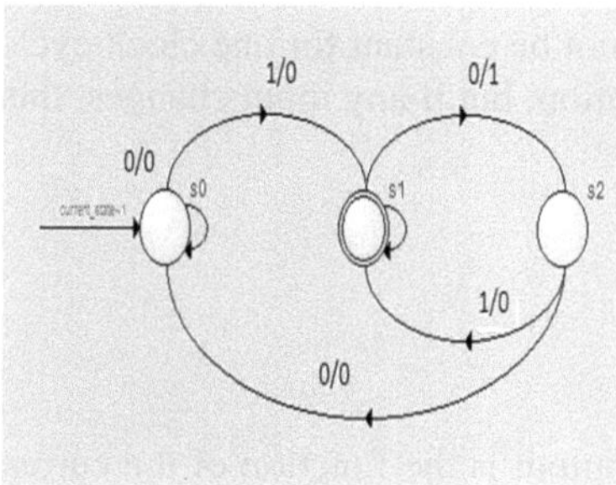

"The true sign of intelligence is not knowledge but imagination.
" --- Albert Einstein

Design the efficient FSM using VHDL using the VHDL constructs. Use the basic VHDL knowledge, apply the intelligence and write efficient VHDL RTL.

Abstract This chapter describes the efficient FSM coding using VHDL constructs. The FSMs are of two types: Moore and Mealy, and this chapter focuses on the RTL design for the Moore and Mealy machines. Even this chapter discusses about the different encoding methods for FSM, and the FSM examples are described using binary, gray, and one-hot encoding method. The examples such as sequence detector and parity checker are useful in the real practical world and are discussed in this chapter. Even this chapter is useful to understand the importance of the multiple process FSM. The key design guidelines for FSM are described with the performance improvement techniques.

Keywords FSM · Moore machine · Mealy machine · Binary encoding · Gray encoding · One-hot encoding · Sequence detector · Glitch-free output · Parity checker · Multiple-process FSM · FSM performance improvement · FSM design guidelines

The original version of this chapter was revised: The incorrect information have been corrected. The erratum to this chapter is available at 10.1007/978-981-10-3296-7_12

V. Taraate, *PLD Based Design with VHDL*,
DOI 10.1007/978-981-10-3296-7_9

9.1 Introduction to FSM

Finite-state machine (FSM) is a source synchronous sequential circuit and can be efficiently described by VHDL. FSMs are used in the design of the sequential circuits, which needs predefined sequence. Even FSMs are used to describe the functionality of the controllers in the ASIC/FPGA based designs. The efficient coding of FSM plays an important role in the design of integrated circuits. FSMs are classified as Moore machine and Mealy Machine.

In the Moore FSM, an output is the function of the current or present state only, and hence, in the Moore FSM, an output is constant for one clock-cycle duration. In the Mealy FSM, an output is the function of the current state and changes in any one of the input, and hence, output may or may not be constant for one clock cycle. Current state is constant for one clock-cycle duration, but if any input changes, then an output also changes irrespective of clock.

9.1.1 Moore Machine

As discussed earlier in the Moore machine, an output is the function of the current state only. Hence, an output is stable or constant for one clock-cycle duration. The representation of Moore machine is shown in Fig. 9.1. As shown in Fig. 9.1, the key blocks of FSM are as follows:

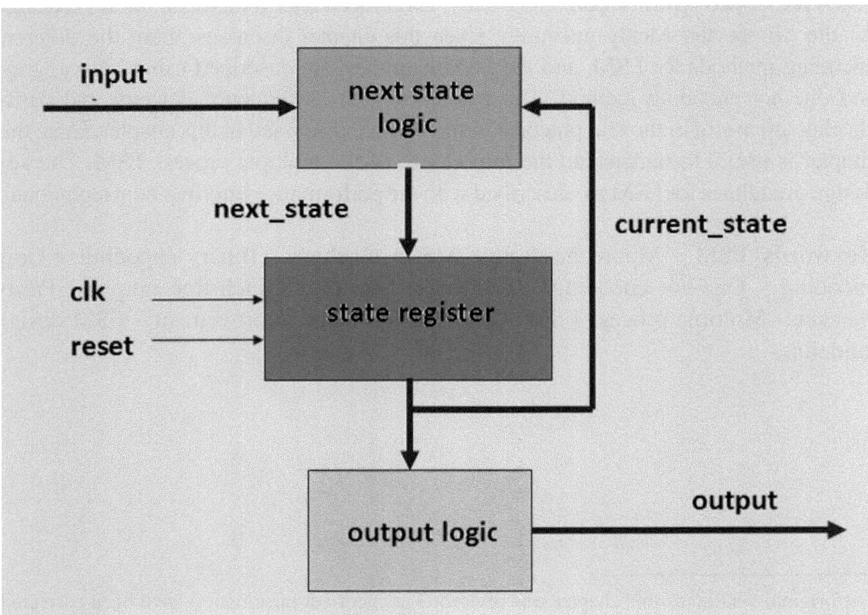

Fig. 9.1 Block diagram of Moore machine

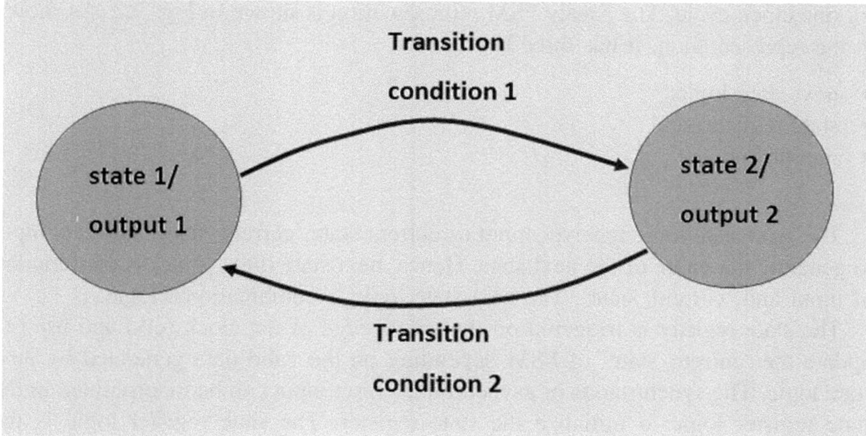

Fig. 9.2 State diagram representation of Moore machine

- next state logic,
- state register, and
- output logic.

The next state logic receives input as current state 'current_state' and data input to generate the next state. Hence, next state (next_state) is the function of input and 'current_state'. The next-state logic is combinational logic.

The state register is triggered on the active edge of the clock (clk) and used to update the 'current_state' of FSM depending on the valid data f. The synchronous or asynchronous reset input can be incorporated in the state register logic to initialize the state register. The state register logic is the sequential block triggered on the active edge of the clock.

Output logic is the combinational logic block, and in the Moore FSM, an output is the function of the current state and constant for one clock cycle.

The state diagram representation of Moore machine is shown in Fig. 9.2. The state diagram has two states: State 1 and State 2. Bubble indicates the state, and the transition from one state to other is indicated by the transition arc. As shown in the state diagram, every state has output and indicated by state 1/output 1 and state 2/output 2. Depending on the changes in the input or transition condition, the state transition occurs.

9.1.2 Mealy Machine

As stated earlier in the Mealy FSM, an output is the function of the current state (current_state) and present input (input). Hence, an output may or may not be stable for one clock cycle. The Mealy FSM representation is shown in Fig. 9.3. As shown in the representation, it has three key blocks:

- next state logic,
- state register, and
- output logic.

The next state logic receives input as current state 'current_state' and data input to generate the value of the next state. Hence, next state (next_state) is the function of input and 'current_state'. The next-state logic is combinational logic.

The state register is triggered on the active edge of the clock (clk) and used to update the 'current_state' of FSM depending on the valid data generated by next state logic. The synchronous or asynchronous reset input can be incorporated in the state register logic to initialize the state register. The state register logic is the sequential block triggered on the active edge of the clock.

Output logic is the combinational logic block, and in the Mealy FSM, an output is function of the current state and input and hence may or may not be constant for one clock cycle.

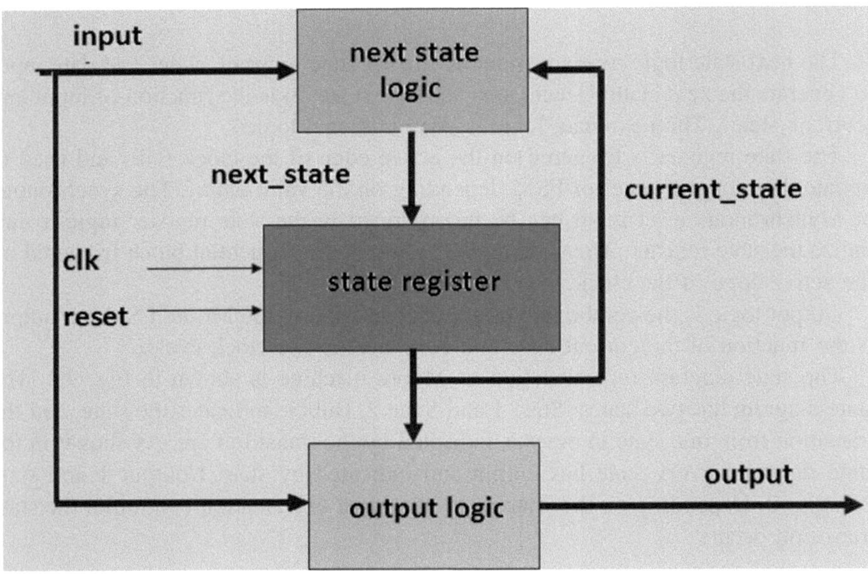

Fig. 9.3 Block diagram of Mealy machine

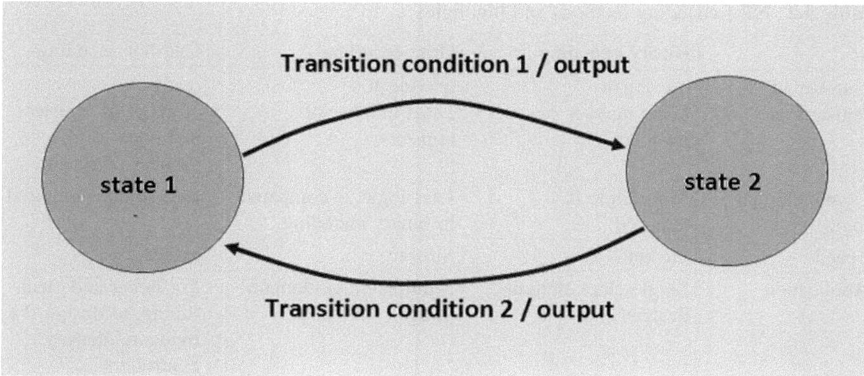

Fig. 9.4 State diagram representation of Mealy machine

The state diagram representation of Mealy machine is shown in Fig. 9.4. The state diagram has two states: State 1 and State 2. Bubble indicates the state, and the transition from one state to other is indicated by the transition arc. As shown in the state diagram state is indicated by State 1 and State 2. Depending on the changes in the input or transition condition, the state transition occurs. As output is function of the current state and input, transition arc shown indicates transition condition/ output.

9.2 FSM Encoding Methods

Depending on the requirement of the design functionality, the FSM can be described by using different encoding styles. The main FSM encoding styles are binary encoding, gray encoding, and one-hot encoding.

For the FSM having 8 states, the state encoding is shown in Table 9.1.

Table 9.1 FSM encoding methods

State	Binary encoding	Gray encoding	One-hot encoding
State 0 (s0)	000	000	00000001
State 1 (s1)	001	001	00000010
State 2 (s2)	010	011	00000100
State 3 (s3)	011	010	00001000
State 4 (s4)	100	110	00010000
State 5 (s5)	101	111	00100000
State 6 (s6)	110	101	01000000
State 7 (s7)	111	100	10000000

Table 9.2 FSM encoding methods and highlights

	Binary encoding	Gray encoding	One-hot encoding
Number of registers	$q = \log_2 n$ Least number of registers	$q = \log_2 n$ Least number of registers	$q = n$ Number of registers and equal to the number of states
Combinational logic	More logic is required	Less logic as compared to binary encoding	Less logic is required
Speed	Slower	Slower	Faster
Application	Single-clock-domain designs	Multiple-clock-domain designs	For better and clean timing, to design the moderate density controllers
Debugging	Difficult to debug	Difficult to debug	Easy to debug

As shown in Table 9.1, the number of flip-flops required for FSM using the binary and gray encoding are same. For eight-state FSM, the numbers of flip-flops required for the binary and gray encoding are equal to 3. Hence, if ‘n’ are number of states and ‘q’ are number of flip-flops, then the relationship between the number of states and number of flip-flops is described as follows: $q = \log_2 n$.

In the one-hot encoding method, as only one bit is logic ‘1’ or hot at a time, the number of flip-flops required for the state machine implementation is same as that of number of states. That is, $q = n$, and hence, these types of machines need more sequential elements as compared to the binary/gray encoding methods.

The key highlights of binary, gray, and one-hot encoding FSM are described in Table 9.2.

The real objective of an RTL design engineer is to design an FSM using one of the encoding styles discussed above. The default encoding style is binary encoding, but due to the following advantages, the one-hot encoding is popular.

- One-hot encoding state machines are faster, and the operation speed is dependent on the number of state transitions.
- Easily synthesized and can be easily described using VHDL to achieve better and clean timing performance.
- Addition and deletion of the states can be easily incorporated without affecting the remaining states.
- Easy to design as the RTL using VHDL can be directly written from the state diagram.
- These kinds of machines are easy to debug.

The subsequent session discusses about the RTL description using VHDL for the state machine depending on the encoding method.

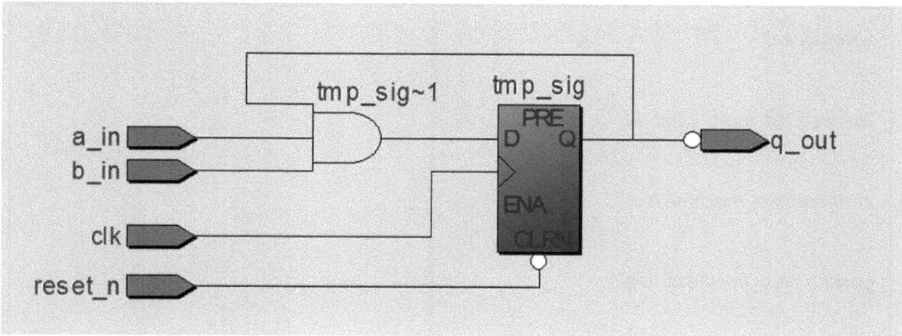

Fig. 9.5 Moore machine sequential circuit

9.3 How to Code Moore FSM Using VHDL?

In the previous few chapters, we have discussed about the RTL design using VHDL constructs. In the practical design scenario, the FSM can be described using single-process block or by using multiple process blocks. Let us consider the design shown Fig. 9.5.

As shown in Fig. 9.5, an output 'q_out' is the function of the 'tmp_sig' that is q_out = not tmp_sig. So let us consider 'tmp_sig' as 'current_state'. The data at 'D' input of register is the function of the changes in the input 'a_in, b_in' and 'current_state (tmp_sig)'. So let us consider the output of AND gate as 'next_state'. The RTL is described using the VHDL constructs and shown in Example 9.1. As output is the function of the 'current_state', only these kind of machines are called as Moore machine.

As shown in Example 9.1, the design functionality is described using the single process block. In the RTL description, it is assumed that next state is the function of the inputs 'a_in, b_in and previous output from register' that is 'tmp_sig'. An output 'q_out' is the function of the 'current_state (tmp_sig)'. But this code has less readability, and it does not give the meaningful information regarding the next-state logic and state register logic. In the above VHDL code, single process is used to describe the next-state logic and state register logic. These kind of coding styles are difficult to debug and inefficient as per as timing, and performance is concern. So it is essential to evolve the efficient technique to describe the Moore FSM. It is recommended to use two- or three-process block FSM to describe the state machines. This improves the readability, and even the debugging also becomes easy. The subsequent sessions discusses about the coding of the Moore FSM using multiple process blocks.

```
library ieee;

use ieee.std_logic_1164.all;

entity moore_machine is

port ( a_in,b_in : in std_logic;

        clk, reset_n : in std_logic;

         q_out : out std_logic);

end moore_machine;

architecture arch_moore_machine of moore_machine is

signal tmp_sig : std_logic;

begin

p1_register:  process (clk, reset_n)

 begin

  if ( reset_n='0') then

    tmp_sig <= '0';

 elsif (clk='1' and clk'event) then

    tmp_sig <= a_in and b_in and tmp_sig;

  end if;

  end process;    q_out<= not tmp_sig;  end arch_moore_machine;
```

- Architecture defines the functionality of design.
- FSM has single state tmp_sig.
- The process is described for the state register logic and sensitive to ‘clk’ and ‘reset_n’.
- For active low ‘reset_n’ the default state is logic ‘0’.
- For rising edge of clock ‘clk’ the ‘current_state’ is assigned as ‘tmp_sig’
- The next_state is function of present input ‘a_in’, ‘b_in’ and ‘curent_state’.
- Output is function of ‘current_state’ only.

Example 9.1 Single-process block VHDL RTL to describe Moore machine

9.3.1 *FSM Design Template for Moore Machine*

Figure 9.6 shows the basic template used to code the Moore FSM. In the practical ASIC design and prototyping using FPGAs, the multiple-process block FSM is used. The process block 'next_state_logic' is used to describe the next-state logic and sensitive to the 'current_state' and input. The process block 'state_register' is sensitive to clock 'clk' and reset 'reset_n' and used to describe the state update depending on the output of the next state logic. The process block 'output_logic' is sensitive to the 'current_state' and describes the output generation. Output logic infers the combinational logic.

9.4 How to Code Mealy FSM Using VHDL?

In the Mealy FSM, an output is the function of the change in the input and current state. In the practical design scenario, the FSM can be described using single process or by using multiple processes. Let us consider the design shown in Fig. 9.7.

As shown in Fig. 9.7, an output 'q_out' is function of the 'tmp_sig' and c_in that is q_out = c_in OR tmp_sig. So let us consider 'tmp_sig' as 'current_state'. The data at 'D' input of register is the function of the changes in the input 'a_in, b_in', and 'current_state (tmp_sig)'. So let us consider the output of AND gate as 'next_state'. The RTL is described using the VHDL constructs and shown in Example 9.2. As output is a function of the 'current_state' and changes in one of the input, these kinds of machines are called as Mealy machine.

As shown in Example 9.2, the design functionality is described using the multiple processes. In the RTL description, it is assumed that next state is function of the inputs 'a_in, b_in, and previous output from register' that is tmp_sig. An output 'q_out' is function of the 'current_state (tmp_sig)' and 'c_in'. But this code has less readability, and it does not give the meaningful information regarding the next state logic and state register logic. In the above VHDL code, single process is used to describe the next state logic and state register logic. These kind of coding styles are difficult to debug and inefficient as per as timing, and performance is concern. So it is essential to evolve the efficient technique to describe the Mealy FSM. It is recommended to use the three-process block FSM to describe the state machines. This improves the readability, and even the debugging also becomes easy. The subsequent sessions discuss about the coding of the Mealy FSM using multiple processes.

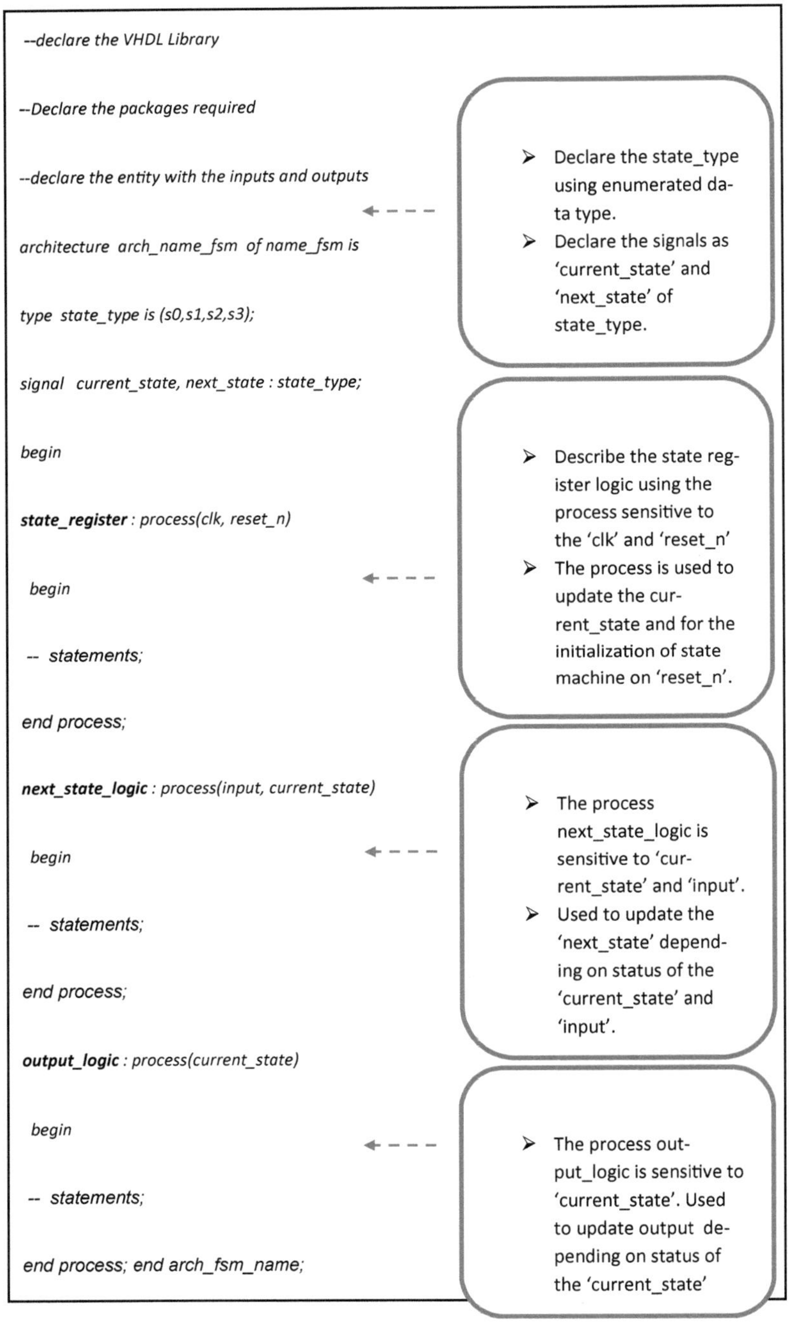

Fig. 9.6 FSM design template for Moore machine

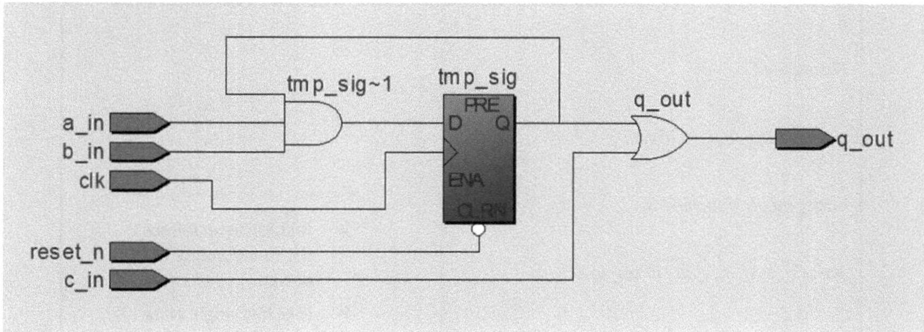

Fig. 9.7 Mealy machine sequential circuit

9.4.1 FSM Design Template for Mealy Machine

Figure 9.8 shows the basic template used to code the Mealy FSM. In the practical ASIC design and prototyping using FPGAs, the multiple-process block FSM is used. The process block 'next_state_logic' is used to describe the next state logic and sensitive to the 'current_state' and input. The process block 'state_register' is sensitive to clock 'clk' and reset 'reset_n' and describes the state update depending on the output generated by the next state logic. The process block 'output_logic' is sensitive to the 'current_state' and 'input'. The next state logic and output logic are combinational processes.

9.5 FSM Examples and VHDL Coding

This section discusses about the FSM examples and efficient coding using VHDL. Most of the FSMs in this section are coded using multiple processes. In some practical scenarios, the same process can be used for the next-state logic and output logic. So the FSM can have two processes: process 'state_register' for updating of the 'current_state' and another process for the 'next_state_logic plus output_logic'. Following are key FSM design guidelines:

- Binary encoding techniques are efficient for a design having 16 or fewer states. As number of states increases, the next state combinational logic performs slower operation.
- One-hot encoding technique is efficient and reliable as compared to the binary encoding due to glitch-free behavior. One-hot encoding requires low-density

```
library ieee;
use ieee.std_logic_1164.all;
entity mealy_machine is
port ( a_in,b_in, c_in : in std_logic;
     clk, reset_n : in std_logic;
     q_out : out std_logic);
end mealy_machine;
architecture arch_mealy_machine of mealy_machine is
signal  tmp_sig : std_logic;
begin
p1_register:  process (clk, reset_n)
 begin
 if ( reset_n='0') then
  tmp_sig <= '0';
 elsif (clk='1' and clk'event) then
  tmp_sig <= a_in and b_in and tmp_sig;
 end if;
 end process; p2_comb_logic: process(tmp_sig, c_in)
 begin
 q_out<= tmp_sig or c_in;
 end process;  end arch_mealy_machine;
```

- Architecture defines the functionality of design.
- FSM has single state tmp_sig.
- The process is described for the state register logic and sensitive to 'clk' and 'reset_n'.
- For active low 'reset_n' the default state is logic '0'.
- For rising edge of clock 'clk' the 'current_state' is assigned as 'tmp_sig'
- The current_state 'tmp_sig' is function of present input 'a_in', 'b_in' and 'curent_state'.

- The process 'p2_comb_logic' is sensitive to the 'current_state' of register that is 'tmp_sig' and input 'c_in'.
- Output is function of the 'tmp_sig' and input 'c_in' hence the FSM is Mealy machine.

Example 9.2 Two-process block VHDL RTL to describe Mealy machine

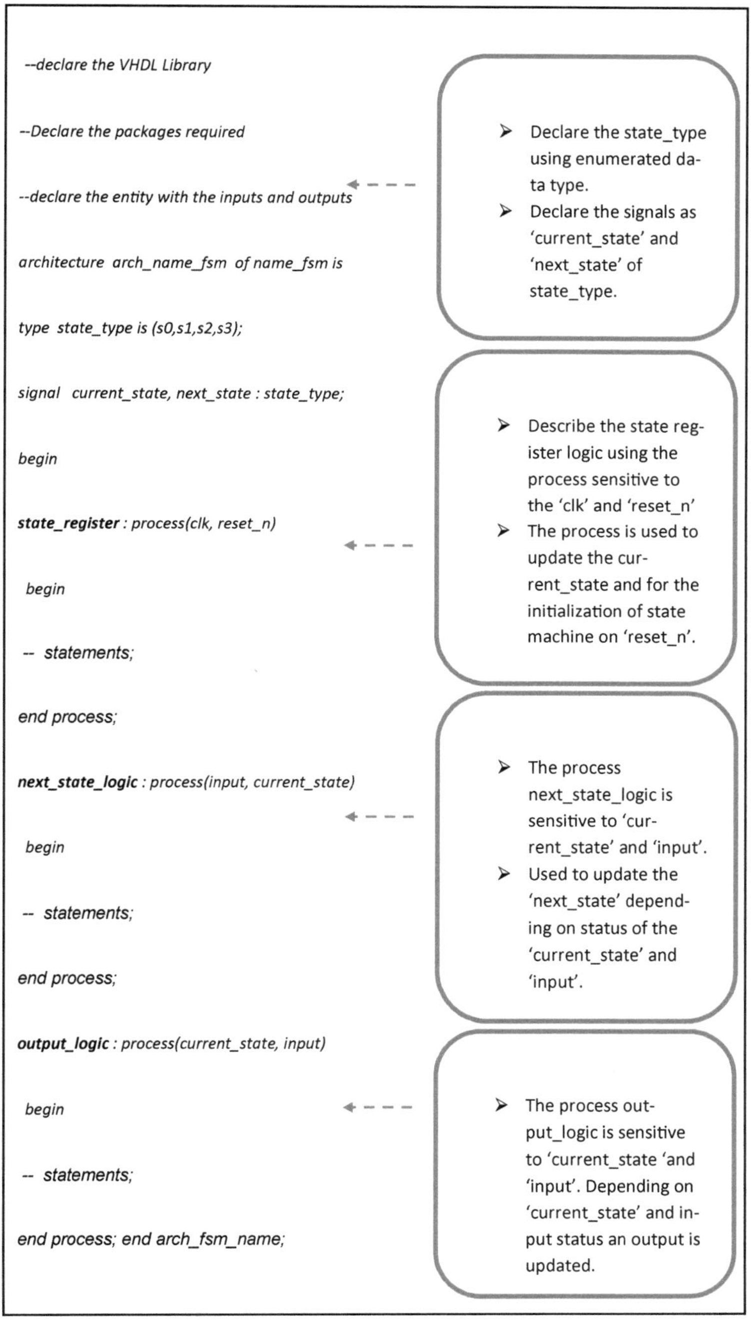

Fig. 9.8 FSM design template for Mealy machine

next-state logic and useful in design of larger FSM blocks. But the main drawback of one-hot encoding is that it uses more registers!

- While designing FSM, designer needs to take care of following key points:
  - Don't leave any undefined states. Initialize the unused states to reset value or use the default statements.
  - Don't implement the FSM with combination of registers and latches. Avoid the unintentional latches in the FSM design to improve the reliability.
  - Model the FSM blocks by using case statements to infer the parallel logic.
  - Separate the next state logic, output combinational logic, and state register logic in different processes to improve the speed of FSM and for better synthesis results.
  - Register FSM output as it preserves the hierarchy.
  - Use the look-ahead mealy machines for better design performance.

9.5.1 Binary Encoding FSM

Consider the following timing sequence, for the timing sequence, it is essential to describe the FSM using binary encoding method. As discussed earlier, binary encoding method uses the less number of registers but has slower speed.

As shown in Fig. 9.9, the design should have four states s0, s1, s2, and s3. The 'reset_in' is active high input, and during normal operation, it should be low. The state transition occurs on the rising edge of clock 'clk' provided that enable input 'enable_in' is active high. The synthesizable RTL using VHDL is shown in Example 9.3.

The synthesis result for Example 9.3 is shown in Fig. 9.10, and it generates two-bit binary counter.

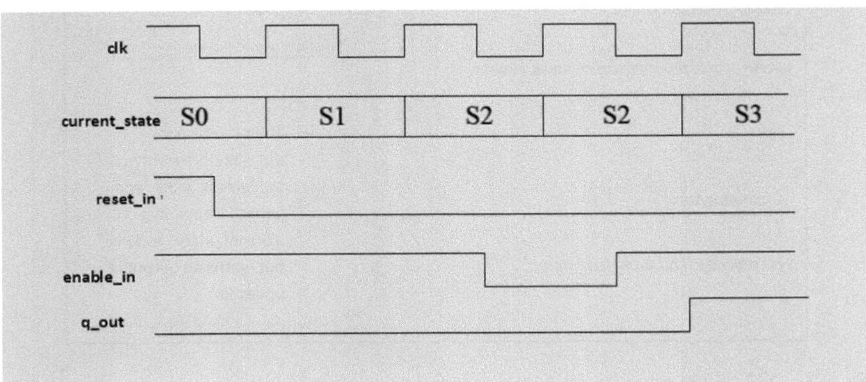

Fig. 9.9 Timing sequence for two-bit counter

```
library ieee;
use ieee.std_logic_1164.all;
entity counter_fsm is
port ( clk, reset_in, enable_in : in std_logic;
          q_out : out std_logic);
end counter_fsm;
architecture arch_counter_fsm of counter_fsm is
type state_type is (s0,s1,s2,s3);
signal current_state, next_state : state_type;
begin
state_register : process(clk, reset_in)
 begin
     if ( reset_in ='1') then
      current_state <= s0;
     elsif ( clk='1' and clk'event) then
      current_state <= next_state;
     end if;
end process;
```

- Architecture defines the functionality of design.
- FSM has four states s0,s1,s2,s3 . The current_state and next_state is defined as of type state_type
- The process is described for the state register logic and sensitive to 'clk'.
- For active low 'reset_n' the default state is 's0'.
- On the rising edge of clock 'clk' the 'next_state' is assigned to 'current_state'.
- The 'reset_in' is asynchronous signal.

Example 9.3 Two-process FSM for the binary counter

```
next_state_logic : process (current_state, enable_in)

begin

 case (current_state) is

    when s0 => if (enable_in='1') then

          next_state<=s1;    else

        next_state<= s0;

       end if;  q_out <= '0';

    when s1 => if (enable_in='1') then

          next_state<=s2;    else

        next_state<= s1;

       end if;   q_out <= '0';

    when s2 => if (enable_in='1') then

          next_state<=s3;    else

        next_state<= s2;

       end if;   q_out <= '0';

    when s3 => if (enable_in='1') then

          next_state<=s0;    else

        next_state<= s3;

       end if;  q_out <= '1';

   end case;    end process;    end arch_counter_fsm;
```

- The next state logic is combinational logic and it is described by the process 'Next_state_logic'.
- Process is sensitive to 'current_state' and 'enable_in'.
- For 'enable_in' value of the next_state is updated.
- The states are 'so,s1,s2,s3' and output is 'q_out'.
- Default state is 's0'.

- The output 'q_out' is function of the 'current_state' only and hence this type of machine is Moore machine.
- The q_out is assigned in the same process.

Example 9.3 (continued)

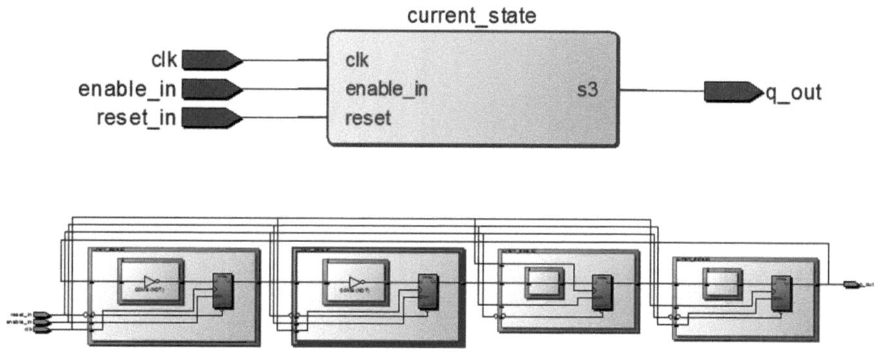

Fig. 9.10 Synthesis result with single output

9.5.2 Binary Counter FSM

The binary counter FSM is described by using the VHDL constructs and shown in Example 9.4. For more information on the binary counters, please refer Chap. 5.

Synthesis result for Example 9.4 is shown in Fig. 9.11. The counter is implemented using two registers and LUTs. LUTs are used to implement the combinational logic.

9.5.3 One-Hot Counter FSM

The one-hot counter FSM is described using VHDL and shown in Example 9.5. Only one output bit is active high or hot at a time, and hence, the desired logic should generate four registers for the four states.

The synthesis result is shown in Fig. 9.12, and as shown, it has four registers and output ‘q_out’ is 4 bit in size.

9.6 Parity Logic Using Moore FSM

In most of the practical design scenario, the Moore machines are used to detect the parity. Consider the one-bit data input ‘d_in’ to the machine, and the state diagram is represented in Fig. 9.13.

```
library ieee;

use ieee.std_logic_1164.all;

entity binary_counter_fsm is

port ( clk, reset_n, enable_in : in std_logic;

         q_out : out std_logic_vector( 1downto 0));

end binary_counter_fsm;

architecture arch_counter_fsm of binary_counter_fsm is

type state_type is (s0,s1,s2,s3);

signal  current_state, next_state : state_type;

begin

state_register : process(clk, reset_n)

 begin

      if ( reset_n ='0') then

         current_state <= s0;

      elsif ( clk='1' and clk'event) then

         current_state <= next_state;

      end if;    end process;
```

- Architecture defines the functionality of design.
- Architecture four states s0,s1,s2,s3 and current_state and next_state is defined as of type state_type
- The process is described for the state register logic.
- For active low 'reset_n' the default state is 's0'.
- For rising edge of clock 'clk' the 'next_state' is assigned to 'current_state'.

Example 9.4 Synthesizable VHDL RTL using binary encoding method

```
next_state_logic : process (current_state, enable_in)

begin

  case (current_state) is

      when s0 => if (enable_in='1') then

                  next_state<=s1;    else

                  next_state<= s0;

              end if;         q_out <= "00";

      when  s1 => if (enable_in='1') then

                  next_state<=s2;    else

             next_state<= s1;

             end if;          q_out <= "01";

      when s2 => if (enable_in='1') then

                  next_state<=s3;  else

             next_state<= s2;

             end if;       q_out <= "10";

      when s3 => if (enable_in='1') then

                 next_state<=s0;  else

            next_state<= s3;

            end if;  q_out <= "11";  end case; end process; end arch_counter_fsm;
```

- The next state logic is combinational logic and it is described by the process 'next_state_logic'.
- Process is sensitive to 'current_state' and 'enable_in'.
- For 'enable_in' is equal to logic '1' the next_state is updated.
- For binary encoding the two bit output is assigned.
- The states are 'so,s1,s2,s3' and output is 'q_out'.
- Output 'q_out' is updated in the same process and function of current_state. Hence this is Moore machine.

Example 9.4 (continued)

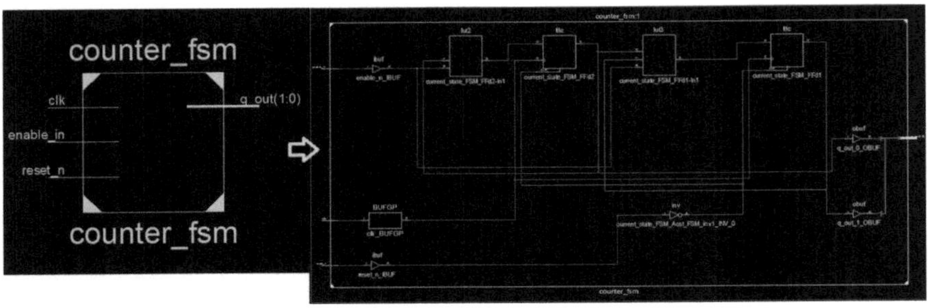

Fig. 9.11 Synthesis result for the binary encoding

The RTL using VHDL for the Moore machine can be described using three-process block FSM and discussed in this section.

9.6.1 Moore Machine: Three-Process Block FSM for Parity Checking

The RTL using VHDL for the Moore machine parity checking logic is described using the three-process FSM in Example 9.6. Process 'state_register' is used to update the 'current_state' value and for the FSM initialization. The process 'next_state_logic' is used to update the 'next_state', and the process 'output_logic' is used to assign output 'parity_out'.

The synthesis result for the parity checking logic for Example 9.6 is shown in Fig. 9.14.

As shown in Fig. 9.14, an output 'parity_out' is the function of the current_state only. As synchronous reset 'reset_n' is used, the reset logic in the data path adds the combinational delay.

9.7 Parity Logic Using Mealy FSM

As discussed earlier in this chapter, in the Mealy machine, output is the function of the current state and change in the input. Hence, output may or may not be stable for one clock-cycle duration. Mealy machine output is prone to glitches, and it is

```
library ieee;
use ieee.std_logic_1164.all;
entity counter_fsm is
port ( clk, reset_n, enable_in : in std_logic;
        q_out : out std_logic_vector(3 downto 0));
end counter_fsm;
architecture arch_counter_fsm of counter_fsm is
type state_type is (s0,s1,s2,s3);
signal  current_state, next_state : state_type;
begin
state_register : process(clk, reset_n)
 begin
     if ( reset_n ='0') then
        current_state <= s0;
     elsif ( clk='1' and clk'event) then
        current_state <= next_state;
     end if;
end process;
```

- Architecture defines the functionality of design.
- Architecture four states s0,s1,s2,s3 and current_state and next_state is defined as of type state_type
- The process is described for the state register logic.
- For active low 'reset_n' the default state is 's0'.
- For rising edge of clock 'clk' the 'next_state' is assigned to 'current_state'.

Example 9.5 Synthesizable VHDL RTL using one-hot encoding

```
next_state_logic : process (current_state, enable_in)

begin

  case (current_state) is

      when s0 => if (enable_in='1') then

                  next_state<=s1;  else

                  next_state<= s0;

               end if;    q_out <= "0001";

      when  s1 => if (enable_in='1') then

                  next_state<=s2;   else

            next_state<= s1;

            end if;      q_out <= "0010";

      when s2 => if (enable_in='1') then

                  next_state<=s3; else

            next_state<= s2;

            end if;     q_out <= "0100";

      when s3 => if (enable_in='1') then

                  next_state<=s0;  else

            next_state<= s3;

            end if;     q_out <= "1000"; end case; end process; end arch_counter_fsm;
```

- The next state logic is combinational logic and it is described by the process 'next_state_logic'.
- Process is sensitive to 'current_state' and 'enable_in'.
- For 'enable_in' is equal to logic '1' the next_state is updated.
- For one hot encoding the four bit output is assigned and only one bit is high or hot at a time.
- The states are 'so,s1,s2,s3' and output is 'q_out'.

Example 9.5 (continued)

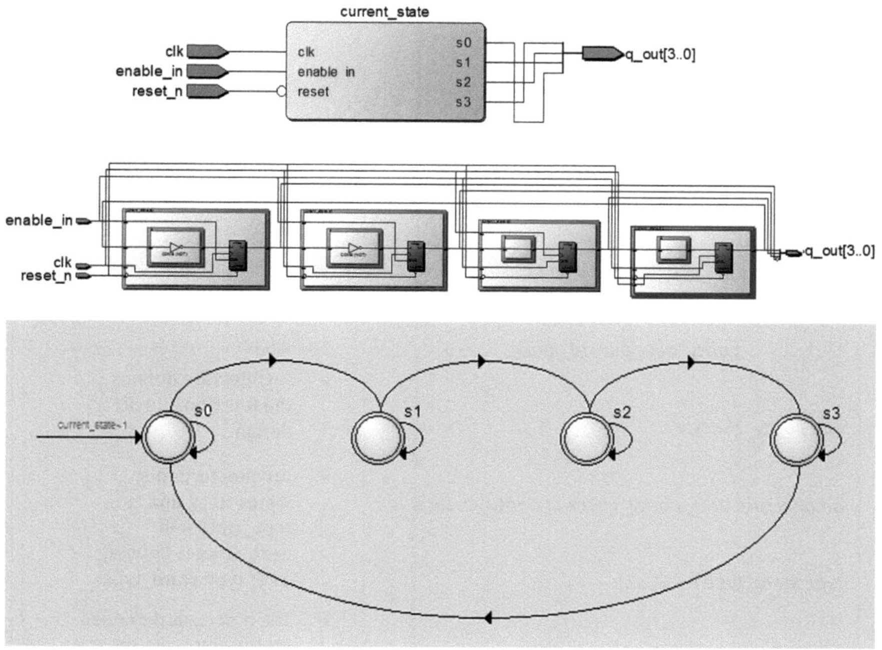

Fig. 9.12 Synthesis result and state diagram for one-hot encoding counter

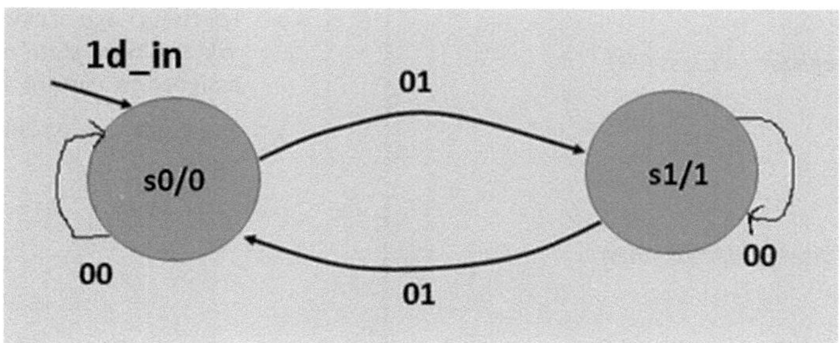

Fig. 9.13 Parity checking logic Moore state machine

essential to use the glitch suppression logic while coding for the Mealy machines. The parity checking logic for the state diagram shown in Fig. 9.15 can be described by using two or three processes.

```
library ieee;
use ieee.std_logic_1164.all;
entity parity_checker is
port ( clk, reset_n, d_in : in std_logic;
        parity_out : out std_logic);
end parity_checker;
architecture arch_parity_check of parity_checker is
type state_type is (s0, s1);
signal current_state, next_state : state_type;
begin
state_register : process (clk)
begin
    if rising_edge(clk) then
        if (reset_n = '0') then
            current_state <= s0;
        else
            current_state <= next_state;
    end if;     end if;   end process;
```

- Architecture defines the functionality of design.
- Architecture four states s0,s1 and current_state and next_state is defined as of type state_type
- The process is described for the state register logic and sensitive to 'clk'.
- For active low 'reset_n' the default state is 's0'.
- For rising edge of clock 'clk' the 'next_state' is assigned to 'current_state'.
- The 'reset_n' is synchronous signal.

Example 9.6 Synthesizable VHDL RTL for parity checking using Moore machine

```
output_logic : process (current_state)

begin

        case (current_state) is

        when S0 => parity_out <= '0';

        when S1 => parity_out<= '1';

        when others => parity_out <= '0';

      end case;

end process;
```

- The output logic is combinational logic and it is described by the process 'output_logic'.
- Process is sensitive to 'current_state' only
- An output is 'parity_out'. And assigned depending on the 'current_state'.
- Default output is '0'.

```
next_state_logic : process (current_state, d_in)

begin

          next_state <= S0;

          case (current_state) is

          when S0 => if (d_in = '1') then

                    next_state <= S1;

                    end if;

          when S1 =>if (d_in = '0') then

                    next_state <= S1;

                    end if;     when others => next_state <= S0;

end case;  end process; end arch_parity_check;
```

- The next state logic is combinational logic and it is described by the process 'next_state_logic'.
- Process is sensitive to 'current_state' and 'd_in'.
- For 'd_in' value, the next_state is updated.
- The states are 'so,s1' and output is 'parity_out'.
- Default state is 's0'.

Example 9.6 (continued)

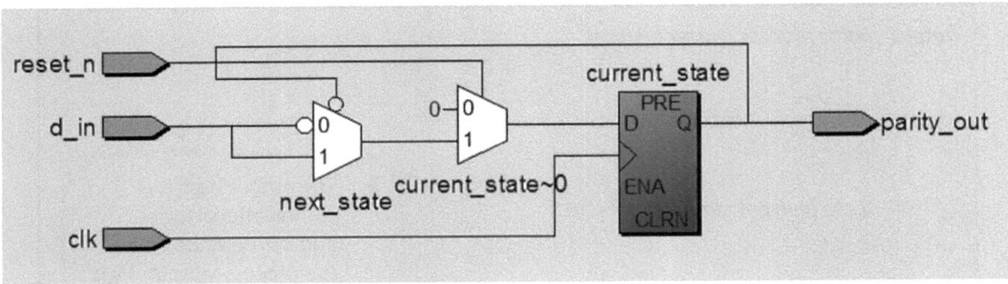

Fig. 9.14 Synthesis result for the Moore machine parity checking logic

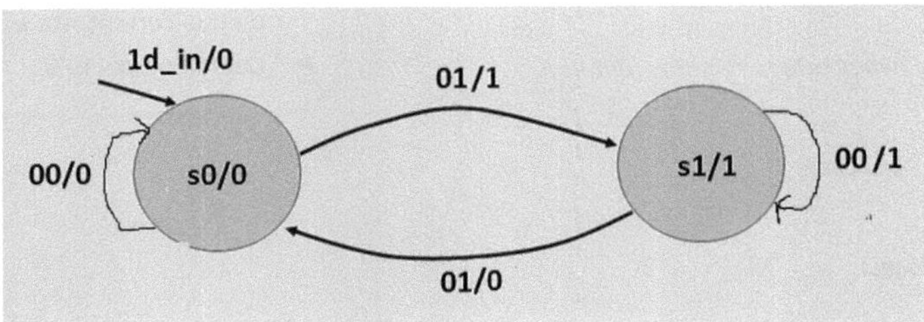

Fig. 9.15 State diagram representation of Mealy machine parity checking logic

9.7.1 Mealy Machine: Two-Process Block FSM for Parity Checking

The FSM for the parity checking logic shown in the Fig. 9.15 is described using two processes in Example 9.7. As shown in the RTL description using VHDL, the process block 'state_register' is used to update the 'current_state' and even to initialize the state machine to the default state. Default state is 's0'. The another procedure block 'next_state_logic' is used to update the 'next_state' and even to assign the output 'parity_out' depending on the status of 'current_state' and 'd_in'.

The synthesis result for the two process of FSM is shown in Fig. 9.16. As shown, an output 'parity_out' is function of the 'current_state' and data input 'd_in'.

9.7.2 Mealy Machine: Three-Process Block FSM for Parity Checker

For better readability and easy debugging, it is recommended to use the three-process FSM. The RTL using VHDL for the Mealy machine parity checker is

```
library ieee;

use ieee.std_logic_1164.all;

entity parity_checker is

port ( clk, reset_n, d_in : in std_logic;

                parity_out : out std_logic);

end parity_checker;

architecture arch_parity_check of parity_checker is

type  state_type is (s0, s1);

signal  current_state, next_state : state_type;

begin

state_register : process (clk)

begin

if rising_edge(clk) then

  if (reset_n = '0') then

   current_state <= s0;

  else

   current_state <= next_state;

end if;  end if; end process;
```

- Architecture defines the functionality of design.
- Architecture four states s0,s1 and current_state and next_state is defined as of type state_type
- The process is described for the state register logic and sensitive to 'clk'.
- For active low 'reset_n' the default state is 's0'.
- For rising edge of clock 'clk' the 'next_state' is assigned to 'current_state'.
- The 'reset_n' is synchronous signal.

Example 9.7 VHDL RTL for Mealy machine parity checking logic using two processes

```
Next_state_logic : process (current_state, d_in)
begin
    parity_out <= '0';
  case (current_state) is
  when s0 =>  if (d_in = '1') then
                parity_out <= '1';
                next_state <= s1;    else
                next_state <= s0;
             end if;
  when s1 => if (d_in= '1') then
               next_state <= s0;    else
               parity_out <= '1';
              next_state <= s1;
             end if;
  when others => next_state <= s0;
end case;
end process;    end arch_parity_check;
```

- The next state logic is combinational logic and it is described by the process 'Next_state_logic'.
- Process is sensitive to 'current_state' and 'd_in'.
- For 'd_in' value, the next_state is updated.
- The states are'so,s1' and output is 'parity_out'.
- Default state is 's0'.

Example 9.7 (continued)

```
library ieee;
use ieee.std_logic_1164.all;
entity parity_checker is
port ( clk, reset_n, d_in : in std_logic;
              parity_out : out std_logic);
end parity_checker;
architecture arch_parity_check of parity_checker is
type  state_type is (s0, s1);
signal  current_state, next_state : state_type;
begin
state_register : process (clk)
begin
if rising_edge(clk) then
  if (reset_n = '0') then
   current_state <= s0;
  else
   current_state <= next_state;
end if; end if; end process;
```

- Architecture defines the functionality of design.
- Architecture four states s0,s1 and current_state and next_state is defined as of type state_type
- The process is described for the state register logic and sensitive to 'clk'.
- For active low 'reset_n' the default state is 's0'.
- For rising edge of clock 'clk' the 'next_state' is assigned to 'current_state'.
- The 'reset_n' is synchronous signal.

Example 9.8 VHDL RTL for Mealy machine parity checking logic using three processes

```
Next_state_logic: process (current_state, d_in)
        Begin    next_state <= S0;
        case (current_state) is
        when S0 => if (d_in = '1') then
                next_state <= S1;    end if;
        when S1 =>if (d_in= '0') then
                next_state <= S1;    end if;
        when others => next_state <= S0;
    end case;    end process;
Output_decode_logic: process (current_state, d_in)
        Begin   parity_out <= '0';
        case (current_state) is
        when S0 => if (d_in = '1') then
                parity_out <= '1';    end if;
        when S1 => if (d_in = '0') then
                parity_out <= '1';  end if;
        when others =>parity_out <= '0';
        end case;
end process;   end arch_parity_check;
```

- The next state logic is combinational logic and it is described by the process 'Next_state_logic'.
- Process is sensitive to 'current_state' and 'd_in'.
- For 'd_in' value the next_state is updated.
- The states are 'so,s1'
- Default state is 's0'.

- The output decode logic is combinational logic and it is described by the process 'Out put_decode_logic'.
- Process is sensitive to 'current_state' and 'd_in'.
- Depending on the status of 'd_in' and 'cur-rent_state' an output 'parity_out' is assigned.
- The FSM described is mealy machine as output is function of 'current_state' and 'd_in'.

Example 9.8 (continued)

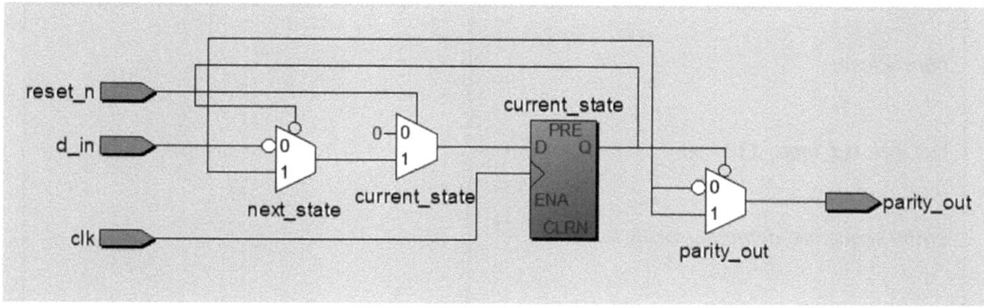

Fig. 9.16 Synthesis result for the Mealy machine parity checking logic

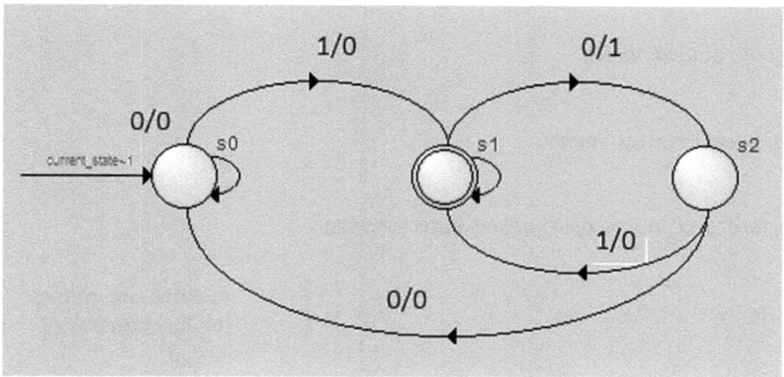

Fig. 9.17 Mealy machine sequence detector state diagram

described using the three-process FSM in the Example 9.8. Process 'state_register' is used to update the 'current_state' value and for the FSM initialization. The process 'next_state_logic' is used to update the 'next_state', and the process 'output_logic' is used to assign output 'parity_out'.

The synthesis result is shown in Fig. 9.16. As shown, the output 'parity_out' is the function of the 'current_state' and data input 'd_in'.

9.8 Sequence Detector Mealy Machine

In most of the practical scenarios, it is essential to design logic circuit to check the required sequence. For example, if we consider the data input 'd_in' with input sequence as '11010100101', from this serial input we wish to detect the sequence '10'.

```
library ieee;

use ieee.std_logic_1164.all;

entity sequence_detector_mealy is

port ( d_in : in std_logic;

reset_n, clk : in std_logic;

    q_out : out std_logic);

end sequence_detector_mealy;

architecture  arch_mealy of sequence_detector_mealy is

type state_type is (s0,s1,s2);

signal  current_state, next_state : state_type;

begin

state_register : process(clk, reset_n)

begin

if (reset_n='0') then

current_state <= s0;

elsif (clk='1' and clk'event)then

current_state <= next_state;

end if;   end process state_register;
```

- Architecture defines the functionality of design.
- FSM has three states s0, s1,s2. The current_state and next_state is defined as of type state_type
- The process is described for the state register logic and sensitive to 'clk' and 'reset_n'.
- For active low 'reset_n' the default state is 's0'.
- For rising edge of clock 'clk' the 'next_state' is assigned to 'current_state'.
- The 'reset_n' is asynchronous signal.

Example 9.9 Synthesizable VHDL RTL for Mealy machine sequence detector

```
comb_fsm: process (d_in, current_state)

begin

case  current_state is

when s0=> if (d_in='0') then

        q_out<='0';       next_state <= s0;

        else

        q_out<='0';     next_state <= s1;

        end if;

when s1=> if (d_in='0') then

        q_out<='1';      next_state <=s2;

        else

        q_out<='0';      next_state <= s1;

        end if;

when s2=> if (d_in='0') then

        q_out<='0';     next_state <=s0;

        else

        q_out<='0';     next_state <= s1;

        end if;  when others => current_state<= s0;

end case;  end process comb_fsm;  end arch_mealy;
```

- The next state and output logic is combinational logic and it is described by the process 'comb_fsm'.
- Process is sensitive to 'current_state' and 'd_in'.
- Depending on 'd_in' value and 'current_state' the next_state is updated.
- The states are 'so,s1,s2' and output is 'q_out'.
- Default state is 's0'.

Example 9.9 (continued)

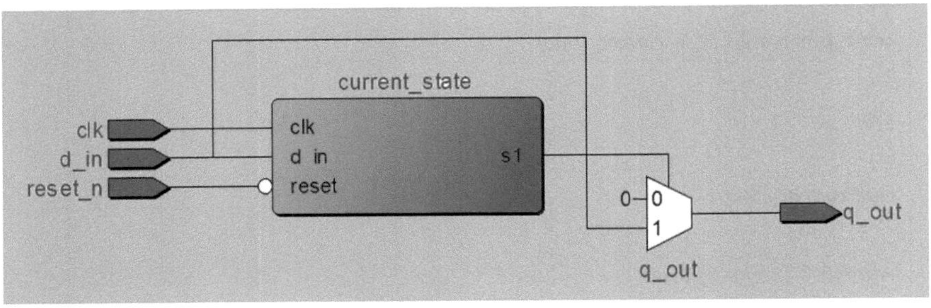

Fig. 9.18 Synthesis result for binary encoded sequence detector

When '10' sequence is detected, then an output 'q_out' should be high. So the output 'q_out' for the above-stated data input should be '00101010010'. To design these kinds of detectors, the FSM using VHDL constructs can be used. The state diagram of sequence detector for the '10' sequence is shown in Fig. 9.17.

As shown in Fig. 9.17, using binary encoding method, the number of states required to detect the sequence '10' are equal to 3 and named as 's0, s1, s2'. The RTL using VHDL for Fig. 9.17 is shown in Example 9.9. The synthesis result is shown in Fig. 9.18.

9.9 One-Hot Encoding Sequence Detector: Moore Machine

For the state diagram shown in Fig. 9.17, the sequence detector using one-hot encoding method is shown in Example 9.10. The synthesis result is shown in Fig. 9.19.

As shown in the above figure, the one-hot encoding state machine uses more register as compared to binary encoding method.

9.10 One-Hot Encoding Sequence Detector: Mealy Machine

To detect the '10' sequence from the input stream 'd_in', the RTL using two-process FSM is described in Example 9.11. The synthesis result is shown in Fig. 9.20.

```
library ieee;
use ieee.std_logic_1164.all;
entity moore_one_hot is
port(d_in,clk,reset_n: in std_logic;
q_out: out std_logic);
end moore_one_hot;
architecture moore_fsm of moore_one_hot is
subtype state_type is std_logic_vector (2 downto 0);
signal state : state_type;
constant s0: state_type:="001";
constant s1: state_type:="010";
constant s2: state_type:="100";
signal current_state, next_state : state_type;
begin
state_register: process (clk, reset_n)
begin
if (reset_n='0') then
current_state <= s0 ;
elsif (clk='1' and clk'event)then
current_state<= next_state;   end if ; end process state_register ;
```

- Architecture defines the functionality of design.
- FSM has three states s0,s1,s2. The current_state and next_state is defined as of type state_type. The encoding method is one-hot.
- The process is described for the state register logic and sensitive to ‘clk’ and ‘reset_n’.
- For active low ‘reset_n’ the default state is ‘s0’.
- For rising edge of clock ‘clk’ the ‘next_state’ is assigned to ‘current_state’.
- The ‘reset_n’ is asynchronous signal.

Example 9.10 Synthesizable RTL for sequence detector using one-hot encoding method

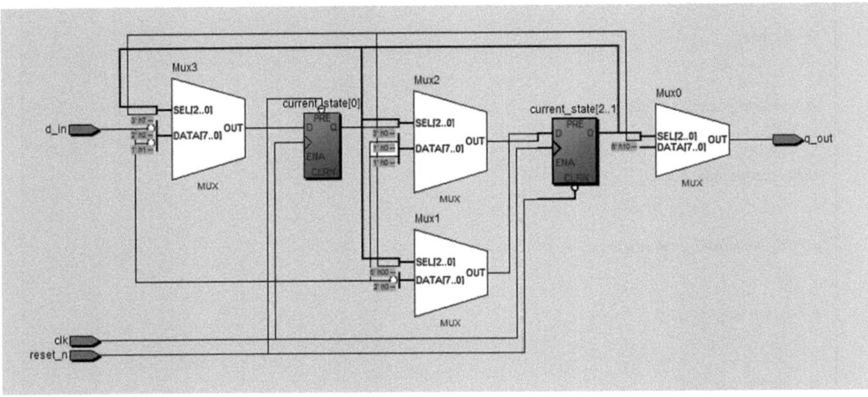

Fig. 9.19 Synthesis result for sequence detector using one-hot encoded Moore FSM

9.11 FSM Optimization

The coded FSM using VHDL can be optimized for the area and performance improvement. Following are the key concepts and can be used during the FSM optimization.

- The design partitioning plays important role while describing the RTL using VHDL. The design is partitioned in such a way that FSM should be individual block. If FSM is used with other logic, then due to poor partitioning, the design performance and area may not be optimum.
- Optimize the state machine by isolating the other logic if the design is not partitioned properly.
- Use the one-hot encoding FSM for better timing results and for glitch-free output.
- If the design consists of the multiple FSMs, then use the separate VHDL description for every FSM.
- Use the FSM compiler for the better design partitioning and to extract the states in the form of state table.
- To code the FSM with glitch free output, ensure that all the outputs are coming out from the flip-flop.
- Use the additional logic circuit as look-ahead output circuit for the Moore machine.
- Use the proper encoding style and try to optimize the register-to-register path delays.
- Improve the overall timing performance of FSM by reducing the combinational delay encountered in the next-state logic. The overall operating frequency of FSM is 1/T, where T is equal to $t_{ctoq} + t_{combo} + t_{su.}$ If t_{combo} is reduced, then the FSM can have improved performance.

```
comb_fsm: process (d_in,current_state)
begin
case current_state is
when s0=> q_out <= '0';
if (d_in='0') then next_state <=s0;
else  next_state <= s1;
end if;
when s1=> q_out <= '0';
if (d_in='0') then  next_state <= s2;
else  next_state <= s1;
end if;
when s2=> q_out <= '1';
if (d_in='0') then next_state <= s0;
else  next_state <= s1;
end if;
when others => q_out <='0'; next_state <= s0;
end case; end process comb_fsm;  end moore_fsm;
```

- The next state and output logic is combinational logic and it is described by the process 'comb_fsm'.
- Process is sensitive to 'current_state' and 'd_in'.
- Depending on 'd_in' value and'current_state' the next_state is updated.
- The states are 'so,s1,s2' and output is 'q_out'.
- Default state is 's0'.

Example 9.10 (continued)

```
library ieee;

use ieee.std_logic_1164.all;

entity mealy_one_hot is

port(d_in,clk,reset_n: in std_logic;

q_out: out std_logic);

end mealy_one_hot;

architecture mealy_fsm of mealy_one_hot is

subtype state_type is std_logic_vector (2 downto 0);

signal state : state_type;

constant s0: state_type:="001";

constant s1: state_type:="010";

constant s2: state_type:="100";

signal current_state, next_state : state_type;

begin

state_register: process (clk, reset_n)

begin

if (reset_n='0') then

current_state <= s0 ;

elsif (clk='1' and clk'event)then

current_state<= next_state;  end if ;  end process state_register ;
```

- ➢ Architecture defines the functionality of design.
- ➢ FSM has three states s0,s1, and s2. The current_state and next_state is defined as of type state_type.The encoding method is one-hot.
- ➢ The process is described for the state register logic and sensitive to 'clk' and 'reset_n'.
- ➢ For active low 'reset_n' the default state is 's0'.
- ➢ For rising edge of clock 'clk' the 'next_state' is assigned to 'current_state'.
- ➢ The 'reset_n' is asynchronous signal.

Example 9.11 Synthesizable RTL for Mealy sequence detector using one-hot encoding method

```
comb_fsm: process (d_in,current_state)

begin

case current_state is

when s0=>  if (d_in='0') then next_state <=s0; q_out <= '0';

               else next_state <= s1; q_out <= '0';

                 end if;

when s1=>  if (d_in='0') then next_state <= s2; q_out <= '1';

                 else next_state <= s1; q_out <= '0';

               end if;

when s2=>  if (d_in='0') then next_state <= s0; q_out <= '0';

               else next_state <= s1; q_out <= '0';

              end if;

when others => q_out <='0';

next_state <= s0;

end case;

end process comb_fsm; end mealy_fsm;
```

- The next state and output logic is combinational logic and it is described by the process 'comb_fsm'.
- Process is sensitive to 'current_state' and 'd_in'.
- Depending on 'd_in' value and 'current_state' the next_state is updated.
- The states are 'so,s1,s2' and output is 'q_out'.
- Default state is 's0'.
- Output is function of 'current_state' and 'd_in' hence the FSM is mealy type.

Example 9.11 (continued)

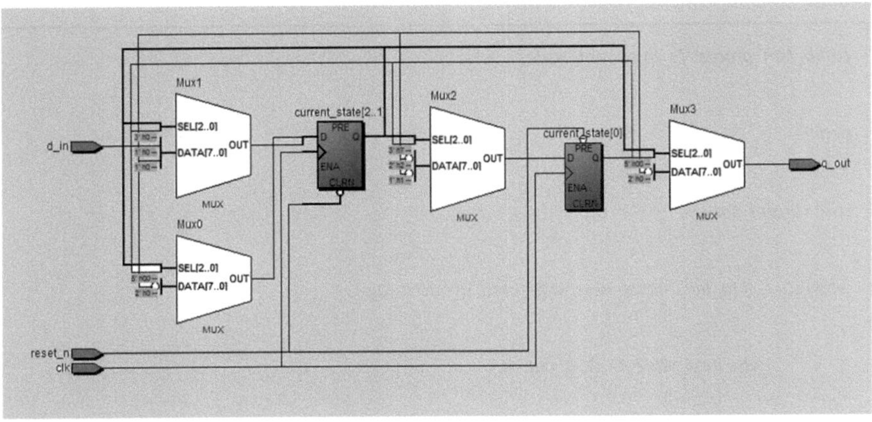

Fig. 9.20 Synthesis result for sequence detector using one-hot encoded Mealy FSM

Chapter 10 discusses about the complex examples, few design performance improvement techniques, and implementation using FPGA.

9.12 Summary

Following are the important points to summarize this chapter:

1. FSM is a source synchronous sequential circuits and of two types Moore and Mealy.
2. FSM encoding methods are binary, gray, and one-hot encoding.
3. Binary and gray encoding methods uses the number of registers equal to $\log_2 n$ where n are number of states.
4. One-hot encoding method uses the number of registers equal to number of states in the machine.
5. The overall FSM speed is dependent on the register-to-register path, and the 'T' is equal to $t_{ctoq} + t_{combo} + t_{su}$.
6. In the Moore FSM, the output is the function of current state only. The look-ahead output circuit can be used for the glitch-less output.
7. In the mealy machine, the output is the function of the current state and input.
8. For better timing, one-hot encoding can be used.

Chapter 10
Synthesis Optimization Using VHDL

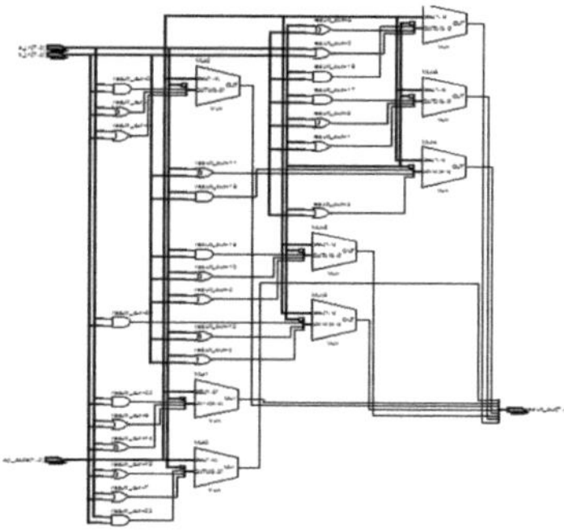

"I have no special talent. I am only passionately curious..." --- Albert Einstein

While writing the VHDL RTL use the synthesis optimization techniques for better performance of the design!

Abstract The PLD-based designs can be described by using concurrent and sequential VHDL constructs. In the practical scenario, the objective is to describe the design functionality by using synthesizable VHDL constructs and that can be accomplished by using important combinational and sequential design guidelines. This chapter focuses on the designs such as ALU, parity checkers, generators, memories, multipliers, and barrel shifters. This chapter also discusses about the synthesis result with the data path and control paths. The synthesis optimization techniques are discussed for the better synthesis outcome and used during RTL design cycle. This chapter is useful for ASIC and FPGA designers to understand the design using VHDL, critical paths and optimizations, and registered inputs and outputs. Even this chapter discusses about the synthesis outcome using Altera and Xilinx PLDs.

Keywords ALU · Logic unit · Arithmetic unit · Data path · Control path · Parity checker · Parity generator · Combinational shifter · Protocol · Registered input · Registered output · Barrel shifter · DSP · Synthesis · PLD · Altera · XILINX · Multiplier · BRAM · Single-port RAM · Dual-port RAM

The original version of this chapter was revised: The incorrect information have been corrected. The erratum to this chapter is available at 10.1007/978-981-10-3296-7_12

V. Taraate, *PLD Based Design with VHDL*,
DOI 10.1007/978-981-10-3296-7_10

As discussed in the previous chapters, VHDL can be efficiently used to code the functionality of the design. The concurrent and sequential constructs discussed in the previous chapters can be used to infer the synthesized logic. In the practical programmable ASIC designs, the design functionality is complex and needs to be described by using the synthesizable VHDL constructs to infer the gate-level netlist and to have the optimal design performance. Most of the programmable ASIC and SOCs uses the processors, buses, arbiters, and protocols (predefined set of rules or transactions). An efficient VHDL coding is an important aspect while describing the functionality of the above blocks. In such scenarios, ASIC designer should use the synthesizable constructs with combinational and sequential design guidelines.

The subsequent section discusses about the efficient designs using VHDL and practical scenarios while describing the processor computational logic, barrel shifters, parity generators, checkers, multipliers, and memories.

10.1 FPGA Design Flow

FPGA design flow includes the following key steps and described in Fig. 10.1:

1. Design entry,
2. Design simulation and synthesis,
3. Design implementation, and
4. Device programming.

These design steps are explained in the following section:

10.1.1 Design Entry

Before the design entry, the design planning need to be done by using the design specifications. The design specifications need to be converted to the architecture and microarchitecture. The design architecture and microarchitecture is design representation of the functionality into small modules to realize the intended functionality. During the architecture design phase, the requirement of memory, speed, and power needs to be estimated. Depending on the requirement, the FPGA device needs to be chosen for the implementation.

Design entry is done by using either Verilog (.v) or VHDL (.vhd) file. After the design entry, the design needs to be simulated for the functional correctness of the design. This is called as functional simulation.

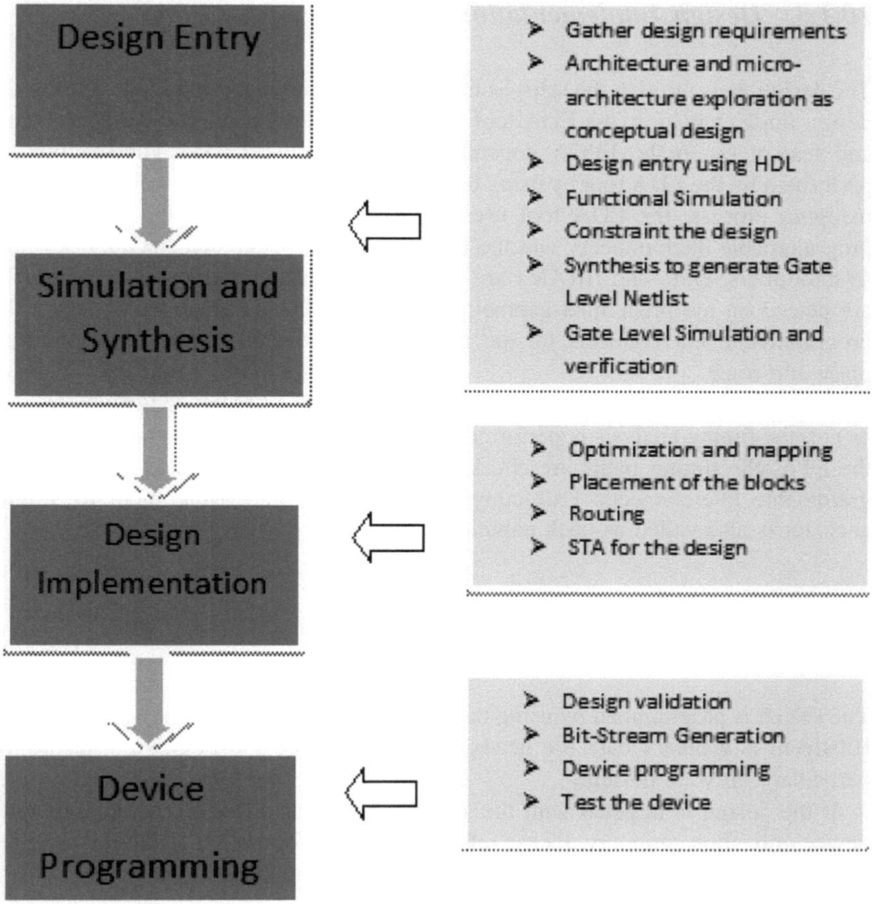

Fig. 10.1 FPGA design flow

10.1.2 Design Simulation and Synthesis

During the functional simulation, the set of inputs are applied to the design to check the functional correctness of the design. Although the timing or area and power issues can crop up during the later design cycle, but designer is at least sure about the functionality of the design.

The major goal of the hardware design engineer is to generate the efficient hardware. The synthesis is the process of converting one level of the design abstraction into the other level. In the logic synthesis, the HDL is converted into the netlist. The netlist is device independent and can be in the standard format like electronic design interchangeable format (EDIF).

10.1.3 Design Implementation

The design goes through the steps as translate, map, and place and route. During the design implementation, the EDA tool translates the design into the required format and map it on to the FPGA depending on the required area. The mapping is performed by the EDA tool by using the actual logic cells or macrocells. During the mapping process, the EDA tool uses the macrocells, configurable logic blocks, programmable interconnects, and the IO blocks. The special dedicated blocks such as multipliers, DSP, and BRAMs are also mapped using vendor tools. The blocks are placed on the predefined geometry inside the FPGA and routed by using the programmable interconnects for the intended functionality. The step is called as place and route.

To check for the design timing performance and whether the constraints are met or not, the timing analysis is performed and it is called as post-layout STA. During the STA, the timing paths are checked with the delays associated with the programmable interconnects. Extracting the RC delays and using them by timing analyzer is also called as back annotation.

10.1.4 Device Programming

The FPGA is programmed by using the vendor-specific or proprietary bitstream file. Bitstream is a binary data file needs to be loaded into the FPGA to execute the particular hardware design.

If the design is targeted with the specific FPGA, then the EDA tool generates device utilization summary. Please refer Appendix B for the XILINX Spartan series devices and Appendix C for the Altera Cyclone II and IV devices.

10.2 Synthesis Optimization Techniques

Before discussion on the synthesis and performance improvement, let us understand the different synthesis techniques used for the optimization. The optimization can be performed at the code level or during the synthesis. The fully optimized design is that which has met the area and timing requirements. The optimization at the RTL level can be achieved by modifying the code to meet the intended functionality. In such type of optimizations, care needs to be taken that the optimized code should have the same simulation results before and after synthesis. But there are few standard techniques used in the real practical scenarios to have better synthesis optimizations and results. Few of such techniques are discussed in this section.

10.2.1 Resource Allocation

This is used for the better synthesis results and this optimization technique uses the sharing of hardware resources.

Consider the VHDL description using process in the following example:

```
Comb_p1: process ( a_in, b_in,c_in,d_in)
begin
if(a_in='1') then
y_out <= b_in+c_in;
else
y_out <= b_in+d_in;
end if;
end process Comb_p1;
```

The above functionality generates two adders one to perform addition of c_in and b_in and another to perform addition of b_in and d_in. It also generates the 2:1 MUX to select one of the outputs of the adder. The synthesis result is shown in Fig. 10.2.

In the above synthesis result, the common input b_in is not shared properly. If the above code is modified using only one adder, then the synthesis optimization results into the better result and minimum area. Figure 10.3 shows the synthesis output.

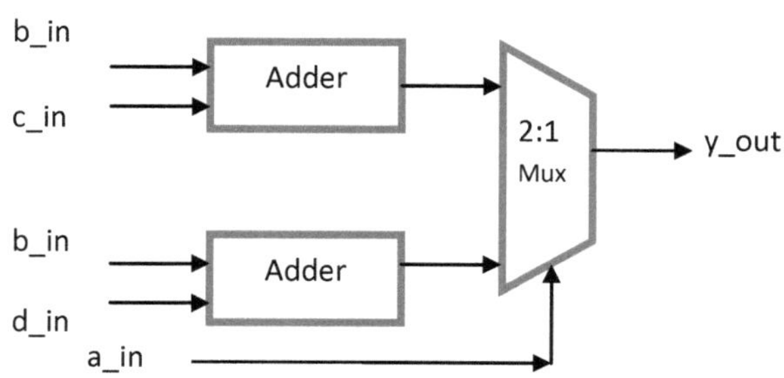

Fig. 10.2 Synthesis result without resource allocation

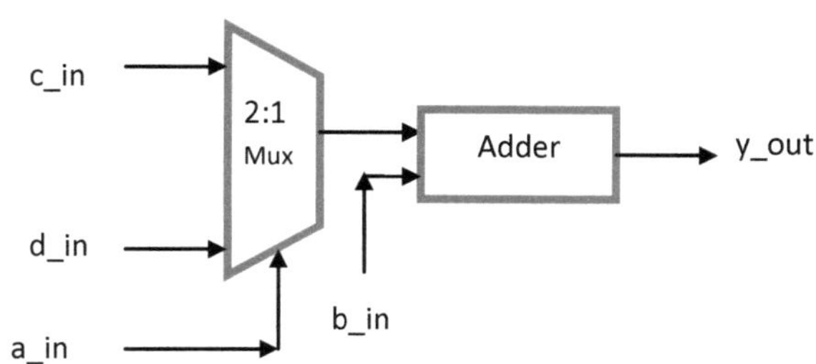

Fig. 10.3 Synthesis result with resource allocation

The modified optimized VHDL code using synthesizable constructs is described in the following example:

```
Comb_p1: process( a_in, b_in, c_in,d_in)
begin
if(a_in='1') then
y_tmp <= c_in;
else
y_tmp <= d_in;
end if;
end process Comb_p1;
```

```
y_out <= b_in + y_tmp;
```

So prior to the sharing of the resources, the area was more but resource sharing technique is effective to reduce the area.

10.2.2 *Common Factors and Subexpressions Used for Optimization*

In most of the RTL designs using VHDL, the RTL engineer uses the expressions or subexpression. In most of the designs, the subexpressions are not reused. If the subexpression-computed are reused, then the synthesizer will be able to perform to synthesis to generate the better results.

Consider the example shown below. In the following example, b_in + c_in is used for the multiple assignments

```
y_tmp <= b_in + c_in;

z_out <= d_in – ( b_in + c_in);
```

Instead of using the z_out <= d_in -(b_in+c_in); the following assignment can give the better logic with minimum resources.

```
z_out <= d_in–y_tmp;
```

Consider another RTL description using VHDL; common factor can be reused while writing an efficient RTL using VHDL.

```
Comb_p1: process(a_in,b_in,c_in,d_in)

begin

if (a_in='1') then

y_out <= b_in and ( c_in + d_in);

else

z_out <= e_in xor (c_in +d_in);

end if;

end process Comb_p1;
```

In the above example, the common factor is (c_in + d_in) and can be reused. The above code can be modified as follows:

```
Comb_p1: process ( a_in, b_in,c_in,d_in)

tmp_add = c_in + d_in;

begin

if (a_in='1') then

y_out <= b_in and ( tmp_add);

else

z_out = e_in xor (tmp_add);

end if;

end process Comb_p1;
```

These minor modifications in the VHDL code can generate more optimized logic.

10.2.3 *Moving the Piece of Code*

In most of the designs using VHDL constructs, the expressions are used in the functional body of for or while loops. These expression values may or may not change during every iteration. Those statements used in the functional body of for or while loops whose value will not change can be handled by using the modifications in the code. The synthesizer during the optimization handles such scenarios, but there are chances of redundant logic generation. This can be avoided by moving the expression outside of the loop. Consider the following design RTL described using VHDL constructs:

```
--The value of y_tmp in the range of 0 to 9

y_tmp <= a_in + b_in;

for y_tmp in  0 to 9 loop;

z_out <= y_tmp-6;

end loop;
```

In the above example, it is assumed that y_out is not assigned with the new value within the loop and the above expression remains constant for every iteration inside the loop. The synthesizer generates the 9 subtractors during the synthesis and this occupies more area. The above VHDL design functionality can be modified to avoid the unnecessary logic.

```
--The value of y_tmp in the range of 0 to 9

 y_tmp <= a_in + b_in;

 tmp<= y_tmp-6;

for y_temp in 0 to 9 loop

z_out <= tmp;

end loop;
```

10.2.4 Constant Folding

Consider the use of constants in the RTL design using VHDL. Instead of writing the code, use the direct computed or required value for the y_out. The piece of code is shown in the following example.

```
integer c_in =3;

y_out <= c_in *3;
```

Instead of using the unnecessary VHDL construct, the better way is to use the value 9 for y_out, and this technique is called as constant folding.

10.2.5 Dead Zone Elimination

The section of the code which is never executed is called as dead zone code. The dead zone code elimination technique needs to be used for the better synthesis results.

The piece of RTL using VHDL is shown in the following example

```
integer c_in=3;
integer b_in =2;

comb_p1: process ( b_in,c_in)
if (b_in >c_in) then
y_out<='1';
else
y_out<='0';
end if;
end process Comb_p1;
```

In the above code, the condition is always false and hence if statement always generates the false output. The synthesizer during the synthesis will perform such kind of optimizations. But if the code is modified, then it will reduce the time during the synthesis.

10.2.6 Use of Parentheses

In the most of the RTL designs using VHDL, if parentheses are used properly, then the synthesis results can be more optimized.

For example, if the assign statement is used in the design without any parentheses, then it generates the logic with more propagation delay.

```
y_out<= a_in + b_in – c_in –d_in;
```

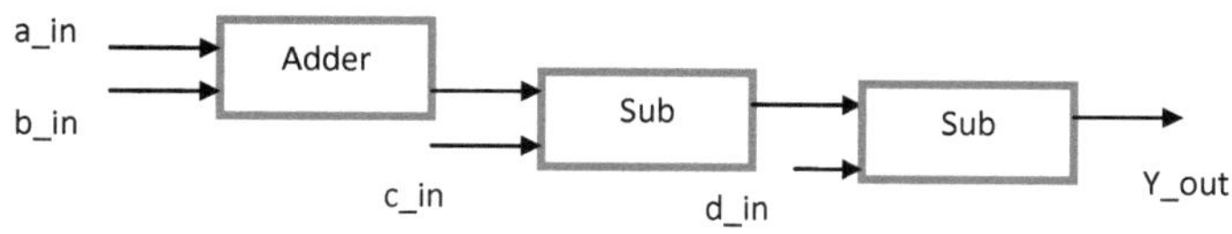

Fig. 10.4 Synthesis result without the use of parentheses

Fig. 10.5 Synthesis result with the use of parentheses

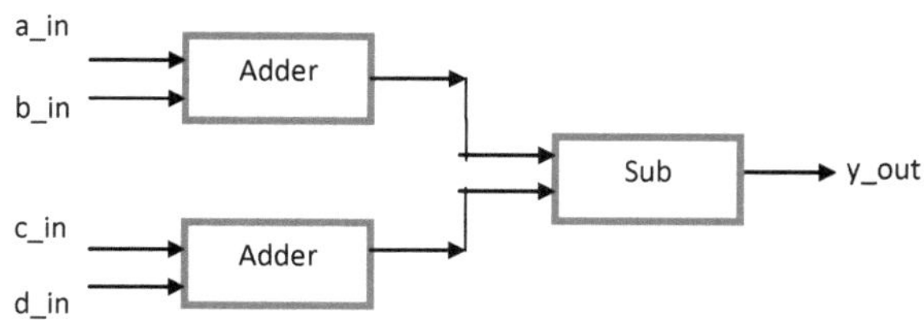

if the above statement is modified as shown below, then it gives the clear timing and data path (Figs. 10.4 and 10.5).

```
y_out<= (a_in+b_in) – (c_in+d_in);
```

10.2.7 Partitioning and Structuring the Design

The design needs to be structured and partitioned for the better synthesis outcome. It is the practical reality that the design which is better partitioned generates better synthesis results and even it reduces the synthesis runtime. The following are the key guidelines recommended for the design partitioning:

1. Partition the design for the design reuse.
2. For the different functionality, use the different module.
3. Use the combinational logic in the same block.
4. Use the separate block or structure logic for the random logic.
5. Partition the design at the top level.
6. Do not use the glue logic at the top level.
7. Use the separate module for state machines; that is, isolating the state machines forms the other logic.
8. Limit the logic size to maximum 10-K gates for every block.
9. Avoid use of the multiple clocks in the same block.
10. Isolate the synchronizers for the multiple-clock-domain designs.

By using the synthesis optimization techniques, the RTL design using VHDL designs by using the ALTERA and XILINX PLDs are discussed in the following sections. The device utilization and the synthesis results are discussed for the better understanding.

10.3 ALU Design

Arithmetic logic unit (ALU) is used in the most of the processors to perform the arithmetic and logical operations. Processor performs one of the operations at a time depending on the operational code (opcode). For 8-bit processors, the ALU is used to perform the operations on two eight-bit operands. Operand is the data on which operation needs to be performed. Similarly for the 16-bit processors, the ALU is used to perform the operations on two 16-bit numbers.

As shown in Fig. 10.6, a ALU architecture is described to perform the operation on two four-bit numbers A (A3 is MSB and A0 is LSB), B (B3 is MSB and B0 is LSB) and carry input C0, A ALU generates an output F (F3 is MSB and F0 is LSB) and an output carry C_{out3}. In the practical design scenario, one-bit ALU can be designed to perform operation on the single bit of data. The operation is performed depending on the opcode bits specified by lines S1 and S0. As shown in the following figure, ALU is designed to perform the execution for the four instructions and the operations are described in the Table 10.1. The functionality is described and it perform one of the operation listed depending on the status of select lines 'S1' and 'S0'. In this example, opcode is 2 bits and is indicated by 'S1' and 'S0'.

10.3.1 Processor Logic Unit and Design

In the practical programmable ASIC design scenario, it is recommended to describe the functionality of design using an efficient VHDL constructs. So at the microarchitecture level, the design is partitioned into multiple modules. The partitioning of design gives the better design understanding and visibility to designer. Consider a scenario to implement the design functionality of an 8-bit ALU, the design is petitioned as separate logic unit and arithmetic unit. Separate arithmetic and logical unit functionality can be described by using efficient VHDL constructs for better readability and better synthesis outcome.

As shown on Fig. 10.7 logical unit need to design to implement the four logical operations, and these logical operations are described in the functional table. The logic unit is designed to perform either AND, OR, XOR or complement operation. Table 10.2 shown below, describes the different logical operations. The complement operation is performed by using adder having one input A_0 and another input logical '1'.

The issue with this type of design is; due to the use of parallel and multiplexing logic the unit performs all the operations at a time. Hence it reduces overall design performance and results into the more area. The data path is from input A0 and B0 to the multiplexer data inputs, and control path is due to the control lines of multiplexers 'S1' and 'S0'. As shown in Fig. 10.7, the processor logic unit performs all the operations at a time and result 'F_0' to 'F_3' is generated depending on the status of the

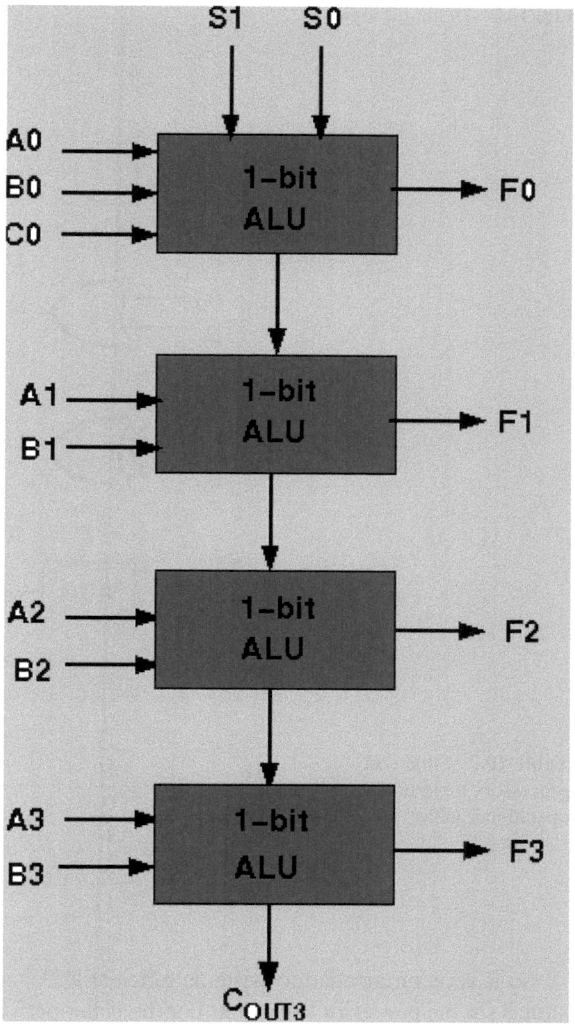

Fig. 10.6 Four-bit ALU architecture

Table 10.1 Four-bit ALU operational table

S1	S0	Operation
0	0	Addition of A, B without carry
0	1	Subtraction of A, B without borrow
1	0	XOR of A, B
1	1	Complement of A

select lines. But this technique is inefficient as it needs more area and power and it does not have the efficient implementation mechanism. If 'S1' and 'S0' are late arriving signals and if this block is used in the register to register path, then there may be possibility of the timing violations. Another important aspect is the concept of resource sharing that is not used in this design.

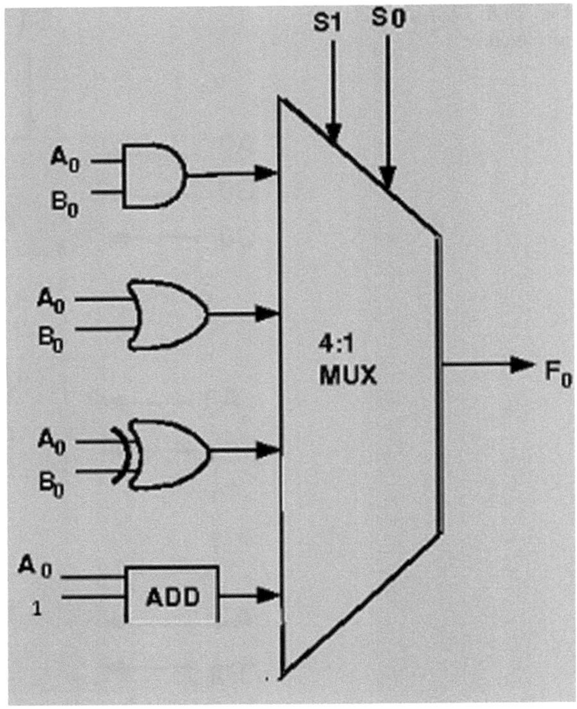

Fig. 10.7 Single-bit logic unit

Table 10.2 Single-bit processors logic unit operational table

S1	S0	Operation
0	0	A0 AND B0
0	1	A0 OR B0
1	0	XOR of A0, B0
1	1	Complement of A0

So it is recommended to write an efficient RTL using synthesizable VHDL constructs for the processor logic unit. For the better performance of the design the 'case' construct and resource sharing technique can be used. The following section describes the RTL using VHDL for the logical unit of the processor to infer the parallel logic.

10.3.1.1 8-bit Logic Unit

Example 10.1 describes the design functionality to perform the operations on two 8-bit binary inputs 'a_in' and 'b_in'. The design functionality is described in Table 10.3. The RTL using VHDL infers the parallel logic with multiplex encoding.

As described in Example 10.1, the functionality is described by using a procedural 'process' block with the 'case' construct. All the case conditions are covered and 'when others' condition is executed to generates output 'result_out' equal

```
library ieee;

use ieee.std_logic_1164.all;

use ieee.std_logic_arith.all;

entity logic_unit is

port ( a_in , b_in: in std_logic_vector (7 downto 0);

        op_code : in std_logic_vector (1 downto 0);

       result_out : out std_logic_vector (7 downto 0));

end entity logic_unit;

architecture arch_logic_unit of logic_unit is

begin

comb_p1 : process ( a_in, b_in, op_code)

    begin

          case op_code is

          when "00" => result_out <= a_in or b_in;

          when "01" => result_out <= a_in xor b_in;

          when "10" => result_out <= a_in and b_in;

          when "11" => result_out <= not a_in ;

         when others => result_out<= "00000000";

          end case;    end process comb_p1;  end architecture arch_logic_unit;
```

- Architecture defines the functionality of design.
- Combinational Process 'comb_p1' is sensitive to the input changes at 'a_in', 'b_in' and 'op_code'.
- Case construct is used to infer the parallel logic.
- Depending on the status of 2-bit 'op_code' the 'result_out' is assigned.
- An output is either 'or', 'xor', 'and', 'not' at a time.

Example 10.1 VHDL RTL for 8-bit ALU using case construct

Table 10.3 Operational table for 8-bit ALU

op_code [1]	op_code[o]	Logic operation
0	0	a_in OR b_in
0	1	a_in XOR b_in
1	0	a_in AND b_in
1	1	Complement of a_in

to ‘00000000’. If op_code is not matching with “00” to “11” then the ‘when others’ clause is executed.

The synthesis result is shown in Fig. 10.8, and it infers the parallel logic using multiplex encoding. For such kind of designs, ‘case’ construct is used instead of

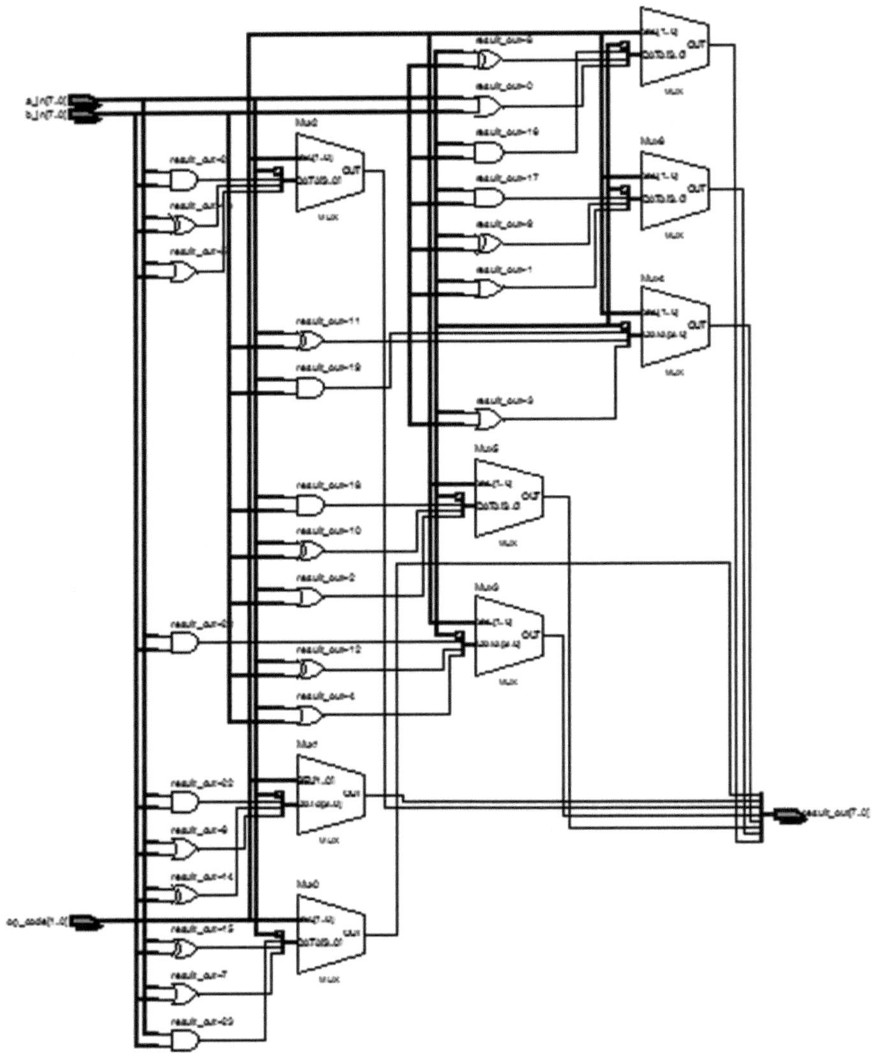

Fig. 10.8 Synthesis result for 8-bit logic unit

using ‘if then else’ construct. As discussed in Chap. 8, the ‘case’ construct infers the parallel logic.

Synthesis result using case construct for the 8-bit processor logic unit is shown in Fig. 10.8. As shown in the above figure, it infers the logic gates with multiplexing logic. In the practical scenario, it is recommended to use the adders as common resources to implement both the logic and arithmetic units.

10.3.1.2 Processor Logic Unit with Registered IO

For the efficient and clean timing analysis, it is recommended to use registered inputs and registered outputs. If all the inputs and outputs are registered that is sampled or captured on the active edge of clock and even if all the outputs are registered and captured on the active edge of clock, then design can give better results and clean register to register timing. The registered inputs and registered outputs can give the clean data path and even the output is glitch or hazard free. For the performance improvement, the pipelining can be used to reduce the data arrival time. Please refer Chap. 5 for the information about the sequential circuit timing.

Example 10.2 uses the registered input and registered output logic. The inputs are sampled or captured on the positive edge of clock ‘clk’ and outputs are launched on the positive edge of ‘clk.’ During the reset condition ‘reset_n = 0’, the processor unit is initialized to logic ‘0’.

The Example 10.2 generates the processor logic unit with all the inputs and outputs registered on positive edge of clock. Readers are requested to assume that every register has an asynchronous reset input ‘reset_n’. The synthesis result is shown in Fig. 10.9.

10.3.2 Arithmetic Unit

The arithmetic unit is used to perform the arithmetic operations such as addition, subtraction, increment, and decrement. The operations are performed on the two different operands. The functional Table 10.4 gives information about the different

Table 10.4 Operational table for the arithmetic unit

op_code[2]	op_code[1]	op_code[o]	Logic operation
0	0	0	Transfer a_in
0	0	1	a_in ADD b_in
0	1	0	a_in ADD b_in with carry input cin_in
0	1	1	a_in SUB b_in
1	0	0	a_in SUB b_in with borrow input cin_in
1	0	1	Increment a_in
1	1	0	Decrement b_in
1	1	1	No operation performed

```
library ieee;

use ieee.std_logic_1164.all;

use ieee.std_logic_arith.all;

entity logic_unit is

port ( a_in : in std_logic_vector (7 downto 0);

       b_in: in std_logic_vector (7 downto 0);

       clk: in std_logic;

       reset_n : in std_logic;

       op_code : in std_logic_vector (1 downto 0);

      result_out : out std_logic_vector (7 downto 0));

end entity logic_unit;

architecture arch_logic_unit of logic_unit is

signal sig_a_in  : std_logic_vector (7 downto 0);

signal sig_b_in : std_logic_vector (7 downto 0);

signal sig_op_code : std_logic_vector (1 downto 0);

begin
```

- Architecture defines the functionality of design.
- Signals are used to hold the intermediate data.
- Signals are declared as 'sig_a_in', 'sig_b_in' and are 8 bit wide and of type std_logic.
- Signal 'sig_op_code' is of type 'std_logic' and is 2-bit wide.
- These are used to assign the values of input on active edge of clock 'clk'.

Example 10.2 VHDL RTL for 8-bit logic unit with registered inputs and outputs

```
reg_p1 : process ( clk, reset_n)
    begin
        if (reset_n='0') then
        sig_a_in <= "00000000";
        sig_b_in <= "00000000";
        sig_op_code <= "00";
        elsif (clk='1' and clk'event) then
        sig_a_in <= a_in;
        sig_b_in <= b_in;
        sig_op_code <= op_code;   end if;      end process reg_p1;
comb_p2 : process ( sig_a_in, sig_b_in, sig_op_code)
    begin
        case sig_op_code is
          when "00" => result_out <= sig_a_in or sig_b_in;
          when "01" => result_out <= sig_a_in xor sig_b_in;
          when "10" => result_out <= sig_a_in and sig_b_in;
          when "11" => result_out <= not sig_a_in ;
        end case;
    end process comb_p2;
end architecture arch_logic_unit;
```

- Sequential process is labeled as 'reg_p1' and sensitive to 'clk' and 'reset_n'.
- For active low 'reset_n' the value of 'sig_a_in', 'sig_b_in' and 'sig_op_code' is assigned to zero.
- For rising edge of the clock the input 'a_in' is assigned to 'sig_a_in'.
- For rising edge of the clock the input 'b_in' assigned to 'sig_b_in'.
- For rising edge of the clock the input 'op_code' is assigned to 'sig_op_code'.

- Combinational Process 'comb_p2' is sensitive to the changes 'sig_a_in', 'sig_b_in' and 'sig_op_code'.
- Case construct is used to infer the parallel logic.
- Depending on the status of 2-bit 'sig_op_code' the 'result_out' is assigned.
- An output is either 'or', 'xor', 'and', 'not' at a time.
- It infers the parallel combinational logic.

Example 10.2 (continued)

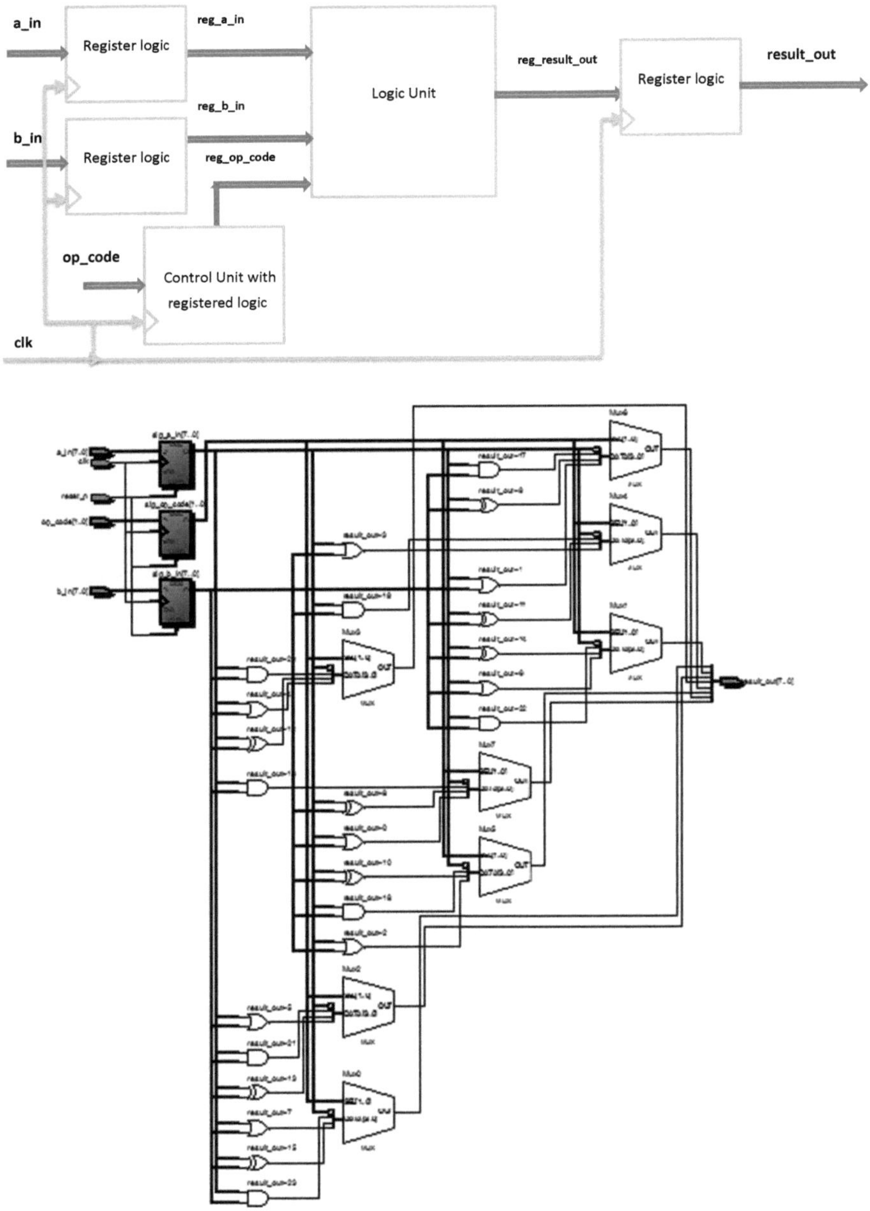

Fig. 10.9 Synthesis result for processor logic unit with registered inputs and outputs

operations need to be performed. The arithmetic unit is described in such a way that it performs only one operation at time. Figure 10.10 describes the block diagram representation of the arithmetic unit (Example 10.3).

```
--arithmetic unit VHDL
library ieee;
use ieee.std_logic_1164.all;
use ieee.std_logic_arith.all;
use ieee.std_logic_unsigned.all;

entity arithmetic_unit is
port ( a_in : in std_logic_vector (7 downto 0);
        b_in : in std_logic_vector (7 downto 0);
       cin_in : in std_logic;
       op_code_in : in std_logic_vector (2 downto 0);
      result_out : out std_logic_vector (7 downto 0);
      co_out : out std_logic );
end entity arithmetic_unit;

architecture arch_arithmetic_unit of arithmetic_unit is
begin

comb_p1 : process ( a_in, b_in, cin_in,op_code_in)
variable tmp_result_out : unsigned ( 8 downto 0);
begin
case op_code_in is
when "000" => tmp_result_out := unsigned ('0' & a_in);
when "001" => tmp_result_out := unsigned ('0' & a_in ) + unsigned ('0' & b_in );
when "010" => tmp_result_out := unsigned ('0' & a_in ) + unsigned ('0' & b_in )+cin_in;
when "011" =>  tmp_result_out := unsigned ('0' & a_in ) - unsigned ('0' & b_in );
when "100" =>  tmp_result_out := unsigned ('0' & a_in ) - unsigned ('0' & b_in )- cin_in;
when "101" =>  tmp_result_out := unsigned ('0' & a_in ) + '1';
when "110" => tmp_result_out := unsigned ('0' & a_in )  -'1';
when others =>  tmp_result_out := "000000000";
end case;
result_out <= std_logic_vector(tmp_result_out(7 downto 0));
co_out <= std_logic(tmp_result_out(8));
end process comb_p1;

end architecture arch_arithmetic_unit;
```

Example 10.3 VHDL RTL for the arithmetic unit

Table 10.5 Signal or pin description of 8-bit ALU

Signal or pin name	Size (bits)	Description
a_in	8	An 8-bit operand
b_in	8	An 8-bit operand
cin_in	1	Carry input to a ALU
op_code_in	4	4-bit opcode for instruction
result_out	8	An 8-bit output from ALU
co_out	1	One-bit output carry from ALU

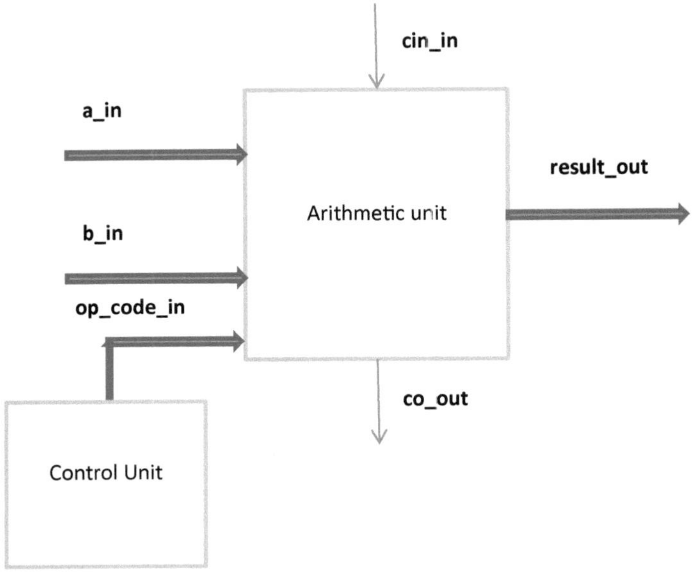

Fig. 10.10 Block diagram of arithmetic unit

The synthesis result for one-bit arithmetic unit is shown in Fig. 10.11. The logic uses the full adder as component to perform the addition and subtraction. Subtraction is performed using 2's complement addition. The synthesized logic also consists of the multiplexer 4:1 to pass the required operand as one of the input of full adder depending on the opcode.

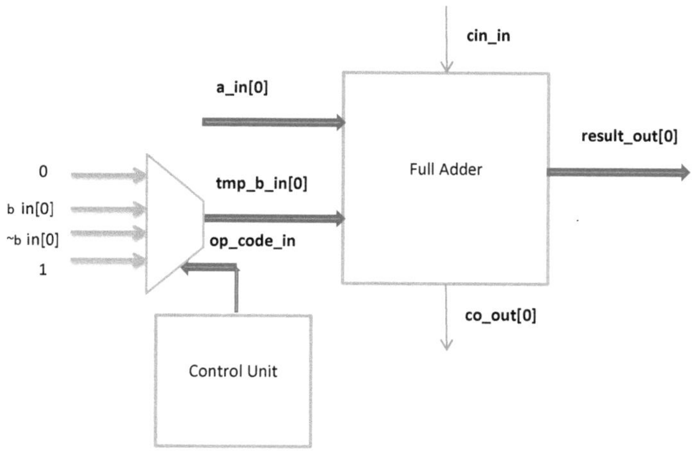

Fig. 10.11 Synthesis result for the one-bit arithmetic unit

Fig. 10.12 ALU top-level diagram

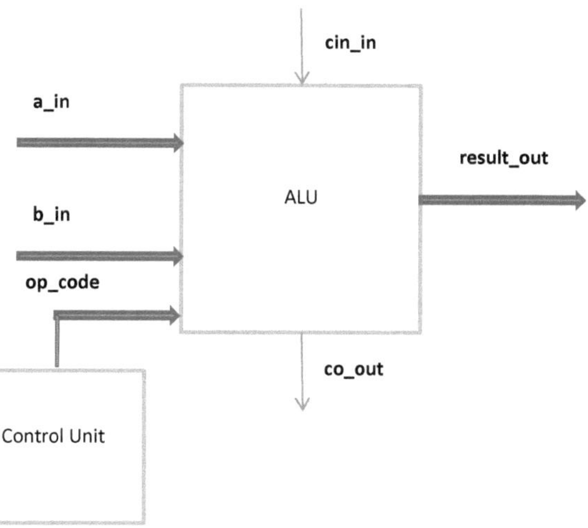

10.3.3 Arithmetic and Logical Unit

Figure 10.12 illustrates the ALU with the associated logic circuit to perform the operation on two 8-bit numbers 'a_in' and 'b_in'. For logic operations, the carry input (cin_in) is ignored and the output 'result_out' is generated depending on the

Table 10.6 Operational table for 8-bit ALU

Operational code	Instruction	Description
0000	Transfer a_in	Generate an output a_in + 0+0
0001	Addition without carry	a_in + b_in +0
0010	Addition with carry	a_in + b_in + 1
0011	Subtract without borrow	a_in –b_in
0100	Subtract with borrow	a_in –b_in-1
0101	Increment a_in by 1	a_in +1
0110	Decrement a_in by 1	a_in -1
1000	a_in OR with b_in	a_in OR b_in
1001	a_in XOR with b_in	a_in XOR b_in
1010	a_in AND with b_in	a_in AND b_in
1011	Complement a_in	Not a_in

operational code of the instruction. Depending of the operational code, ALU is used to perform either arithmetic or logical operation. During arithmetic operations if result is more than 8 bits, then carry output 'co_out' is set to logical '1' that indicates carry propagation outside to MSB.

Table 10.6 describes the number of instructions need to be performed by the ALU. As shown in the table ALU performs 7 arithmetic operations and 4 logical operations. The pin or signal description is shown in Table 10.5.

An efficient RTL using VHDL to infer the parallel logic is described in Example 10.4. For the 'op_code_in = 0', it performs the arithmetic operation, and when 'op_code_in = 1', it performs the logic operation.

The synthesis result for the 8-bit ALU is shown in Fig. 10.13. As shown in the figure, it consists of the parallel logic for the arithmetic operations and logic operations. Using the multiplexer at the output side, either arithmetic or logical operation result can be selected. The logic does not use the concept of resource sharing and area and power optimization. This RTL can be modified by using the concept of resource sharing for the better synthesis result.

```
--8-bit arithmetic logic unit VHDL
library ieee;
use ieee.std_logic_1164.all;
use ieee.std_logic_arith.all;
use ieee.std_logic_unsigned.all;

entity arithmetic_logic_unit is
port ( a_in : in std_logic_vector (7 downto 0);
b_in : in std_logic_vector (7 downto 0);
cin_in : in std_logic;
op_code_in : in std_logic_vector (3 downto 0);
result_out : out std_logic_vector (7 downto 0);
co_out : out std_logic );
end entity arithmetic_logic_unit;

architecture arch_arithmetic_unit of arithmetic_logic_unit is
begin
comb_p1 : process ( a_in, b_in, cin_in,op_code_in)
variable tmp_result_out : unsigned ( 8 downto 0);
begin

if(op_code_in(3)='0') then
case op_code_in(2 downto 0) is
when "000" => tmp_result_out := unsigned ('0' & a_in);
when "001" => tmp_result_out := unsigned ('0' & a_in ) + unsigned ('0' & b_in );
when "010" => tmp_result_out := unsigned ('0' & a_in ) + unsigned ('0' & b_in )+cin_in;
when "011" => tmp_result_out := unsigned ('0' & a_in ) -unsigned ('0' & b_in );
when "100" => tmp_result_out := unsigned ('0' & a_in ) -unsigned ('0' & b_in )-cin_in;
when "101" => tmp_result_out := unsigned ('0' & a_in ) + '1';
when "110" => tmp_result_out := unsigned ('0' & a_in ) -'1';
when others => tmp_result_out := "000000000";
```

Example 10.4 VHDL RTL for 8-bit ALU

```
end case;

else
case op_code_in  (2 downto 0) is
when "000" => tmp_result_out :=  unsigned ( '0' & (a_in  OR  b_in)) ;
when "001" => tmp_result_out :=  unsigned ( '0' & (a_in  XOR  b_in)) ;
when "010" => tmp_result_out := unsigned ( '0' & (a_in  AND  b_in)) ;
when "011" => tmp_result_out := unsigned ( '0' & NOT (a_in )) ;
when others => tmp_result_out := "000000000";
end case;
end if;

result_out <= std_logic_vector(tmp_result_out(7 downto 0));
co_out <= std_logic(tmp_result_out(8));
end process comb_p1;
end architecture arch_arithmetic_unit;
```

Example 10.4 (continued)

10.4 Barrel Shifters

In most of the DSP applications, the combinational shifters are used to perform the shifting operations on the data input. The combinational shifters are called as barrel shifter. The advantage of barrel shifter is that it performs the shifting operation depending on the required number of shifts depending on the control inputs without use of any clocking logic. Most of the barrel shifters are designed by using the multiplexer logic.

Example 10.5 is RTL using VHDL and has 8-bit input ‘d_in’, three-bit control input ‘c_in’, and an 8-bit output ‘q_out’ (Fig. 10.14).

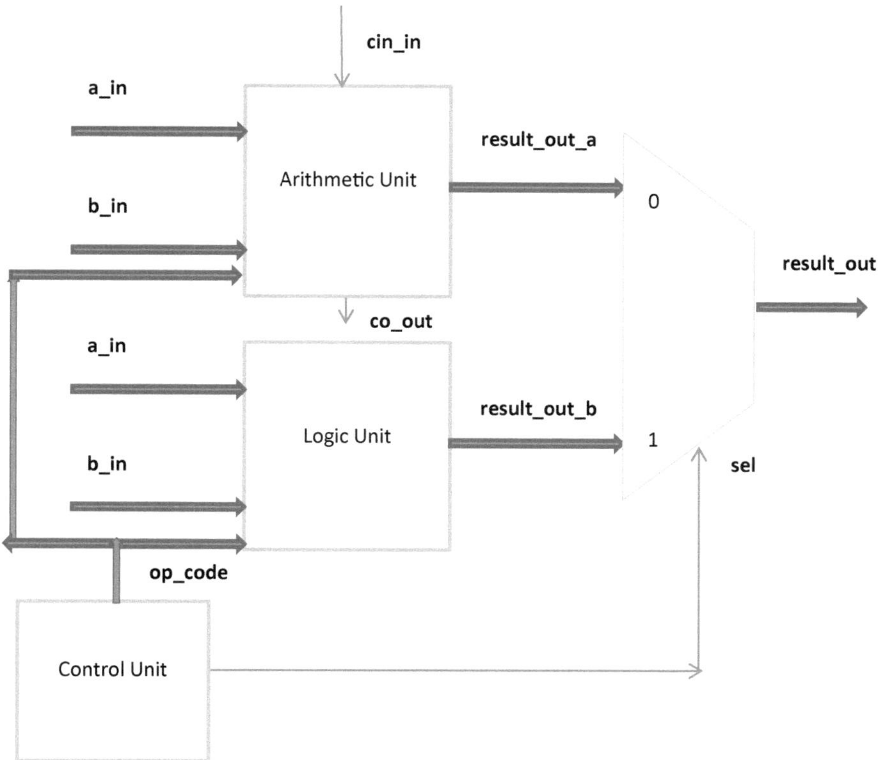

Fig. 10.13 Synthesis result for the 8-bit ALU

10.5 Parity Checkers and Generators

In most of the programmable ASIC and SOC designs, RTL using VHDL constructs is used to describe the protocol behavior. The requirement and objective is functional correctness of the design and even to have timing and cycle accurate models. In most of the practical applications, the parity needs to be detected as even parity or odd parity. For example if the even number of 1's is there in any string, then the parity is treated as even parity, and if odd number of 1's are there in the string, then parity will be treated as odd parity. This section focuses on the parity generator and checker.

```
library ieee;

use ieee.std_logic_1164.all;

use ieee.std_logic_arith.all;

use ieee.std_logic_unsigned.all;

entity barrel_shifter is

 port (

  d_in       : in  std_logic_vector(7 downto 0);  --data input vector

    shift_lr : in  std_logic;                  -- 0=>left_operation
    --1=>right_operation

  shift_value   : in  std_logic_vector(2 downto 0);  -- shift value
```

Example 10.5 VHDL RTL for barrel shifter

10.5.1 Parity Checker

Efficient RTL using VHDL construct for the parity checker is described in Example 10.6. As described in the RTL, the even or odd parity is checked, and at output 'y_out', even parity is indicated by logic '0' and odd parity is indicated by logic '1'.

The syntheis result is shown in Fig. 10.15 and it is combinational logic and implemented using XOR gate.

10.5.2 Parity Generator

The RTL using VHDL for the 8-bit parity generator is described by using efficient constructs and shown in Example 10.7

```
  clk       : in  std_logic;                      -- clock  input signal
  reset_n   : in  std_logic;                      --  active low reset signal
  enable_in : in  std_logic;                      -- used for parallel load
  y_out     : out std_logic_vector(7 downto 0));  -- shifted data output
end barrel_shifter;
architecture arch_barrel_shifter of barrel_shifter is
begin
sequ_p1: process (clk,reset_n,shift_value,shift_lr)
variable tmp_x,tmp_y : std_logic_vector(7 downto 0);
variable ctrl_0,ctrl_1,ctrl_2 : std_logic_vector(1 downto 0);
 begin  -- process p1
ctrl_0:=shift_value(0) & shift_lr;
ctrl_1:=shift_value(1) & shift_lr;
ctrl_2:=shift_value(2) & shift_lr;
```

Example 10.5 (continued)

```
if(reset_n = '0') then
y_out<="00000000";
elsif(clk'event and clk = '1') then
if (enable_in='0')then
  assert(false) report "data load is disabled" severity warning;
elsif(shift_lr='1')then
  assert(false) report "shift data right" severity warning;
elsif(shift_lr='0')then
  assert(false) report "shift data left" severity warning;
 end if;
if (enable_in='1') then
case ctrl_0 is
  when "00"|"01" =>tmp_x:=d_in ;
```

Example 10.5 (continued)

```
when "10" =>tmp_x:=d_in(6 downto 0) & d_in(7);  --shift the data input
    left by 1 bit

when "11" =>tmp_x:=d_in(0) & d_in(7 downto 1);  --shift the data input
    right by 1 bit

when others => null;

end case;

case ctrl_1 is

when "00"|"01" =>tmp_y:=tmp_x;

when "10" =>tmp_y:=tmp_x(5 downto 0) & tmp_x(7 downto 6);  --shift
    data input to left by 2 bits

when "11" =>tmp_y:=tmp_x(1 downto 0) & tmp_x(7 downto 2);  --shift
    data input to right by 2 bits

when others => null;

end case;

case ctrl_2 is

when "00"|"01" =>y_out<=tmp_y ;
```

Example 10.5 (continued)

```
when "10"|"11" =>y_out<= tmp_y(3 downto 0) & tmp_y(7 downto 4); --
   shift to righ or left by 4 bits

when others => null;

end case;

end if;

end if;

end process sequ_p1;

end arch_barrel_shifter;
```

Example 10.5 (continued)

The synthesis result is shown in Fig. 10.16. The synthesized logic consists of XOR logic, and it is a purely combinational design. For the input string of the 7 bits, the output is 8 bits.

10.6 Memories

Depending on the design requirements, the distributed RAM or BRAMs can be used while prototyping. The memories can have synchronous or asynchronous read–write capabilities. The single-port and dual-port BRAMs are discussed in this section.

10.6.1 Single-Port RAM

Example 10.8 is the VHDL description of the distributed RAM with the asynchronous read. Depending on the design requirements, the distributed or BRAM can be modeled using VHDL.

The synthesis outcome of single-port RAM with asynchronous read using Altera Quartus II for MAXII device is shown in Fig. 10.17.

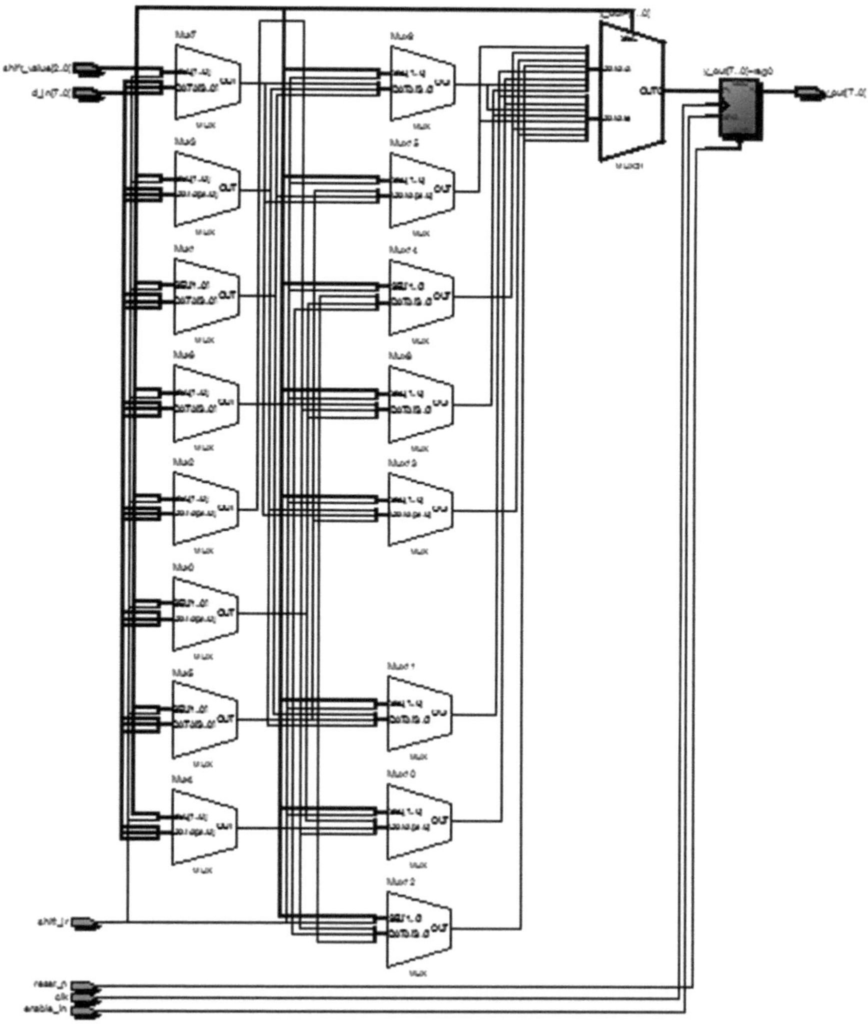

Fig. 10.14 Synthesis result of barrel shifter

Another type of single-port BRAM with read-first mode is described using VHDL and shown in Example 10.9.

The synthesis outcome of single-port RAM with read-first mode using Altera Quartus II for MAXII device is shown in Fig. 10.18.

```
library ieee;

use ieee.std_logic_1164.all;

entity parity_checker is

 port (

   a0_in : in  std_logic;

   a1_in : in  std_logic;

   a2_in : in  std_logic;

   a3_in : in  std_logic;
  y_out  : out std_logic);

end parity_checker;

architecture arch_parity_checker of parity_checker is

signal sig_tmp_1,sig_tmp_2 : std_logic;

begin

      sig_tmp_1 <= a0_in xor a1_in;

      sig_tmp_2 <= a2_in xor sig_tmp_1;

      y_out <= sig_tmp_2 xor a3_in;

end arch_parity_checker;
```

Example 10.6 VHDL RTL for the parity checker

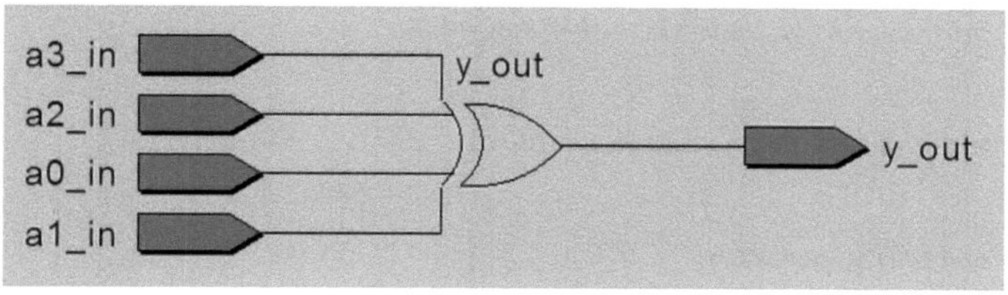

Fig. 10.15 Synthesized logic for parity checker

```
--8 Bit parity Generator

library ieee;

use ieee.std_logic_1164.all;

use ieee.std_logic_arith.all;

use ieee.std_logic_unsigned.all;

entity parity_generator is

generic(n:integer:=7);
```

Example 10.7 VHDL RTL for 8-bit parity generator

Another type of single-port BRAM with write-first mode is described using VHDL and shown in Example 10.10.

The synthesis outcome of single-port RAM with write-first mode using Altera Quartus II for MAXII device is shown in Fig. 10.19. Reader can target these single port RAM VHDL codes on the different Altera devices (Cyclone Iv, Cyclone II).

```
port(a_in:in std_logic_vector(n-1 downto 0);

y_out:out std_logic_vector(n downto 0));

end parity_generator;

architecture arch_parity_gen of parity_generctor is

begin

comb_p1: process(a_in)

        variable tmp_1:std_logic;

        variable tmp_2:std_logic_vector(y_out'range);

        begin

                tmp_1:='0';

                for i in a_in'range loop

                        tmp_1:=tmp_1 xor a_in(i);

                        tmp_2(i):=a_in(i);

                end loop;

                        tmp_2(y_out'high):=tmp_1;

                        y_out<=tmp_2;

end process comb_p1;

end arch_parity_gen;
```

Example 10.7 (continued)

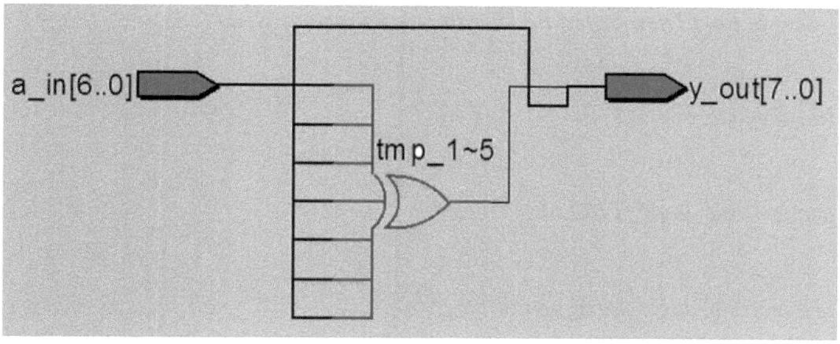

Fig. 10.16 Synthesized 8-bit parity generator

10.6.2 Dual-Port RAM

Example 10.11 is the VHDL description of the simple dual-port BRAM with single clock.

The synthesis outcome of dual-port RAM using Altera Quartus II for MAXII device is shown in Fig. 10.20.

For the dual-port RAM with two clocks, the RTL using VHDL is described and shown in the Example 10.12.

The synthesis outcome of dual-port RAM with two clocks using Altera Quartus II for MAXII device is shown in Fig. 10.21.

If the VHDL RTL is synthesized by using the XILINX ISE, then the device utilization for the dual-port RAM is shown in Table 10.7.

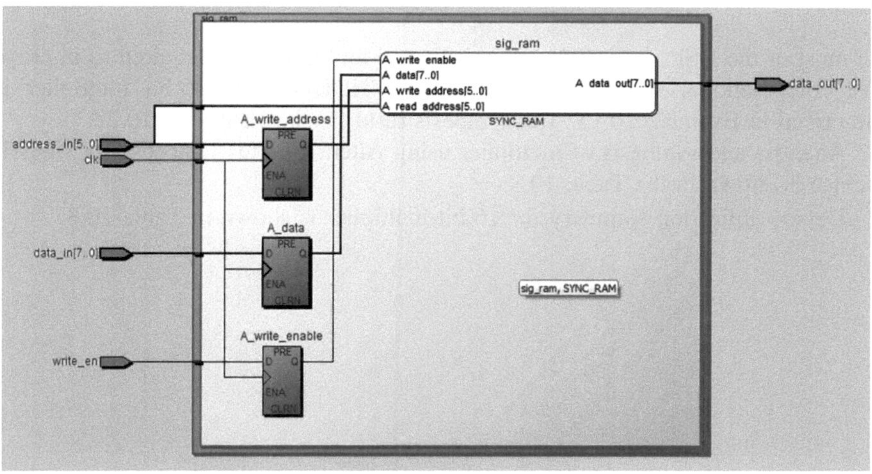

Fig. 10.17 Synthesized single-port RAM with asynchronous read

```
--Single Port Distributed RAM with Asynchronous Read

library ieee;

use ieee.std_logic_1164.all;

use ieee.std_logic_unsigned.all;

entity ram_single_port is

port(

    clk : in std_logic;

    write_en : in std_logic;

    address_in : in std_logic_vector(5 downto 0);
```

Example 10.8 VHDL RTL for single-port RAM with asynchronous read

10.7 Multipliers

In most of the DSP applications, the dedicated multipliers are required to improve the computational speed. The RTL using VHDL for the 16-bit multiplier is described in Example 10.13. The synthesis result is shown in Fig. 10.22.

Analysis and synthesis of multiplier using Altera Quartus II license for MAXII device is shown in the Table 10.8.

Device utilization summary for 16-bit multiplier is shown in Table 10.8.

```
        data_in : in std_logic_vector(7 downto 0);
        data_out : out std_logic_vector(7 downto 0)
    );
end ram_single_port;
architecture arch_ram of ram_single_port is
type ram_m_type is array (63 downto 0) of std_logic_vector(7 downto 0);
signal sig_ram : ram_m_type;
begin
sequ_p1: process(clk)
        begin
        if (clk'event and clk = '1') then
          if (write_en = '1') then
            sig_ram(conv_integer(address_in)) <= data_in;
          end if;
      end if;
end process sequ_p1;
data_out <= sig_ram(conv_integer(address_in));
end arch_ram;
```

Example 10.8 (continued)

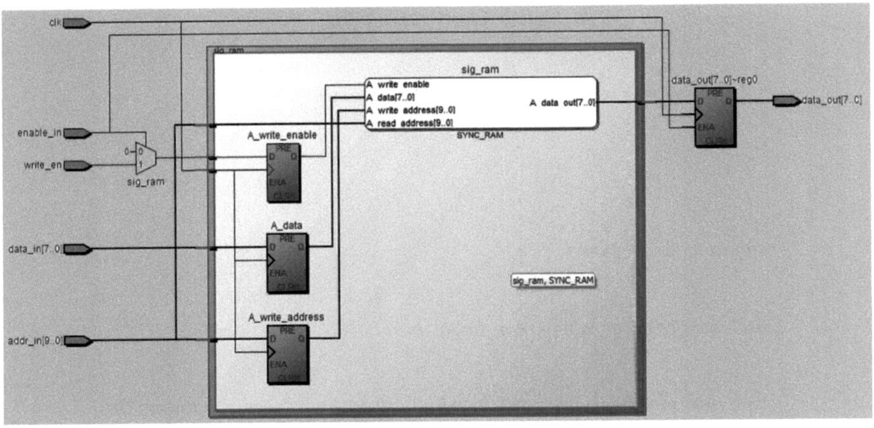

Fig. 10.18 Synthesized single-port BRAM with read-first mode

```
-- Single-Port Block RAM with Read-First Mode

library ieee;

use ieee.std_logic_1164.all;

use ieee.std_logic_unsigned.all;

entity ram_sp_read_first is

port(

        clk : in std_logic;

        write_en : in std_logic;

        enable_in : in std_logic;

        addr_in : in std_logic_vector(9 downto 0);

        data_in : in std_logic_vector(7 downto 0);

        data_out : out std_logic_vector(7 downto 0)
```

Example 10.9 VHDL RTL for single-port BRAM with read-first mode

```
  );

end ram_sp_read_first;

architecture arch_ram_sp of ram_sp_read_first is

type ram__m_type is array (1023 downto 0) of std_logic_vector(7
    downto 0);

signal sig_ram : ram_m_type;

begin

sequ_p1: process(clk)

begin

    if (clk'event and clk = '1)' then

        if( enable_in = '1' )then

            if( write_en = '1' )then

                sig_ram(conv_integer(addr_in)) <= data_in;

            end if;
        data_out <= sig_ram(conv_integer(addr_in));

    end if;

  end if;

end process sequ_p1;

end arch_ram_sp;
```

Example 10.9 (continued)

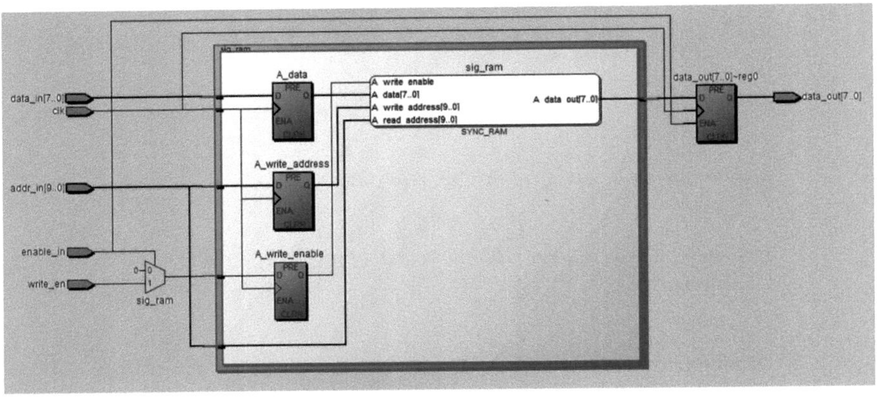

Fig. 10.19 Synthesized single-port RAM with write-first mode

```
-- Single port BRAM with the write first

library ieee;

use ieee.std_logic_1164.all;

use ieee.std_logic_unsigned.all;

entity ram_sp_write_first is

port(

     clk : in std_logic;

     write_en : in std_logic;

     enable_in : in std_logic;

     addr_in : in std_logic_vector(9 downto 0);

     data_in : in std_logic_vector(7 downto 0);

     data_out : out std_logic_vector(7 downto 0)

  );

end ram_sp_write_first;
```

Example 10.10 VHDL RTL for single-port BRAM with write-first mode

```
architecture arch_ram_sp of ram_sp_write_first is

type ram_m_type is array (1023 downto 0) of std_logic_vector(7
     downto 0);

signal sig_ram : ram_m_type;

begin

sequ_p1: process(clk)

    begin

    if (clk'event and clk = '1') then

       if (enable_in = '1' ) then

        if (write_en = '1' ) then

          sig_ram(conv_integer(addr_in)) <= data_in;

          data_out <= data_in;

       end if;

     data_out <= sig_ram(conv_integer(addr_in));

    end if;

  end if;

end process sequ_p1;

end arch_ram_sp;
```

Example 10.10 (continued)

```
-- Simple Dual-Port Block RAM with single Clock

library IEEE;

use IEEE.std_logic_1164.all;

use IEEE.std_logic_unsigned.all;

entity dual_port_ram is

port(

      clk : in std_logic;

      enable_a_in : in std_logic;

      enable_b_in : in std_logic;

      write_en : in std_logic;

      addr_a_in : in std_logic_vector(9 downto 0);

      addr_b_in : in std_logic_vector(9 downto 0);

      data_a_in : in std_logic_vector(7 downto 0);

      data_b_out : out std_logic_vector(7 downto 0)
```

Example 10.11 VHDL RTL for dual-port RAM

```
);

end dual_port_ram;

architecture arch_dual_port_ram of dual_port_ram is

type ram_m_type is array (1023 downto 0) of std_logic_vector(7
downto 0);

shared variable sig_ram : ram_m_type;

begin

sequ_p1: process(clk)

    begin

    if (clk'event and clk = '1') then

     if (enable_a_in = '1' ) then

       if (write_en = '1') then

          sig_ram(conv_integer(addr_a_in)) := data_a_in;

       end if;
```

Example 10.11 (continued)

```
        end if;
    end if;
end process sequ_p1;
sequ_p2: process(clk)
    begin
    if (clk'event and clk = '1') then
        if (enable_b_in = '1' ) then
            data_b_out <= sig_ram(conv_integer(addr_b_in));
        end if;
    end if;
end process sequ_p2;
end arch_dual_port_ram;
```

Example 10.11 (continued)

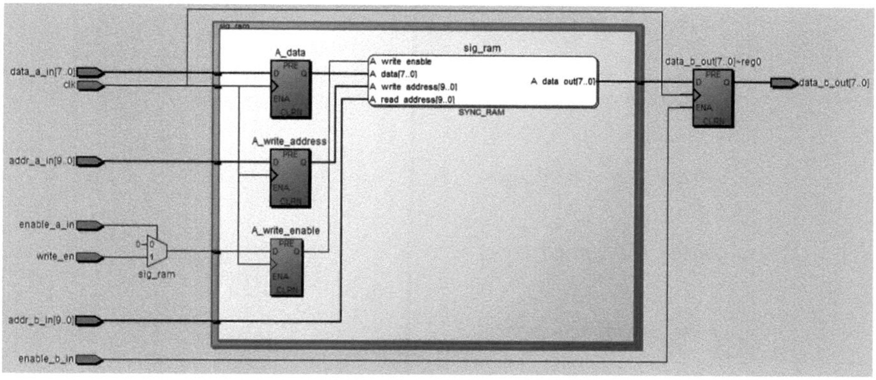

Fig. 10.20 Synthesized dual-port RAM

```
--Dual port RAM with two clocks

library ieee;

use ieee.std_logic_1164.all;

use ieee.std_logic_arith.all;

use ieee.std_logic_unsigned.all;

entity dual_port_ram is

port(
```

Example 10.12 VHDL RTL for dual-port BRAM with two clocks

```
clk_a : in std_logic;
clk_b : in std_logic;
enable_a_in : in std_logic;
enable_b_in : in std_logic;
write_en : in std_logic;
addr_a_in : in std_logic_vector(9 downto 0);
addr_b_in : in std_logic_vector(9 downto 0);
data_a_in : in std_logic_vector(7 downto 0);
data_b_out : out std_logic_vector(7 downto 0)
);
end dual_port_ram;
architecture arch_dual_port_ram of dual_port_ram is
type ram_m_type is array (1023 downto 0) of std_logic_vector(7 downto
0);
```

Example 10.12 (continued)

```
shared variable sig_ram : ram_m_type;

begin

port_a: process(clk_a)

begin

    if (clk_a'event and clk_a = '1' )then

      if (enable_a_in = '1') then

        if (write_en = '1') then

          sig_ram(conv_integer(addr_a_in)) := data_a_in;

        end if;

      end if;

   end if;

end process port_a;

port_b: process(clk_b)

begin
```

Example 10.12 (continued)

```
        if (clk_b'event and clk_b = '1') then

          if enable_b_in = '1' then

            data_b_out <= sig_ram(conv_integer(addr_b_in));

          end if;

        end if;

end process port_b;

end arch_dual_port_ram;
```

Example 10.12 (continued)

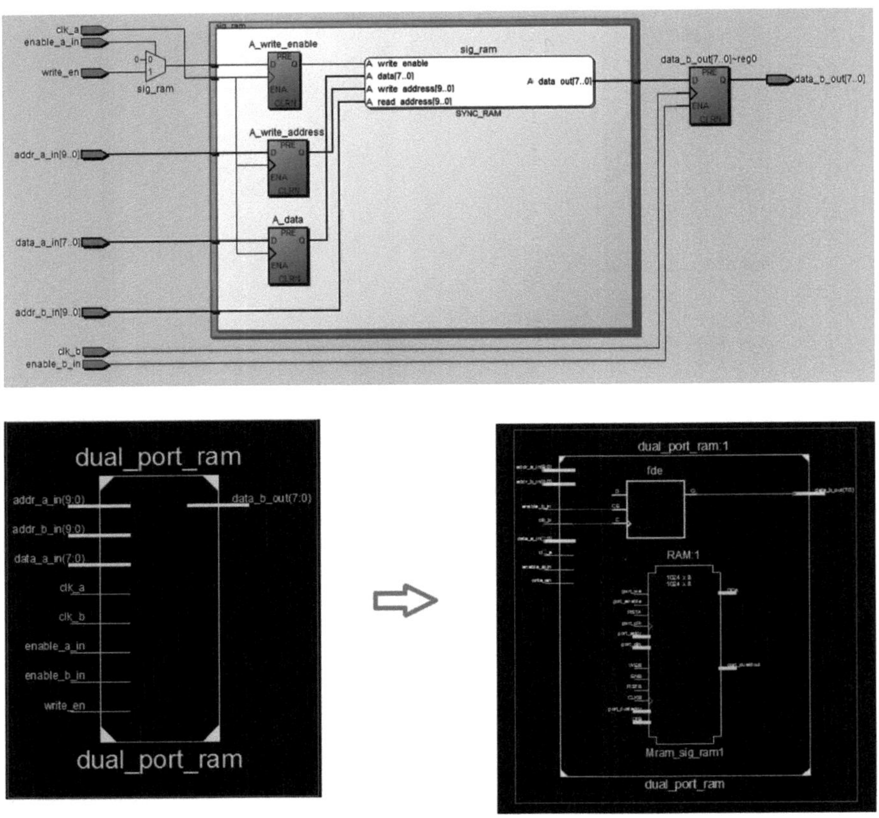

Fig. 10.21 Synthesized dual-port BRAM with two clocks

Table 10.7 Device utilization for XILINX FPGA device: XC3S100e-5vq100

Device utilization summary (estimated values)			
Logic utilization	Used	Available	Utilization (%)
Number of slices	1	960	0
Number of 4 input LUTs	1	1920	0
Number of bonded IOBs	41	66	62
Number of BRAMs	1	4	25
Number of GCLKs	2	24	8

```
--VHDL RTL for 16 bit multiplier

library ieee;

use ieee.std_logic_1164.all;

use ieee.std_logic_arith.all;

entity multiplier is

generic (data_size :integer := 16;

      data_level:integer:=4);

port (

      clk : in std_logic;

      a_in : in std_logic_vector (data_size-1 downto 0);

      b_in : in std_logic_vector (data_size-1 downto 0);

      y_out : out std_logic_vector (2*data_size-1 downto 0));

end multiplier;

architecture arch_multiplier of multiplier is
```

Example 10.13 VHDL RTL for 16-bit multiplier

```
type register_levels is array (data_level-1 downto 0) of unsigned
      (2*data_size-1 downto 0);

signal register_bank :register_levels;

signal sig_a, sig_b : unsigned (data_size-1 downto 0);

begin

y_out <= std_logic_vector (register_bank (data_level-1));

seq_mul: process ( clk)

begin

      if( clk'event and clk = '1')  then

            sig_a <= unsigned(a_in);

            sig_b <= unsigned(b_in);

            register_bank (0) <= sig_a * sig_b;

      for i in 1 to data_level-1 loop

          register_bank (i) <= register_bank (i-1);
      end loop;

    end if;

end process seq_mul;

end arch_multiplier;
```

Example 10.13 (continued)

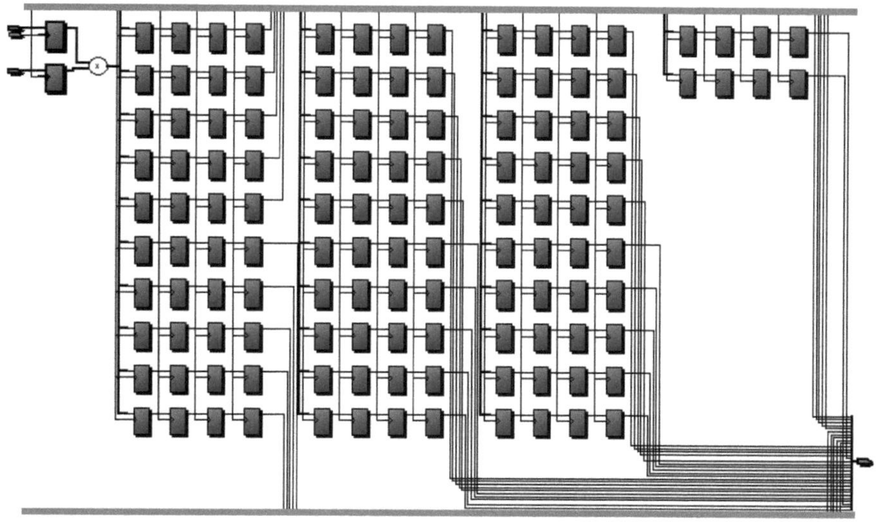

Fig. 10.22 Synthesis result for the 16-bit multiplier

Analysis & Synthesis Status	***Successful - Mon May16 21:48:55 2016***
Quartus II 32-bit Version	***13.0.0 Build 156 04/24/2013 Revision***
Name Mult_design	
Top-level Entity Name	***multiplier***
Family	***MAX II***
Total logic elements	***472***
Total pins	***65***
Total virtual pins	***0***
UFM blocks	***0 / 1 (0 %)***

Table 10.8 Resource usage summary for multiplier 16 bits using Altera Quartus II

	Analysis and synthesis resource usage summary	Usage
	Resource	
1	Total logic elements	472
1	combinational with no register	312
2	register only	126
3	combinational with a register	**34**
2		
3	Logic element usage by number of LUT inputs	
1	4-input functions	120
2	3-input functions	168
3	2-input functions;	37
4	1-input functions	19
5	0-input functions	2
4		
5	Logic elements by mode	
1	normal mode	297
2	arithmetic mode	175
3	qfbk mode	0
4	register cascade mode	0
5	synchronous clear/load mode	0
6	asynchronous clear/load mode	0
6		
7	Total registers	160
8	Total logic cells in carry chains	185
9	I/O pins	65
10	Maximum fan-out node	clk
11	Maximum fan-out	160
12	Total fan-out	1395
13	Average fan-out	2.60

10.8 Summary

The following are the key points to summarize this chapter:

1. The design partitioning can give the good and clear visibility of the data and control paths for the programmable ASIC design.
2. The RTL using VHDL for the complex design should have the separate functionality for the data paths and control paths.
3. Use the resource sharing concepts while coding for the logic unit. All the logical operations can be performed by using full adder component with additional combinational logic.

4. Parity generators are used to generate an even or an odd parity for the data input string.
5. Barrel shifters are combinational shifters and designed by using MUX-based logic.
6. Memories can be of distributed or BRAM type and inferred depending on the design requirement.
7. For less storage, distributed RAM can be inferred using the LUTs.
8. For the complex designs with large memory requirements, BRAMs can be inferred using dedicated block RAM resources of FPGA.
9. Multipliers are used as dedicated resource to perform the multiplication to realize DSP functions.

References

1. www.springer.com http://www.springer.com/us/book/9788132227892
2. www.xilinx.com "ISE Design Suite 14.6"
3. www.altera.com "Altera Quartus II 13.0 (32-bit)

Chapter 11
Design Implementation Using Xilinx Vivado

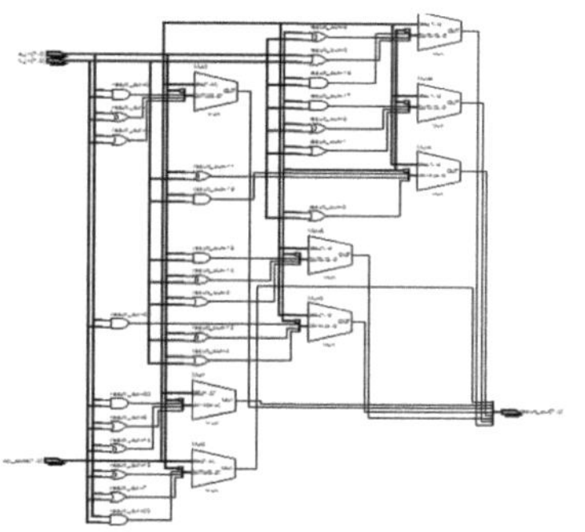

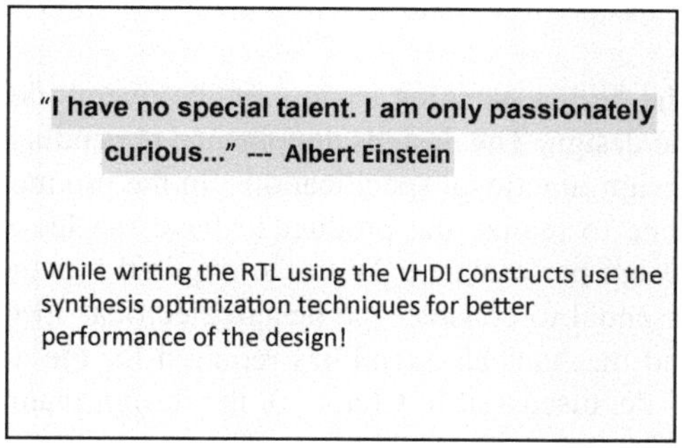

Abstract The PLD-based designs can be implemented by using the FPGA and by using the vendor-specific EDA tool chain. The chapter discusses about the design implementation using XILINX Vivado. The design flow using XILINX Vivado to perform the design simulation, synthesis, and implementation is discussed with the case study. Even this chapter discusses about the FIFO depth calculations and FIFO design.

Keywords FIFO · Vivado · Design planning · Verification · Translate · Map · Place and route · Simulation · Synthesis · Timing simulation · Bitstream · Device programming · IO planning · IO assignment · Constraints · Debugging

As discussed in the previous chapters, VHDL is efficiently used to code the functionality of the design. The design using VHDL can be implemented using the vendor-specific EDA tools. The subsequent section discusses about the design implementation using the XILINX Vivado.

V. Taraate, *PLD Based Design with VHDL*,
DOI 10.1007/978-981-10-3296-7_11

11.1 Design Implementation Case_Study Using Xilinx Vivado

The subsequent section discusses about the design implementation using XILINX Vivado. For better understanding of the design flow the combinational design with 8 inputs and 8 outputs is realized using XILINX Vivado. The design flow using the Xilinx Vivado is shown in Fig. 11.1.

11.1.1 Design Planning

The design planning stage uses the overall design specification while planning for the design. The designs are planned depending on the end application of the design, design functional specifications. In the product design cycle, the design planning is done to realize the product to have the lesser area, lesser power, and maximum design performance. With the required functional specifications of the design, it is essential to consider the design electrical specifications, environmental conditions, and mechanical assemblies required for the design.

As discussed in Chap. 10, the design planning stage starts with the architecture development for the design. Architecture for the design is block-level representation of the functionality and it gives information about the design intent. Depending on the functionality requirements, the architecture document can be created and can be evolved during this stage. The architecture of any design should give information about the processing logic, external memories, interfaces, and internal storage required. The overall data and control path information is described at the higher level in the architecture document. Even the architecture document should be able

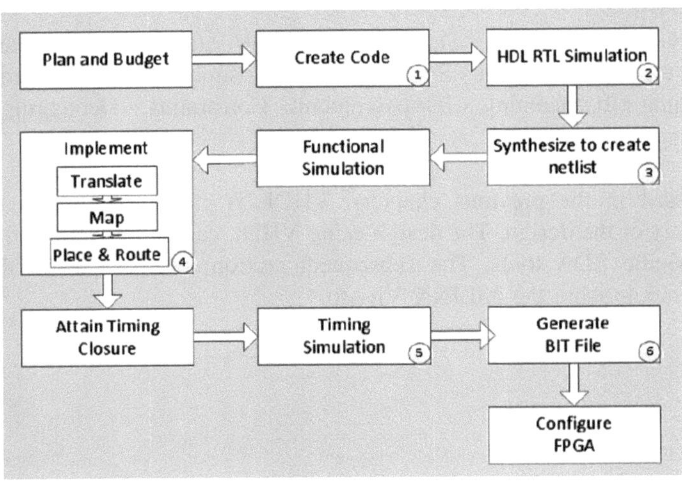

Fig. 11.1 Design flow [2]

to give information about the overall data rate for the design, operating frequency, area requirement, and power requirements. The architecture document also needs to focus on the requirement of third-part IPs, processors, the timing requirements, memories, and latency. The document should give the clarity about the overall area estimations and the target FPGA requirements for the design.

The architecture document is used in the later stages to plan the design. It acts as the source document while developing the microarchitecture of the design. The microarchitecture document can be created by using the architecture document. This document consists of the sub-blocks for every functional unit described in the architecture. For example, consider processor logic; the microarchitecture document should give information about the functional blocks of the processors, parallel and sequential processing algorithms, internal registers required, pipelining stages, buffer requirements, instructions and their decoding, communication, and interface mechanism with other functional blocks. Even the microarchitecture document should be able to specify the overall timing requirements for the individual functional blocks. For the SOC designs, the document should give the clarity on the hardware and software partitioning and their dependability. This document gives the information about the data and control flow for the individual functional blocks and hence used as reference document in the later stages of the design.

The RTL design using VHDL or Verilog uses the microarchitecture document as reference document. For the complex designs, better design partitioning plays an important role to realize the design with lesser area, lesser power, and improved design performance. The RTL designer should have clarity about the target technology for which design need to be implemented. The area, speed, and power improvement techniques with the coding and design guidelines need to be used while writing the RTL using VHDL or Verilog. The individual functional blocks with the correct functional intent can be integrated during this stage to realize the design.

The functional correctness of the design is checked by using the verification techniques. During the design verification, the objective is to check for the bugs to have the functional correctness of the design. As the complexity of the design increases, the design verification time and budgeting are additional overhead, but efforts during this stage can detect the functional bugs. For complex designs, the overall verification planning using the sophisticated self-checking testbenches can boost the design performance. In the practical scenario, the verification planning and verification architecture is used to improve the overall coverage for the design. Before the synthesis, the functional correctness of design is checked without using any delays. This kind of verification is called as presynthesis verification.

The fully functional RTL design is one of the input by the synthesis tool. Other inputs used by synthesis tool are ASIC libraries, design constraints. For the design using the PLD, the target FPGA device family information is used by the vendor-specific EDA tool. The synthesis outcome is gate-level netlist, and it is lower level abstraction of the HDL.

Consider the design shown in Example 11.1. To implement the design using the Xilinx Vivado, use the following steps:

```
library ieee;
use ieee.std_logic_1164.all;
use ieee.std_logic_arith.all;
use ieee.std_logic_unsigned.all;

entity comb_design is

 port (
  d_in      : in  std_logic_vector(7 downto 0);
  y_out     : out std_logic_vector(7 downto 0) );

end comb_design;

architecture arch_comb_design of comb_design is

Begin

  y_out(0) <= not d_in (0);

        y_out(1) <=( not d_in (2) ) d_in(1);

        y_out(2) <= (d_in (2) and d_in(3)) or (( not d_in (2) ) and
d_in(1));

        y_out(3) <= (d_in (2) and d_in(3));

        y_out(4) <= d_in(4);

        y_out(5) <= d_in(5);

        y_out(6) <= d_in (6);

        y_out(7) <= d_in(7);

        end arch_comb_design;
```

Example 11.1 VHDL RTL for comb_design

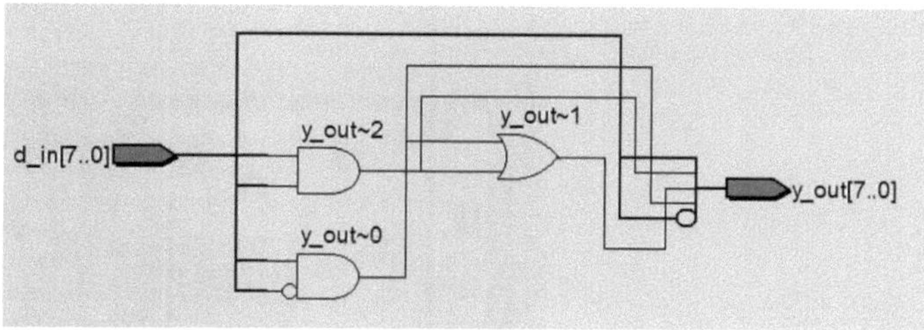

Fig. 11.2 Synthesis result for comb_design

1. Create the Vivado project and input the source file,
2. Simulate the design using Xsim simulator,
3. Perform the design synthesis,
4. Implement the design using Vivado,
5. Perform the timing simulation, and
6. Using Nexys 4 board, perform the functionality verification.

The design shown in the Example is implemented and verified using Xilinx Vivado.

Add the source VHDL file comb_design.vhd using the Xilinx Vivado, and perform the RTL analysis on the added source file. The result for the synthesis without the IO assignment is shown in Fig. 11.2.

11.1.2 IO Planning and IO Constraints

Use the IO planning layout shown in Fig. 11.3.

To perform the IO planning, use the following auxiliary view after clicking on IO planning (Fig. 11.4).

Fig. 11.3 IO planning layout [2]

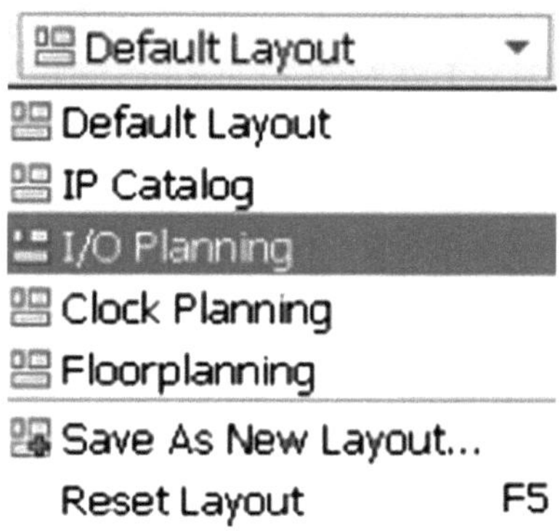

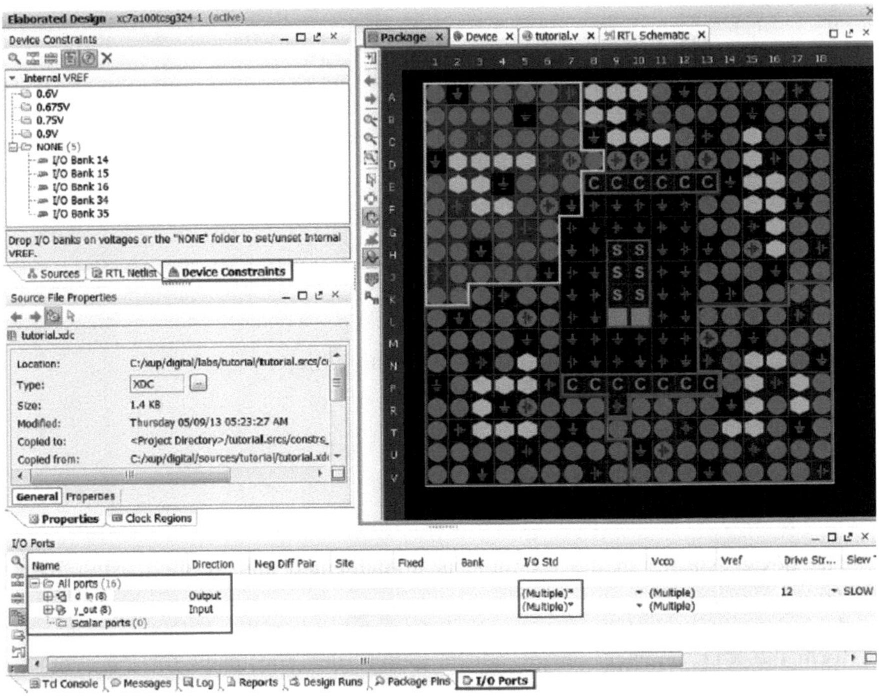

Fig. 11.4 IO planning auxiliary view [2]

In the auxiliary view, the package is displayed, and after selection of the device constraints, the IO ports are displayed in the console area. With multiple IO standards, the design inputs and outputs are listed in the IO tab area.

In the IO tab area, click on the (+) box for inputs (d_in) and output (y_out) (Fig. 11.5).

Now you can see the IO standards. For the d_in(6 downto 0) and y_out(6 downto 0), the IO standard LVCMOS33 is used, and for the d_in(7) and y_out(7), the default IO standard LVCMOS18 is used. Depending on the IO requirements, one of the IO standards can be chosen. Now to change the IO standard for the y_out (7) to LVCMOS33, use the following Fig. 11.6.

By using the tcl commands, also IO standards can be assigned. Use the following commands.

```
set_property package_pin V5 [get_ports {y_out[7]}]
set_property iostandard LVCMOS33 [get_ports [list
{y_out[7]}]]
```

Even by using the IO port properties, the IO standards can be assigned. After assignment of IO standards, save the constraints in the comb_design.xdc file.

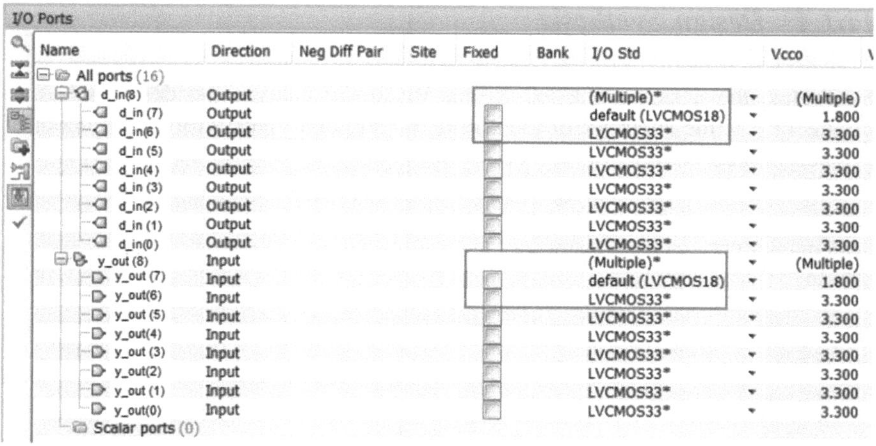

Fig. 11.5 IO standards [2]

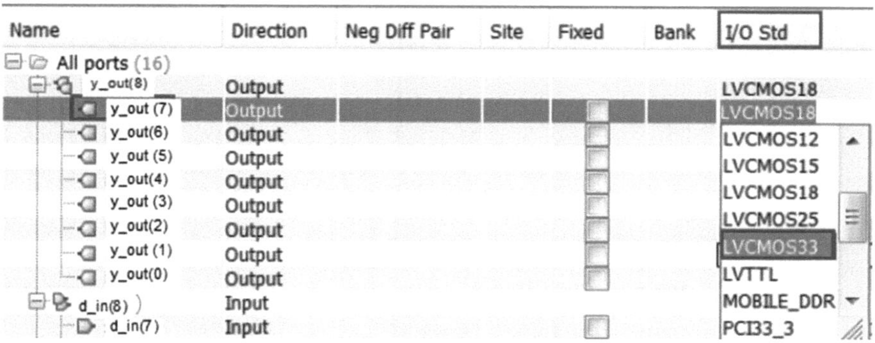

Fig. 11.6 Selection for IO standard [2]

11.1.3 Functional Simulation of the Design

Carry out the functional simulation of the design using Xsim simulator. The simulation results are shown in Fig. 11.7. Functional simulation is carried out buy writing the testbench using VHDL constructs.

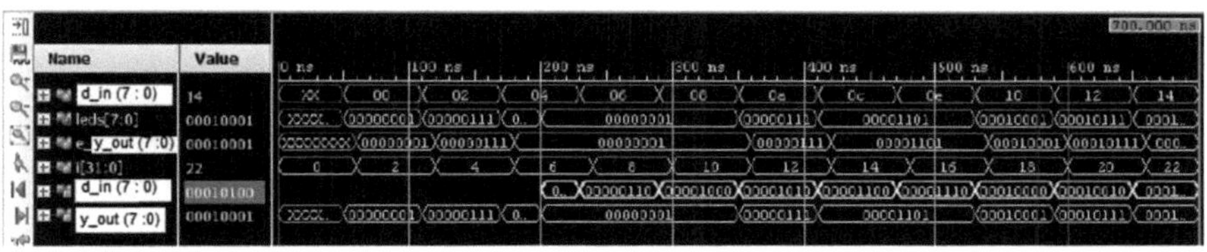

Fig. 11.7 Functional simulation result [2]

11.1.4 Design Synthesis

Synthesize the design using the Xilinx Vivado to analyze the design summary. Figure 11.8 is the snapshot of the Xilinx Vivado and gives information about the synthesis-phase completion.

Click on the Table tab to get the device utilization. The device utilization for the comb_design is shown in Fig. 11.9. As the design is basic combinational logic, it uses three LUTs and 16 IOs only.

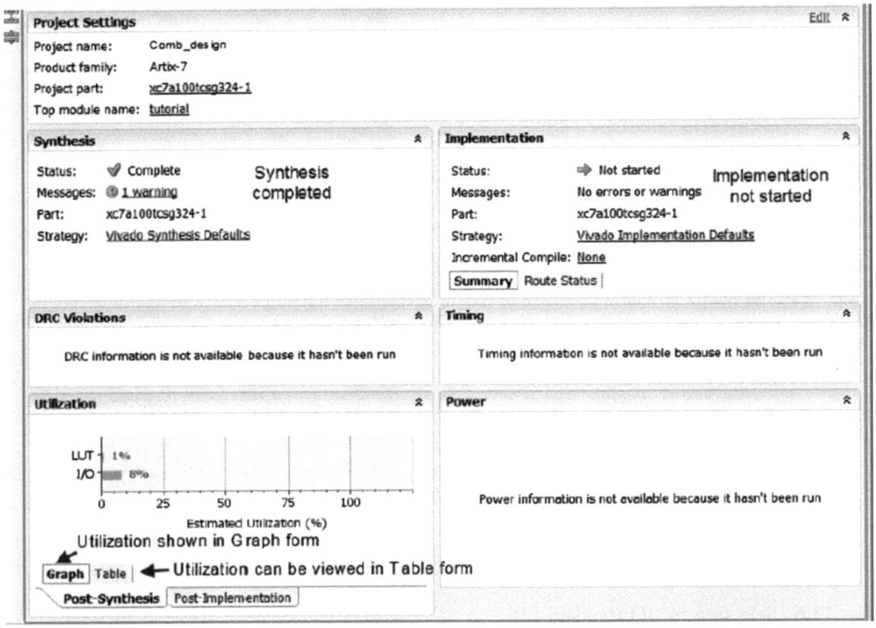

Fig. 11.8 Synthesis-completed window [2]

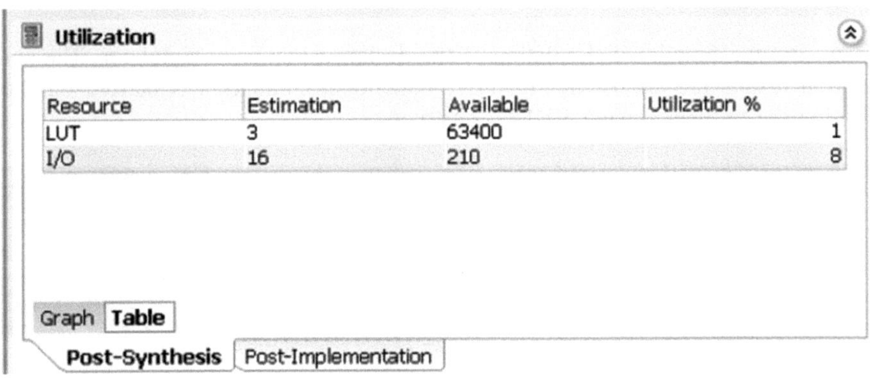

Fig. 11.9 Device utilization [2]

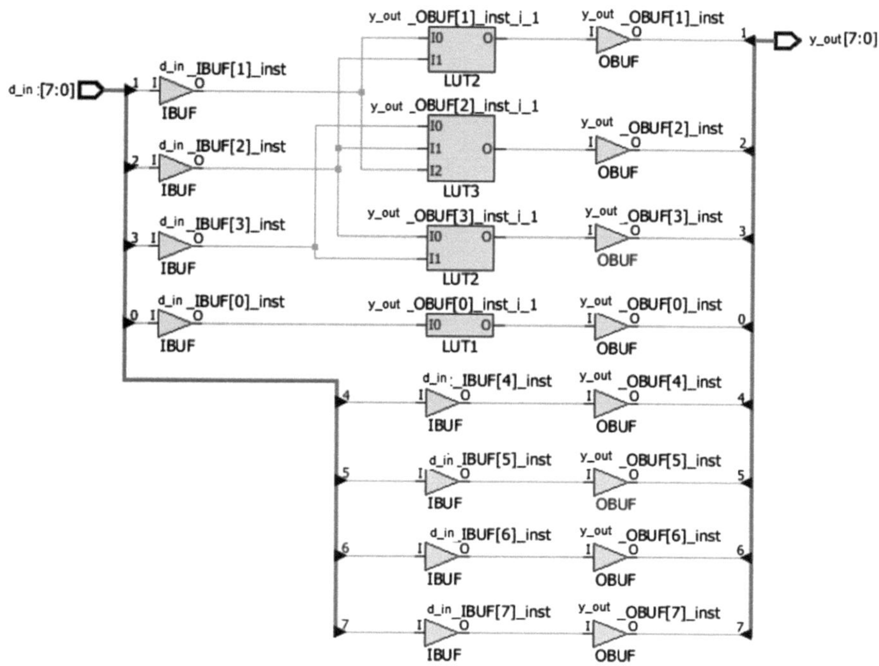

Fig. 11.10 Netlist view [2]

The synthesis result can be in the form of gate level netlist and for the above design the synthesis result is shown in Fig. 11.10. As shown, the IO buffers are automatically added by the tool in the input and output path. LUTs are used to map the gates.

11.2 Design Implementation

The design is implemented using Vivado by clicking on the 'run implementation' which is in the implementation task. The design implementation is performed by Vivado using the synthesis output file, and after design implementation, the implementation result can be viewed in the schematic form by clicking on the 'open implemented design.' Figure 11.11 shows the implemented design.

To check the project summary, close the implemented design view and select the project summary tab. Select the post-implementation tab under the timing and utilization window. Figure 11.12 shows the post-implementation status, and the device utilization is only 3 LUTs with 16 IOs. As the design is combinational, there are no any timing constraints provided for the design.

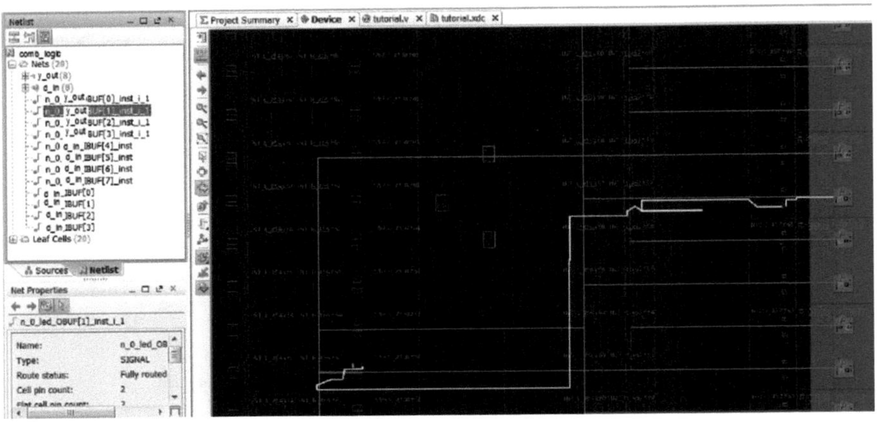

Fig. 11.11 Implemented design [2]

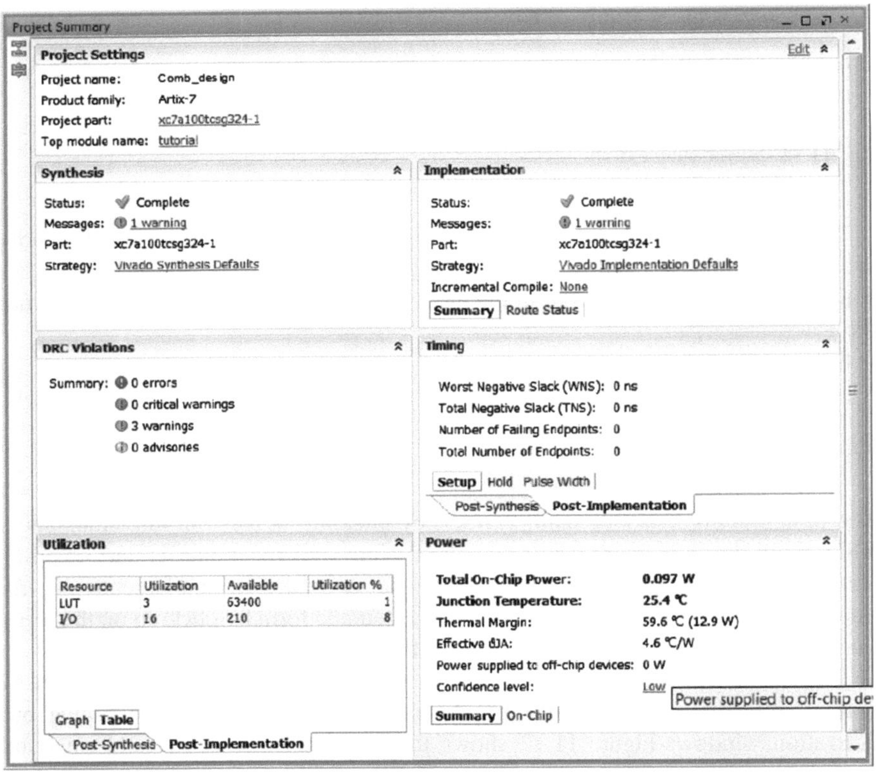

Fig. 11.12 Post-implementation summary [2]

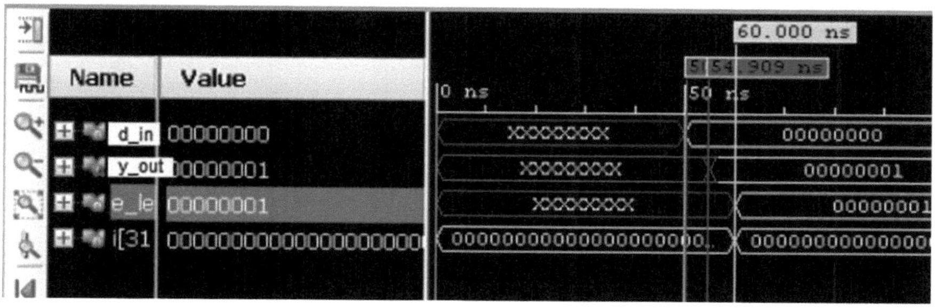

Fig. 11.13 Timing simulation result [2]

11.2.1 Timing Simulation

Perform the timing simulation by using the Vivado. Use the run simulation > run post-implementation timing simulation. Use the comb_design.tb tas top-level module. The result of timing simulation is shown in Fig. 11.13.

11.3 FPGA Board Bring-up

Create the bitstream file and verify the design functionality.

1. Click on the 'Generate Bitstream' under the program and debug tasks.
2. This will be generated by using the implemented design output. The bitstream file 'comb_design.bit' is generated under 'impl_1' directory.
3. Check for the board setting and power on status of the board. The Nexys 4 board is shown below (Fig. 11.14).
4. Click on 'Open new hardware target' link. The link is shown in Fig. 11.15.
5. Click 'Next' to see the Vivado CSE server name form.

Fig. 11.14 Nexys 4 board [2]

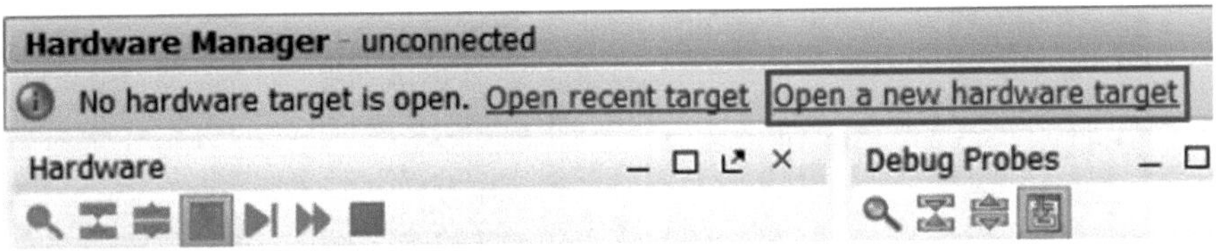

Fig. 11.15 New hardware tab [2]

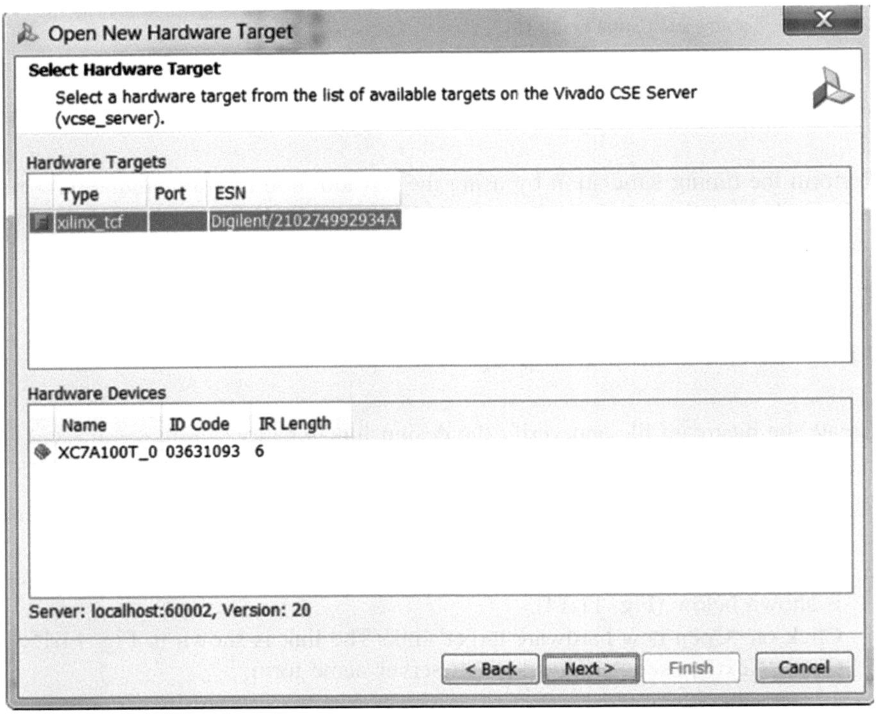

Fig. 11.16 Hardware device-detected window [2]

6. Click 'Next' with local host port selected. The JTAG cable will be searched to detect the Xilinx_tcf. This shows the hardware device detected in the chain. Figure 11.16 shows the detected hardware device.
7. Click 'Next' till Finish and this will give the status of hardware session from unconnected to the server name. The device is highlighted and indicates that it is not programmed.
8. Now select the device and use comb_design.bit file as programming file (Fig. 11.17).

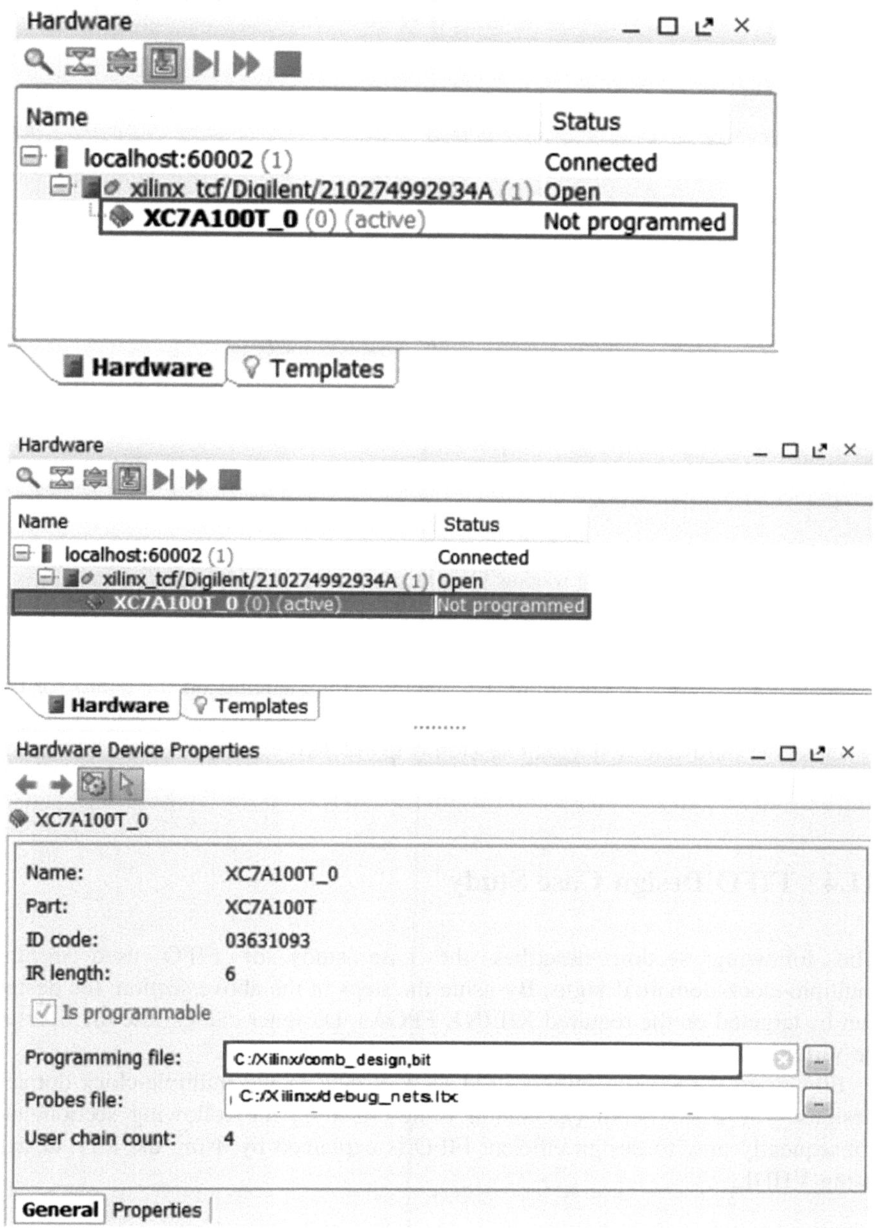

Fig. 11.17 Bitstream programming [2]

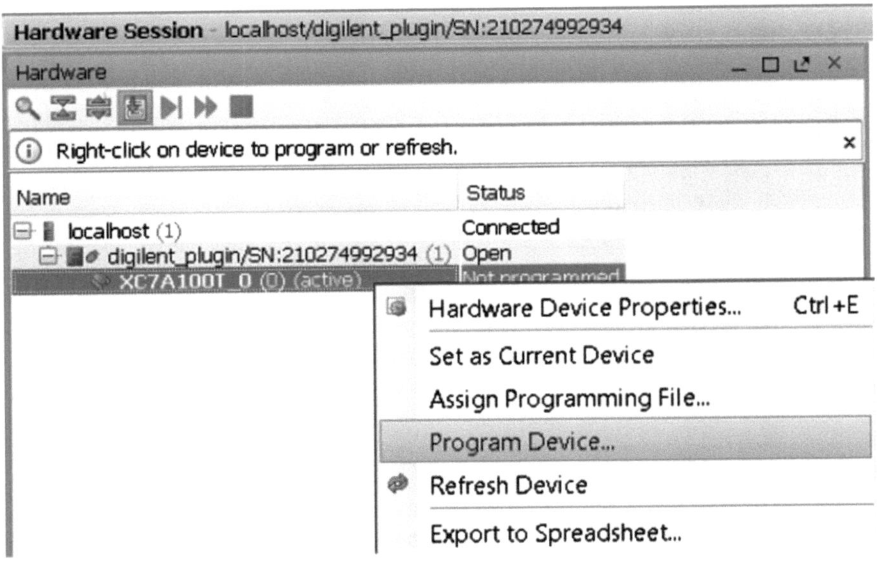

Fig. 11.18 Device programming window [2]

9. Now select the Program Device, use the right click to configure FPGA. The snapshot is shown in the figure. By changing the switches on the board for the respective inputs, verify the output. By using File > Close Hardware Manager close the hardware debugging session (Fig. 11.18).

11.4 FIFO Design Case Study

The following section describes the case study of FIFO used in the multiple-clock-domain designs. By using the steps in the above section, the design can be targeted on the required XILINX FPGAs. Designer can choose the Spartan or Virtex series FPGAs required for the suitable applications.

FIFOs are the storage buffers used to pass data in the multiple-clock-domain designs. The FIFO depth calculation is described in the following section and subsequently how to design efficient FIFO is explained by using the RTL design using VHDL.

11.4.1 Asynchronous FIFO Depth Calculations

Scenario I: Clock domain 1 is faster as compared to clock domain 2; that is, f1 is greater than f2 without any idle cycle between write and read.

Consider the design where f1 = 100 MHz and f2 = 50 MHz and the burst of data transfer from clock domain one to clock domain 2 is 100 without idle cycles that is consecutive write and read cycles.

The depth of FIFO can be calculated as :

1. **Find time required to write one data** :

   Twrite = 1/100 MHz = 10 nsec
2. **Find out time required to write burst of data** :

   Tburst_write= Twrite * Burst length = 10nsec * 100 = 1micro-second
3. **Find time required to read one data :**

   Tread = 1/50 MHz = 20 nsec
4. **Find out number of data read in duration of Tburst_write :**

   No of reads = 1000 nsec/20 nsec = 50
5. **The depth of FIFO :**

   Depth = Burst length – No of reads = 100-50 = 50

Scenario II: Clock domain 1 is faster as compared to clock domain 2; that is, f1 is greater than f2 with idle cycles between writes and reads.

Consider the design where f1 = 100 MHz and f2 = 50 MHz and the burst of data transfer from clock domain one to clock domain 2 is 100 with idle cycles . Number of idle cycles between two successive writes = 1 and number of idle cycle between two successive reads =3

The depth of FIFO can be calculated as :

1. ***Find time required to write one data*** :

As between two successive writes the idle cycle is one therefore for every two cycles one data is written

Twrite = 2 * (1/100 MHz) = 20 nsec

2. ***Find out time required to write burst of data*** :

Tburst_write= Twrite * Burst length = 20nsec * 100 = 2 micro-second

3. **Find time required to read one data** :

As between two successive reads the idle cycle is three therefore for every four cycles one data is read

Tread = 4 * (1/50 MHz) =80 nsec

4. **Find out number of data read in duration of Tburst_write** :

No of reads = 2000 nsec/80 nsec = 25

5. **The depth of FIFO :**

Depth = Burst length – No of reads = 100-25 = 75

Scenario III: Clock domain 1 is slower as compared to clock domain 2; that is, f1 is less than f2 with idle cycles between two successive writes and two successive reads.

Consider the design where f1 = 50 MHz and f2 = 80 MHz and the burst of data transfer from clock domain one to clock domain 2 is 100 with idle cycles . Number of idle cycles between two successive writes = 1 and number of idle cycle between two successive reads =3

The depth of FIFO can be calculated as :

1. ***Find time required to write one data :***

As between two successive writes the idle cycle is one therefore for every two cycles one data is written

Twrite = 2 * (1/50 MHz) = 40 nsec

2. ***Find out time required to write burst of data :***

Tburst_write= Twrite * Burst length = 40nsec * 100 = 4 micro-second

3. ***Find time required to read one data :***

As between two successive reads the idle cycle is three therefore for every four cycles one data is read

Tread = 4 * (1/80 MHz) =50 nsec

4. ***Find out number of data read in duration of Tburst_write :***

No of reads = 4000 nsec/50 nsec = 80

5. ***The depth of FIFO :***

Depth = Burst length – No of reads = 100-80 = 20

Scenario IV: Clock domain 1 is the frequency equal to clock domain 2; that is, f1 is equal to f2 and idle cycles between two successive reads and writes

Consider the design where f1 = 100 MHz and f2 = 100 MHz and the burst of data transfer from clock domain one to clock domain 2 is 100 with idle cycles . Number of idle cycles between two successive writes = 1 and number of idle cycle between two successive reads =3

The depth of FIFO can be calculated as :

1. ***Find time required to write one data :***

As between two successive writes the idle cycle is one therefore for every two cycles one data is written

Twrite = 2 * (1/100 MHz) = 20 nsec

2. ***Find out time required to write burst of data :***

Tburst_write= Twrite * Burst length = 20nsec * 100 = 2 micro-second

3. ***Find time required to read one data :***

As between two successive reads the idle cycle is three therefore for every four cycles one data is read

Tread = 4 * (1/100 MHz) =40 nsec

4. ***Find out number of data read in duration of Tburst_write :***

No of reads = 2000 nsec/40 nsec = 50

5. ***The depth of FIFO :***

Depth = Burst length – No of reads = 100-50 = 50

11.4.2 FIFO Design Using VHDL

The FIFO design uses the dual-port RAM as component and the RTL description using the VHDL is shown in Example 11.2. Please refer Chap. 10 for the dual-port RAM implementation.

The design can be implemented by using the XILINX and ALTERA PLDs.

```
library ieee;
use ieee.std_logic_1164.all;
use ieee.std_logic_unsigned.all;
entity fifo _RTL is
    generic(
        DATA_WIDTH:natural:=8;
        ADDRESS_WIDTH:natural:=2
    );
    port(
        reset_n      :in std_logic;
        clk      :in std_logic;
```

Example 11.2 VHDL RTL for FIFO

```
        data_in  :in std_logic_vector(DATA_WIDTH-1 downto 0);

        data_out :out std_logic_vector(DATA_WIDTH-1 downto 0);

        read_en    :in std_logic;

        write_en    :in std_logic;

        fifo_empty    :out std_logic;

        fifo_full    :out std_logic

    );

end entity fifo_RTL;

architecture arch_RTL_fifo  of fifo_RTL is

    signal read_ptr:std_logic_vector(ADDRESS_WIDTH-1 downto
    0):=(others=>'0');

    signal write_ptr:std_logic_vector(ADDRESS_WIDTH-1 downto
    0):=(others=>'0');

    signal count:std_logic_vector(ADDRESS_WIDTH downto
    0):=(others=>'0');
```

Example 11.2 (continued)

```
signal valid_read:std_logic:='0';

signal valid_write:std_logic:='0';

signal empty:std_logic:='1';

signal full:std_logic:='0';

constant MAX:std_logic_vector(ADDRESS_WIDTH downto
0):=('1',others=>'0');

constant MIN:std_logic_vector(ADDRESS_WIDTH downto
0):=(others=>'0');

component dual_port_ram

   generic(

            DATA_WIDTH:natural:=8;

            ADDRESS_WIDTH:natural:=4

   );

   port(
```

Example 11.2 (continued)

```
            reset_n      :in std_logic;

            read_clk   :in std_logic;

            write_clk   :in std_logic;

            data_in  :in std_logic_vector(DATA_WIDTH-1 downto 0);

    data_out :out std_logic_vector(DATA_WIDTH-1 downto 0);

    read_addr  :in std_logic_vector(ADDRESS_WIDTH-1 downto 0);

    write_addr  :in std_logic_vector(ADDRESS_WIDTH-1 downto 0);

            read_en    :in std_logic;

            write_en    :in std_logic

    );

end component dual_port_ram;

begin

memory:dual_port_ram

    generic map(
```

Example 11.2 (continued)

```
        DATA_WIDTH=>DATA_WIDTH,

        ADDRESS_WIDTH=>ADDRESS_WIDTH

)

port map(

        reset_n=>reset_n,

        read_clk=>clk,

        write_clk=>clk,

        data_in=>data_in,

        data_out=>data_out,


        read_addr=>read_ptr(ADDRESS_WIDTH-1 downto 0),

        write_addr=>write_ptr(ADDRESS_WIDTH-1 downto 0),

        read_en=>valid_read,

        write_en=>valid_write
```

Example 11.2 (continued)

```
  );

valid_read<='1' when (read_en='1' and empty='0') else '0';

valid_write<='1' when (write_en='1' and full='0') else '0';

empty<='1' when count=MIN else '0';

full<='1' when count=MAX else'0';

  fun_p2:process(reset_n,clk) is

        begin

              if (reset_n='0') then

                    read_ptr<=(others=>'0');

                    write_ptr<=(others=>'0');

                    count<=(others=>'0');

              elsif rising_edge(clk) then

                          if (valid_read='1') then

                                read_ptr<=read_ptr+1;
```

Example 11.2 (continued)

```
            if (valid_write='1') then
                count<=count;
            else
                count<=count-1;
            end if;
        end if;
        if (valid_write='1') then
            write_ptr<=write_ptr+1;
            if (valid_read='1') then
                count<=count;
            else
                count<=count+1;
            end if;
        end if;
```

Example 11.2 (continued)

```
                end if;

          end process fun_p2;

   fifo_empty<=empty;

   fifo_full<=full;

end architecture arch_RTL_fifo
```

Example 11.2 (continued)

11.5 Summary

The following are the key points to summarize this chapter:

1. The RTL description using VHDL for the complex design should have the separate functionality for the data paths and control paths.
2. Use the FIFO for passing the data from one of the clock domains to another clock domain.
3. Design can be implemented on XILINX and Altera PLDs depending on the available resources.
4. The combinational logic is mapped into the LUT.
5. The timing simulation of design is post-layout simulation which includes the delays.
6. In the prelayout simulation, delays are not included.

References

1. www.springer.com http://www.springer.com/us/book/9788132227892.
2. www.xilinx.com "XILINX Vivado Design guide".

Erratum to: PLD Based Design with VHDL

Vaibhav Taraate

Erratum to:
V. Taraate, *PLD Based Design with VHDL*, DOI 10.1007/978-981-10-3296-7

The original version of the book was inadvertently published without the following corrections:

1. In Front matter, p. viii, Preface: explanation for Chaps. 9 and 10 are modified as per TOC.
2. In Chap. 1, p. 2, sixth line of second paragraph in Sect. 1.1: "**V**ery-**H**igh-**S**peed **I**ntegrate **C**ircuit **H**ardware **D**escription **L**anguage (**V**$_{\mathbf{HSIC}}$**HDL**)" has been modified to "**V**$_{\mathbf{HSIC}}$**HDL**: **V**ery **H**igh **S**peed **I**ntegrated **C**ircuit **H**ardware **D**escription **L**anguage".
3. In Chap. 1, p. 16, last line: "end struct;" is changed to "end arch_struct;" in program "Example 1.1 Structural style for the half adder"; p. 17, the last line: "end behav;" is changed to "end arch_behav;" in program "Example 1.2 Behavior style of the VHDL code for half adder".
4. In Chap. 2, p. 28 and p. 30: "Example 2.2 Synthesizable VHDL code for two-input OR logic" and "Example 2.3 Synthesizable VHDL code for NOR logic" are replaced with the new programing codes, respectively.

The updated original online version for this book can be found at
http://dx.doi.org/10.1007/978-981-10-3296-7

V. Taraate (✉)
Pune, Maharashtra, India
e-mail: vaibhavt@gmail.com

V. Taraate, *PLD Based Design with VHDL*,
DOI 10.1007/978-981-10-3296-7_12

5. In Chap. 4, p. 100: “Example 4.6 Synthesizable RTL for binary-to-excess-3 code converter” is replaced with the new programming code; the typo error in cross links for “Examples 4.13, 4.14, 4.16, and 4.18” are changed to “Examples 4.12, 4.13, 4.14 and 4.15” respectively.
6. In Chap. 7, p. 227: “if(clk= ‘1’ and clk’event) then” is updated with “if(clk= ‘0’ and clk’event) then”.
7. In Chap. 8, p. 250, Example 8.8: the condition “when “1100” => y_out <= “00010”” is updated with “when “1100” => y_out <= “0001000000000000””.
8. In Chap. 9, p. 277, Table 9.1: s6 state value for binary encoding “101” is updated with “110”.
9. In Chap. 10, p. 333: “Example 10.3 VHDL RTL for the arithmetic unit” is updated with new programing code.
10. The cover type is changed from “Soft Cover” to “Hard Cover”.

The erratum chapters and the book have been updated with the changes.

Appendix A
Key Differences VHDL 87 and VHDL 93

The key differences in the syntax of VHDL 87 and VHDL 93 are listed in this appendix.

1. **Alias**

In VHDL 87, aliases are declared for the object, but in VHDL 93, aliases are declared for the objects, subprograms, types, and operators and even for the named entities. In VHDL 93, aliases cannot be declared for the entities with the loop parameters, labels, and generate parameters.

2. **Attributes**

In VHDL 93, the following attributes are added:

- ASCENDING
- DRIVING
- DRIVING_VALUE
- IMAGE
- INSTANCE_NAME
- PATH_NAME
- SIMPLE_NAME
- VALUE

3. **Bit-String literals**

In VHDL 87, bit-string literals are of type ***Bit_Vector***. For example, if signal 'tmp_sig' is declared as std_logic_vector (0–7), then using VHDL 87 the assignment can be

```
tmp_sig <= to_stdlogicvector(x"B1A2");
```

But using VHDL 93, the above signal assignment generates error. So for VHDL 87 and VHDL 93, the following can work

V. Taraate, *PLD Based Design with VHDL*,
DOI 10.1007/978-981-10-3296-7

```
tmp_sig <= to_stdlogicvector(Bit_vector(x"B1A2"));
```

4. **Character Set**

The character set in VHDL 87 is 128 characters, but in VHDL 93, the character set is of 256 characters.

5. **Direct Instantiations**

In VHDL 87, component declaration is required, but in VHDL 93, it is possible to exclude the component declaration and it is possible to instantiate an entity or configuration declaration. VHDL 87 does not allow any international characters even in comments. But using VHDL 93, many standard EDA tools support the international characters in the comments.

6. **Delayed concurrent statements**

In VHDL 87, it is not possible to have all concurrent statements active during simulation. In VHDL 93, it is possible to have all the concurrent statements active during simulation as postponed.

7. **Extended Identifiers**

Extended identifiers with the backslash '\' are supported in VHDL 93. Extended identifiers always start with the '\' and are case sensitive. The extended identifiers may include the reserved words and spaces.

8. **Files**

File handling is very different in VHDL 93 as compared to VHDL 87. The predefined subprograms such as File_Open and File_Close are not supported in VHDL 87. VHDL 93 supports Impure for the functions using files outside the local scope. VHDL 87 does not support the Impure.

File parameters for the subprogram do not have mode as In and Out in VHDL 93.

9. **Generate**

The generate statement in VHDL 87 does not support the declaration. By using VHDL 93, the declaration is possible.

10. **Impure functions**

The function ***Now*** that returns the current simulation time is impure function in VHDL 93. An impure function works using the parameters and returns the different values for the identical input parameters. Function calling an impure function must be declared as ***Impure.*** The procedure not working by using only parameters can be declared as ***Impure.***

11. **Port associations**

In VHDL 87, actual parameter must be of signal type. VHDL 93 allows the use of the constant value as input port parameter. VHDL 93 allows ***slice*** as the formal parameter. Even VHDL 93 allows the type conversion functions and direct type conversions between the formal and actual parameters.

12. **Report**

By using VHDL 87, it is possible to use ***Report*** statement with Assert. ***Report*** statement is new in VHDL 93.

13. **Shared Variables**

VHDL 87 does not allow shared variable, but VHDL 93 allows the use of the shared variables in the concurrent declarations.

14. **Signal delay**

By using VHDL 93, it is possible to describe inertial delay by using ***Inertial.*** By using VHDL 93, it is possible to use the ***Reject*** to combine the ***inertial*** and ***transport***.

By using VHDL 87, additional signal is required to have the similar functionality.

For example:

In VHDL 87 the signal delay can be expressed by using

```
tmp_sig <= d_in after 3 ns;
y_out <=  transport tmp_sig after 5 ns;
```

In VHDL 93 the delay assignment is declared as

```
y_out <= reject 3 ns inertial d_in after 8 ns;
```

15. **Syntax**

VHDL 87 syntax is allowed in VHDL 93, and the following are the differences in the declaration using VHDL 87 and VHDL 93.

S. No.	VHDL 87	VHDL 93
1	end arch_name;	end architecture arch_name;
2	end entity_name;	end entity entity_name;
3	end conf_name;	end configuration conf_name;
4	end component;	end component comp_name;

(continued)

(continued)

S. No.	VHDL 87	VHDL 93
5	end fun_name;	end function fun_name;
6	end proc_name;	end procedure proc_name;
7	end record;	end record rec_name;
8	end pck_name	end package pck_name;

16. **Statement declaration**

The statement declaration using VHDL 87 and VHDL 93 differs, and the declaration style is shown below.

S. No.	VHDL 87	VHDL 93
1	component : comp_name	component : comp_name is
2	blk_name : block	blk_name : block is
3	proc_name: process	proc_name: process is

Appendix B
Xilinx Spartan Devices

- **XILINX SPARTAN 3 DEVICES**

V. Taraate, *PLD Based Design with VHDL*,
DOI 10.1007/978-981-10-3296-7

Device	System gates	Equivalent	CLB array (One CLB = Four Slices)			Distributed RAM Bits (K = 1024)	Block RAM Bits (K = 1024)	Dedicated Multipliers	DCMs	Max. User I/O	Maximum Differential I/O Pairs
		Logic cells[(1)]	Rows	Columns	Total CLBs						
XC3S50[(2)]	50K	1,728	16	12	192	12K	72K	4	2	124	56
XC3S200[(2)]	2G0K	4,320	24-	20	480	30K	216K	12	4	173	76
XC3S400[(2)]	400K	8,064	32	28	896	56K	280K	16	4	264	116
XC3S1000[(2)]	1M	17,280	48	40	1,920	120K	432K	24	4	391	175
XC3S1500	1.5M	29,952	64	52	3,328	208K	576K	32	4	487	221
XC3S2000	2M	46,080	80	64	5,120	320K	720K	40	4	565	270
XC3S4000	4M	62,208	96	72	6,912	432K	1,728K	96	4	633	300
XC3S5000	5M	74,880	1104	80	8,320	520K	1,872K	104	4	633	300

Notes

1. Logic cell 1 [U Look-Up Table (LUI) plus a ‘D’ flip-flop. ‘Equivalent Logic Cells’ equals ‘Total CLBs’ × 8 logic cells/CLB × 1.125 effectiveness
2. These devices are available in Xilinx Automotive versions as described in DS314: *Spartan-3 Automotive XA FPGA Family*

- **Spartan 3 Family Architecture**

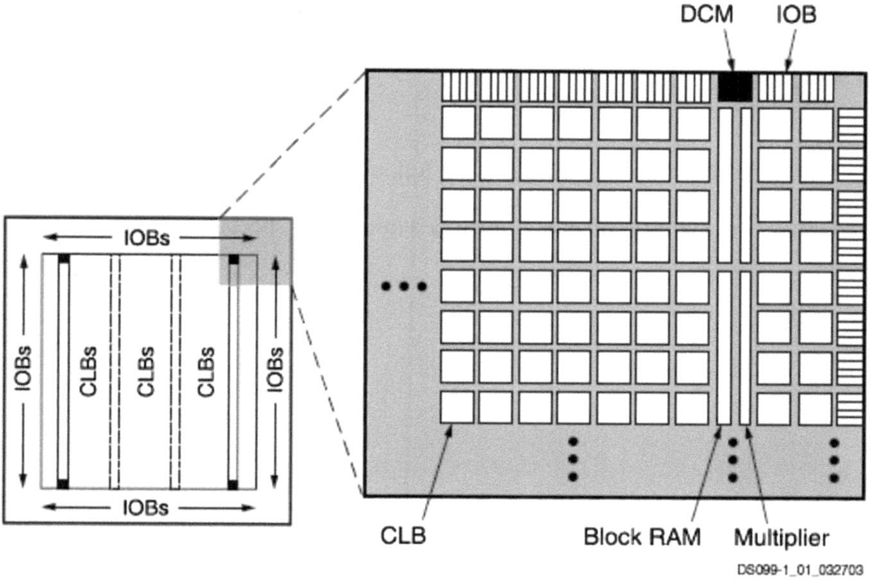

Notes:

1. The two additional block RAM columns of the XC3S4000 and XC3S5000 devices are shown with dashed lines. The XC3S50 has only the block RAM column on the far left.

- **Xilinx Spartan 3 Package information for Part no. XC3S400-4PQ208C**

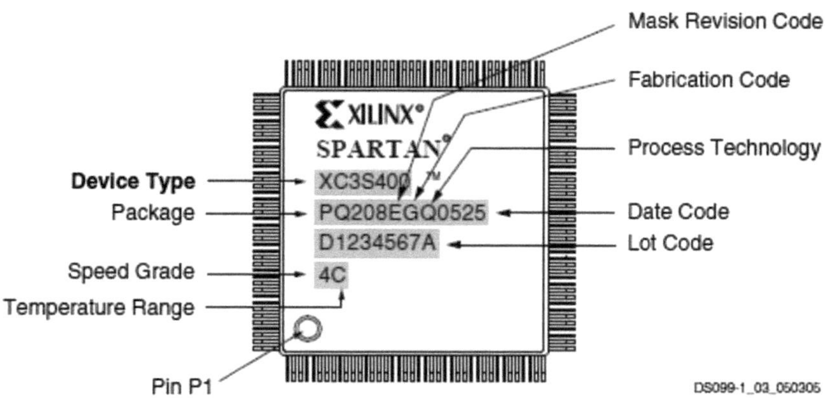

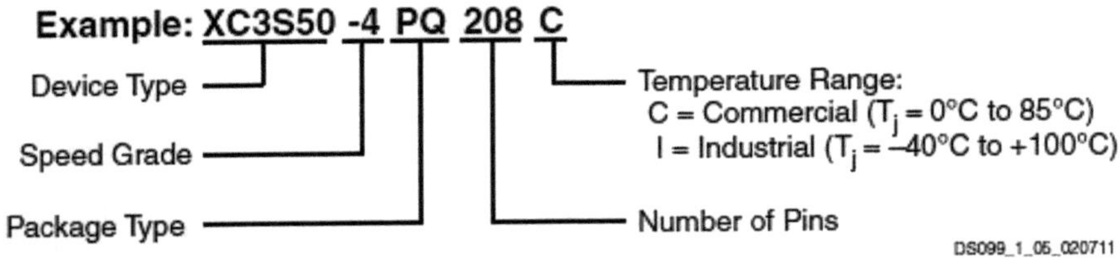

For more information, please use the following link:

http://www.xilinx.com/support/documentation/data_sheets/ds099.pdf

- **Xilinx FPGA Spartan 3E Devices**

Device	System gates	Equivalent logic cells	CLB Array (One CLB = Four Slices)				Distributed RAM bits[1]	Block RAM bits[1]	Dedicated multipliers	DCMs	Maximum user I/O	Maximum differential I/O pairs
			Rows	Columns	Total CLBs	Total slices						
XC3S100E	100K	2,160	22	16	240	960	15K	72 K	4	2	108	40
XC3S250E	250K	5,508	34	26	612	2,448	38K	216K	12	4	172	68
XC3S500E	500K	10,476	46	34	1,164	4,856	73K	360K	20	4	232	92
XC3S1200E	1200K	19,612	60	46	2,168	8,872	136K	504K	28	8	304	124
XC3S1600E	1600K	33,192	76	58	3,688	14,762	231K	648K	36	8	376	156

Notes

1. By convention, 1 Kb is equivalent to 1,024 bits

- **XILINX SPARTAN 3E Architecture**

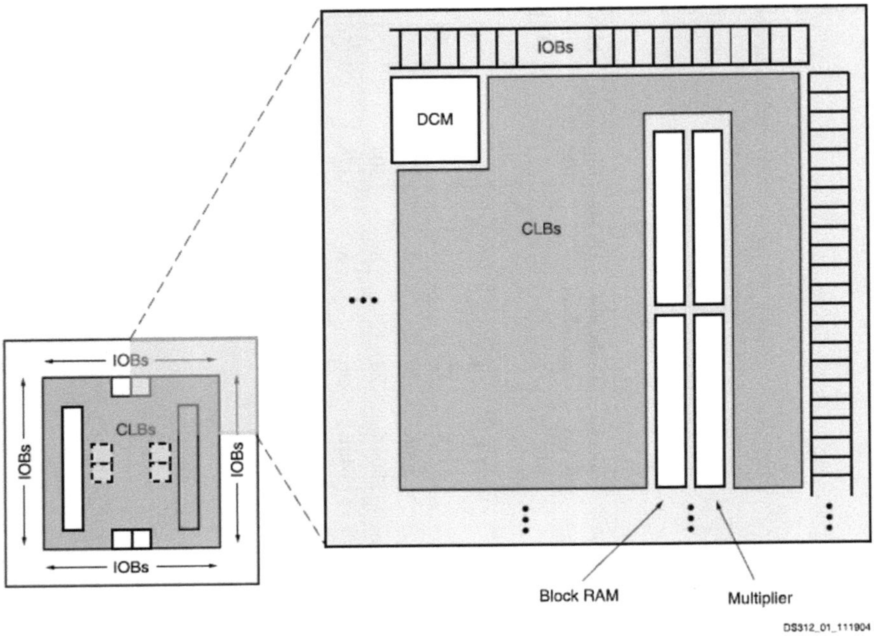

- **XILINX Spartan 3E package information**

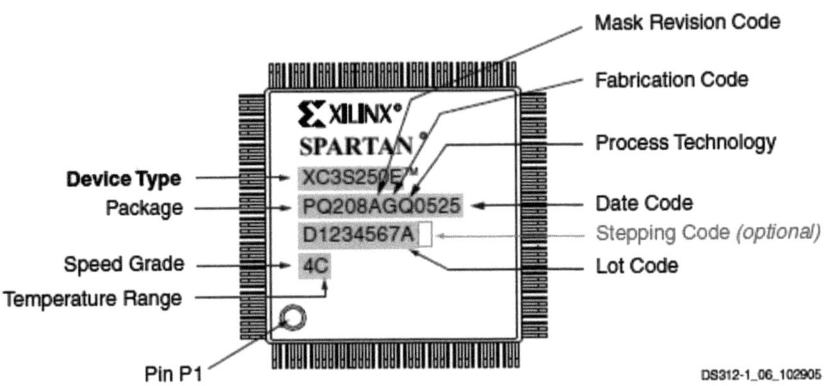

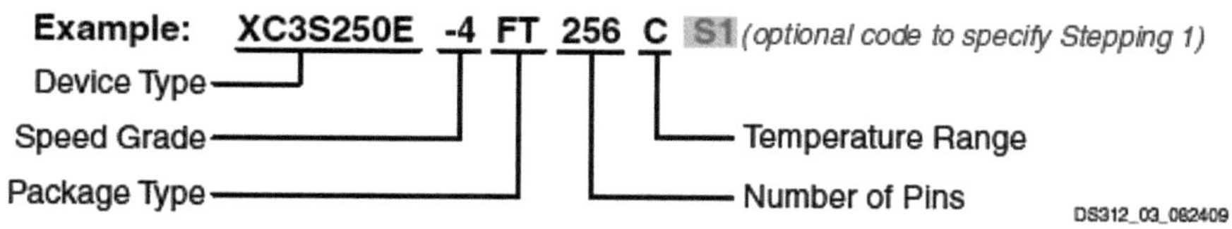

For more information, please use the following link:

http://www.xilinx.com/support/documentation/data_sheets/ds312.pdf

Appendix C
Altera (Intel FPGA) Cyclone IV Devices

- **Cyclone IV GX FPGA Devices**

V. Taraate, *PLD Based Design with VHDL*,
DOI 10.1007/978-981-10-3296-7

Cyclone IV GX FPGA Devices

		Maximum resource count for cyclone IV GK FPGAs (1.2 V)						
		EP4CGX15	EP4CGX22	EP4CGX30	EP4CGX50	EP4CGX75	EP4CGX110	EP4CGX150
Resources	LEs (K)	14	21	29	50	74	109	150
	M9K memory blocks	60	B4	120	278	462	666	720
	Embedded memory (Kb)	540	756	1,030	2,502	4,158	5,490	6,480
	18 × 18 multipliers	0	40	80	140	108	280	360
Architectural features	Global clock networks	20	20	20	30	30	30	30
	PLLs	3	4	4	8	9	9	8
I/O features	I/O voltage levels supported (V)	1.2, 1.5, 1.8, 25, 13						
	I/O standards supported	LVTTL, LVCMOS, PCI, PCI-X, LVDS, mini-LVDS, RSDS, LVPECL. SSTL-18 (1 and II), SSTL-15 (I and II), SSTL-2 (I and II). HSTL-1B (I and II), HSTL-15 (I and II), HSTL-12 (I and II), Differential SSTL-18 (I and II), Differential SSTL-15 (I and II), Differential SSTL-2 (I and II), Differential HSTL-18 (I and II), Differential HSTL-15 (I and II), Differential HSTL-12 (I and II), Differential HSUL-12						
	Emulated LVDS channels	9	40	40	73	73	139	139
	LVDS channels. B40 Mbps (receive/transmit)	7/7	14/14	14/14	49/49	49/49	59/59	59/59
	Transceiver count[1] (2.5 Gbps/3.125 Gbps)	2/0	2, 0/4, 0	4, 0/0, 4[2]	0, 8	0, 8	0, 8	0, 8
	PCIe hard IP blocks (Gen1)				1			
External memory interfaces	Memory devices supported	DDR2, DDR, SDR						

- **Cyclone IV E FPGA Features**

		Maximum resource count for cyclone IV E FPGAs								
		EP4CE6	EP4CE10	EP4CE15	EP4CE22	EP4CE30	EP4CE40	EP4CE55	EP4CE75	EP4CE115
Resources	LEs (K)	6	10	115	22	29	40	56	75	114
	M9K memory blocks	30	46	56	66	66	126	260	305	432
	Embedded memory (Kb)	270	414	504	594	594	1,134	2,340	2,745	3,888
	18 × 18 multipliers	15	23	56	66	66	116	154	200	266
Architectural features	Global clock networks	10	10	20	20	20	20	20	20	20
	PLLs	2	2	4	4	4	4	4	4	4
I/O features	I/O voltage levels supported (V)	1.2, 1.5, 1.8, 2.5, 3.3								
	I/O standards supported	LVTTL, LVCMOS, PCI, PCI-X, LVDS, mini-LVDS, RSDS, LVPECL, SSTL-18 (1 and II), SSTL-15 (I and II), SSTL-2 (I and II), HSTL-18 (I and II), HSTL-15 (I and II), HSTL-12 (I and II), Differential SSTL-18 (I and II), Differential SSTL-15 (I and II), Differential SSTL-2 (I and II), Differential HSTL-18 (I and II), Differential HSTL-15 (I and II), Differential HSTL-12 (I and II), Differential HSUL-12								
	LVDS channels	66	66	137	52	224	224	163	178	230
External memory interfaces	Memory devices supported	DDR2, DDR, SDR								

- **ALTERA (Intel FPGA) Cyclone II Architecture**

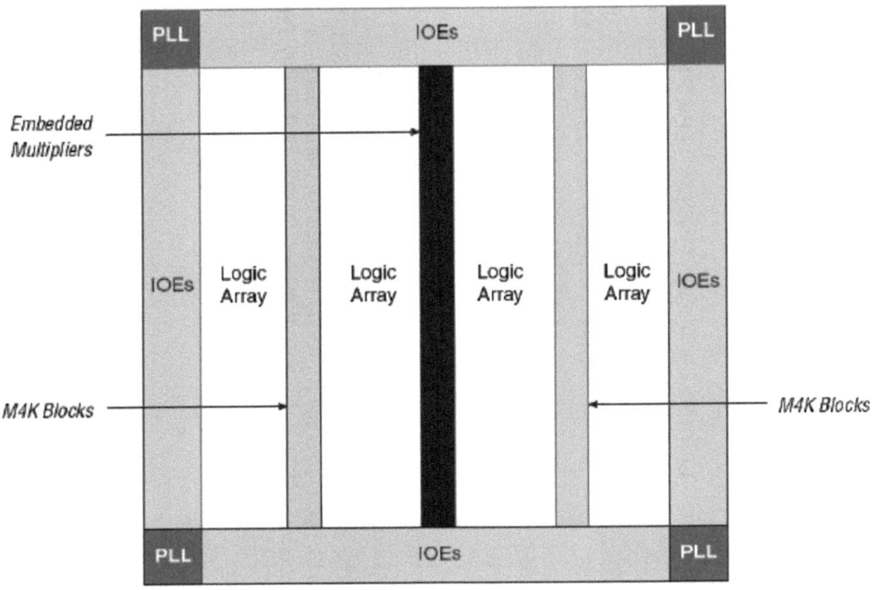

- **ALTERA (Intel FPGA) Cyclone II FPGA Features** (Table 1.1)

Table 1.1 Cyclone II FPGA family features

Feature	EP2C5	EP2C8	EP2C20	EP2C35	EP2C50	EP2C70
LEs	4,608	8,256	18,752	33,216	50,528	63,416
M4K RAM blocks (4 Kbits plus 512 parity bits)	26	36	52	105	129	250
Total RAM bits	119,808	165,888	239,616	483,840	594,432	1,152,000
Embedded multipliers *(1)*	13	I8	26	35	86	150
PLLs	2	2	4	4	4	4
Maximum user I/O pins	158	182	315	475	450	622

Note

(1)This is the total number of 18 × 18 multipliers. For the total number of 9 × 9 multipliers per device, multiply the total number of 18 × 18 multipliers by 2

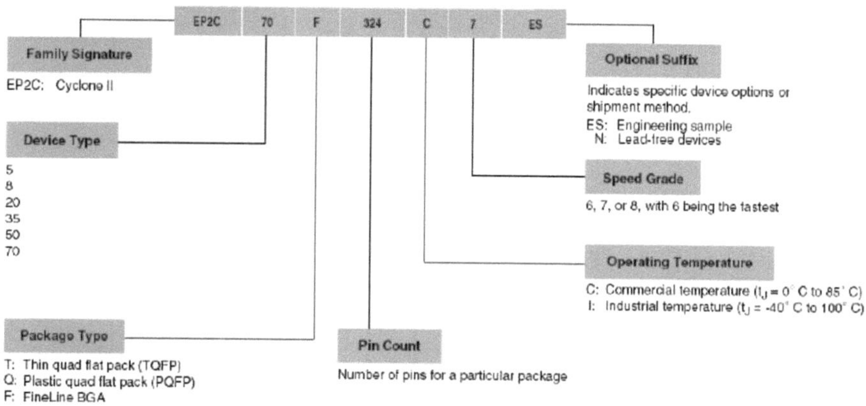

For more information, please use the following link:

http://www.altera.com

Appendix D
VHDL Design Units

VHDL design units are classified as primary design units and secondary design units.

Primary design units are as follows:

- Entity
- Package
- Configuration

Secondary design units are as follows:

- Architecture or multiple architectures
- Package body declarations

The syntax of the VHDL design units and constructs are listed below for the quick reference.

Primary Design Units

1. **Entity Declaration**

Entity is used to define the input and output interfaces for the given design, and it consists of the port declaration and generic clauses. The environment in which entity is used may consist of the following declarations:

- Type
- Subprogram
- Alias
- File
- Constants
- Signals

V. Taraate, *PLD Based Design with VHDL*,
DOI 10.1007/978-981-10-3296-7

```
entity entity_name is

        generic (generic_list);
        port (port_list);

        subprogram_declaration
        | subprogram_body
        | type_declaration
        | subtype_declaration
        | constant_declaration
        | signal_declaration
        |shared_variable_declaration
        | file_declaration
        | alias_declaration
        | attribute_declaration
        | attribute_specification
        | disconnection_specification
        | use_clause
        | group_template_declaration
        | group_declaration

begin

        concurrent_assertion_statement
        | passive_concurrent_procedure_call
        | passive_process_statement

end entity_name ;
```

2. **Package Declaration**

Packages are used to define the input and output interfaces of common elements which are visible to other designs. Packages consist of the following declarations:

- Subprogram
- Attributes
- Aliases
- Types
- Files
- Components

package *package_name is*

subprogram_declaration
| subprogram_body
| type_declaration
| subtype_declaration
| constant_declaration
| shared_variable_declaration
| file_declaration
| alias_declaration
| use_clause
| group_template_declaration
| group_declaration

end package package_name ;

3. **Configuration Declaration**

In the case of the VHDL complex designs which consist of multiple entities or components, configuration statement is used. Configuration statement is used to select the required components from the IEEE library. It may contain the following:

- Component configuration
- Block configuration
- Generate statement
- Attribute specifications

configuration *conf_name of entity_name is*

use_clause
| attribute_specification
| group_declaration

for
architecture_name
block statement label
|generate statement label
(discret range | static expression)
|use clause
(block configuration | component configuration)

end for;

end configuration conf_name ;

Secondary Design Units

1. Package Body Declaration

Package body consists of the functional information of the procedures and functions. The functional information may be visible to many other designs.

```
Package body package_name is

    subprogram_declaration
    | subprogram_body
    | type_declaration
    | subtype_declaration
    | constant_declaration
    | shared_variable_declaration
    | file_declaration
    | alias_declaration
    | use_clause
    | group_template_declaration
    | group_declaration

end package body package_name ;
```

2. Architecture Declaration

Architecture is used to describe the functionality of the design. The input and output relations are described by the architecture. Single entity can have more than one architecture. Architecture can have the different modules such as processes, subprograms (function and procedure calls), and block statements,

```
architecture arch_name of entity_name is

architecture_declarative_part

begin
    concurrent_statements;

end architecture arch_name ;
```

Libraries

Library is used to store the previously compiled or analyzed information. By using the library clause, the information is available to the design units. Library can contain one or more than one package. By using the 'use' clause, the information of library element can be accessible. Even the required package element or all the elements of the packages can be accessible

library library_name;
use library_name.package_name.(selected_elements | all);

Index

V. Taraate, *PLD Based Design with VHDL*,
DOI 10.1007/978-981-10-3296-7

MIX
Papier aus verantwortungsvollen Quellen
Paper from responsible sources
FSC® C105338

If you have any concerns about our products,
you can contact us on
ProductSafety@springernature.com

In case Publisher is established outside the EU,
the EU authorized representative is:
Springer Nature Customer Service Center GmbH
Europaplatz 3, 69115 Heidelberg, Germany

Printed by Libri Plureos GmbH
in Hamburg, Germany